AF355958

Application of Plant-Based Molecules and Materials in Cosmetics

Application of Plant-Based Molecules and Materials in Cosmetics

Guest Editor

Paulraj Mosae Selvakumar

Basel • Beijing • Wuhan • Barcelona • Belgrade • Novi Sad • Cluj • Manchester

Guest Editor
Paulraj Mosae Selvakumar
Science and Math Program
Asian University for Women
Chittagong
Bangladesh

Editorial Office
MDPI AG
Grosspeteranlage 5
4052 Basel, Switzerland

This is a reprint of the Special Issue, published open access by the journal *Cosmetics* (ISSN 2079-9284), freely accessible at: https://www.mdpi.com/journal/cosmetics/special_issues/ZYU9N33CY9.

For citation purposes, cite each article independently as indicated on the article page online and as indicated below:

Lastname, A.A.; Lastname, B.B. Article Title. *Journal Name* **Year**, *Volume Number*, Page Range.

ISBN 978-3-7258-2991-0 (Hbk)
ISBN 978-3-7258-2992-7 (PDF)
https://doi.org/10.3390/books978-3-7258-2992-7

Cover image courtesy of Paurlaj Mosae Selvakumar

Contents

About the Editor

Paulraj Mosae Selvakumar

Dr. Paulraj Mosae Selvakumar is presently working as an Associate Professor and the Director of the Environmental Sciences Program at the Asian University for Women in Chittagong, Bangladesh. He earned his Bachelor's and Master's degrees in Chemistry from Manonmaniam Sundaranar University in Tamil Nadu, India. He completed his Ph.D. at Bhavnagar University and the Central Salt and Marine Chemical Research Institute. His professional career began as a Scientist in the cheminformatics division at Jubilant Biosys Ltd., Bangalore. Later, he served as an Assistant Professor in Chemistry at Karunya University, Coimbatore, Tamil Nadu. He was also a DAAD research fellow at the Technical University of Kaiserslautern, Germany. His research interests include Palmyraculture—focusing on the plantation and utilization of the Palmyra palm for sustainable development—waste-to-wealth initiatives, the synthesis of small bioactive molecules, nanomaterials, supramolecular architectures, and the development of sensor molecules for detecting anions, cations, and small molecules in water, as well as water treatment technologies. He is also the founder of the Green Bangle Movement, a tree plantation initiative that aims to empower women in environmental conservation and sustainable development through the popularization of ecofeminism and ecopreneurship in Asia. He has received several awards/fellowships and recognitions, including a Senior Research Fellowship (SRF) in Chemistry from CSIR, India, and a DAAD fellowship from DAAD, Germany, and the DST Fast Track for Young Scientists. He is a member of the International Union for Conservation of Nature (IUCN), the Chemical Research Society of India (CRSI), the Material Research Society of India (MRSI), and the Luminescent Society of India (LSI). As an educator, he has developed curricula and pedagogies for both undergraduate and postgraduate programs, covering a wide range of subjects in Chemistry and Environmental Sciences.

 *cosmetics*

Editorial

Application of Plant-Based Molecules and Materials in Cosmetics

Paulraj Mosae Selvakumar

Environmental Sciences Program, Green Bangle Movement, Asian University for Women, Chittagong 4000, Bangladesh; p.selvakumar@auw.edu.bd

Citation: Selvakumar, P.M. Application of Plant-Based Molecules and Materials in Cosmetics. *Cosmetics* **2024**, *11*, 211. https://doi.org/10.3390/cosmetics11060211

Received: 27 November 2024
Accepted: 2 December 2024
Published: 4 December 2024

With the growing demand for products that are gentle on the skin and manufactured by eco-friendly means, the field of cosmetics is witnessing a remarkable shift toward natural ingredients. This Special Issue of Cosmetics delves into the fascinating realm of plant-based molecules and materials, navigating their potential to transform the cosmetic industry.

Since ancient times, plants have been considered a source of nourishment, healing, and beauty. Traditional botanical medicines have been employed to alleviate ailments for centuries, and in recent years, the therapeutic effects of plant-derived active molecules and nanomaterials have been scientifically recognized and amplified [1,2]. Gradually, the adoption of plant-based molecules and materials in the field of cosmetics has garnered significant attention, with increasing numbers of studies exploring their properties and effectiveness [3,4]. This shift towards natural cosmetics aligns with the current trend of "clean" beauty products that foster healthy skin conditions by excluding harmful chemicals, animal extracts, and synthetic ingredients [5]. The rapid expansion of the global cosmetic industry has created a broad market for plant-based cosmetic ingredients that can have multiple beneficial effects, such as depigmentation, brightening, and anti-aging [6].

A total of 10 manuscripts, comprising 7 research articles, a communication, and 2 reviews, are included in the Special Issue "Application of Plant-Based Molecules and Materials in Cosmetics", where each underwent a rigorous and systematic review process by the Cosmetics editorial team. These contributions are briefly summarized below:

Wan-Teng Lin et al. investigated the potential of an essential oil prepared from *Glossogyne tenuifolia* as a natural skin-whitening agent. The study (1) demonstrates the oil's ability to inhibit melanin biosynthesis by targeting the MITF signaling pathway, establishing its potential as an innovative skin-lightening product. The study (2) by Alfredo Martínez-Gutiérrez et al. explored the combined effects of apigenin and phloretin on skin aging and hyperpigmentation. The research recommends this combination for the effective regulation of melanogenesis by inducing autophagy in melanocytes, which offers a promising approach for treating skin hyperpigmentation. Laziz Bouzidi et al. investigated the potential of blending Astrocaryum pulp oil and kernel fat to produce multifunctional cosmetic ingredients. The study (3) proposed the blend to achieve a rich assortment of bioactive compounds, texture, and physical properties that makes it suitable for various cosmetic applications. Laura Rubio et al. conducted a comprehensive analysis (4) of several essential oils and natural extracts, marking potential phytomarkers and allergens. These findings underscore the safety and efficacy parameters crucial for natural cosmetic formulations. Faten Mohamed Ibrahim et al. presented research (5) on a novel nanoemulgel formulation that incorporates lemon peel extract as a non-toxic, antimicrobial, and alcohol-free hand sanitizer. The formulation projected promising antimicrobial activity against bacterial and fungal attacks. The paper (6) by Ugnė Žlabienė et al. explored the potential of patchouli extract and allantoin-based anti-dandruff shampoo formulations. The study showed that this combination is able to improve the technological properties of the shampoo while potentially resolving dandruff concerns. Cloé Boira et al. investigated the potential of *Centella asiatica* extract to reduce the appearance of stretch marks in their paper (7). The findings demonstrated its ability to accelerate fibroblast proliferation and promote skin repair, suggesting its capacity as a natural treatment for stretch marks. Andrea Cavagnino et al. explored the use of adaptogenic plants, such as *Lactobacillus plantarum*, *Withania somnifera*, and *Terminalia ferdinandiana*, to support skin resilience. The study (8) suggested that these plant extracts can protect skin from environmental stressors and aid in skin regeneration. Mokgadi Ursula Makgobole et al. reviewed the traditional use of medicinal plants in West Africa for the treatment of skin diseases. The review (9) discussed the potential of these plant species as sources of bioactive compounds with antimicrobial and anti-inflammatory properties, which can be employed in cosmetic formulations. Sunehra Sayanhika et al. explored the potential of Asian Palmyra palm, which is the official tree of Tamil Nadu Government, as a cosmetic ingredient. This review (10) dissected the range of skin benefits from the plant's molecules, including antioxidant and antimicrobial properties. The bioactive compounds present in the different parts of the palmyra plant make it a promising ingredient for various cosmetic formulations.

In summary, this Special Issue provides an extensive overview of the recent research advancements in the field of plant-based cosmetics. The studies presented demonstrate the multidirectional potential of natural ingredients to address skincare concerns, offering safer and more sustainable alternatives to synthetic cosmetics. As research continues to explore the uncharted potential of plant-based molecules and materials, we anticipate more of these sophisticated and effectual cosmetic formulations in the near future.

Conflicts of Interest: The author declares no conflicts of interest.

List of Contributions

1. Lin, W.-T.; Chen, Y.-J.; Kuo, H.-N.; Yu, C.-Y.; Abomughaid, M.M.; Senthil Kumar, K.J. *Glossogyne tenuifolia* Essential Oil Prevents Forskolin-Induced Melanin Biosynthesis via Altering MITF Signaling Cascade. *Cosmetics* **2024**, *11*, 142. https://doi.org/10.3390/cosmetics11040142.
2. Martínez-Gutiérrez, A.; Sendros, J.; Noya, T.; González, M.C. Apigenin and Phloretin Combination for Skin Aging and Hyperpigmentation Regulation. *Cosmetics* **2024**, *11*, 128. https://doi.org/10.3390/cosmetics11040128.
3. Bouzidi, L.; Deonarine, S.; Soodoo, N.; Emery, R.J.N.; Martic, S.; Narine, S.S. Extending the Physical Functionality of Bioactive Blends of *Astrocaryum* Pulp and Kernel Oils from Guyana. *Cosmetics* **2024**, *11*, 107. https://doi.org/10.3390/cosmetics11040107.
4. Rubio, L.; Pita, A.; Garcia-Jares, C.; Lores, M. Natural Extracts and Essential Oils as Ingredients in Cosmetics: Search for Potential Phytomarkers and Allergen Survey. *Cosmetics* **2024**, *11*, 84. https://doi.org/10.3390/cosmetics11030084.
5. Ibrahim, F.M.; Shalaby, E.S.; El-Liethy, M.A.; Abd-Elmaksoud, S.; Mohammed, R.S.; Shalaby, S.I.; Rodrigues, C.V.; Pintado, M.; Habbasha, E.S.E. Formulation and Characterization of Non-Toxic, Antimicrobial, and Alcohol-Free Hand Sanitizer Nanoemulgel Based on Lemon Peel Extract. *Cosmetics* **2024**, *11*, 59. https://doi.org/10.3390/cosmetics11020059.
6. Žlabienė, U.; Bartkutė, E.; Bernatonienė, J. Unlocking the Potential: A Comprehensive Analysis of the Technological Properties and Consumer Perception of Shampoo Enriched with Patchouli Extract and Allantoin. *Cosmetics* **2024**, *11*, 53. https://doi.org/10.3390/cosmetics11020053.
7. Boira, C.; Meunier, M.; Bracq, M.; Scandolera, A.; Reynaud, R. The Natural *Centella asiatica* Extract Acts as a Stretch Mark Eraser: A Biological Evaluation. *Cosmetics* **2024**, *11*, 15. https://doi.org/10.3390/cosmetics11010015.
8. Cavagnino, A.; Breton, L.; Ruaux, C.; Grossgold, C.; Levoy, S.; Abdayem, R.; Roumiguiere, R.; Cheilian, S.; Bouchara, A.; Baraibar, M.A.; et al. Adaptogen Technology for Skin Resilience Benefits. *Cosmetics* **2023**, *10*, 155. https://doi.org/10.3390/cosmetics10060155.

9. Makgobole, M.U.; Mpofana, N.; Ajao, A.A.-n. Medicinal Plants for Dermatological Diseases: Ethnopharmacological Significance of Botanicals from West Africa in Skin Care. *Cosmetics* **2023**, *10*, 167. https://doi.org/10.3390/cosmetics10060167.
10. Sayanhika, S.; Selvakumar, P.M. An Insight into the Cosmetic and Dermatologic Applications of the Molecules of Palmyra Palm. *Cosmetics* **2024**, *11*, 196. https://doi.org/10.3390/cosmetics11060196.

References

1. Mariselvam, R.; Ignacimuthu, S.; Ranjitsingh, A.; Mosae, S.P. An insight into leaf secretions of Asian Palmyra palm: A wound healing material from nature. *Mater. Today Proc.* **2021**, *47*, 733–738. [CrossRef]
2. Thevamirtha, C.; Balasubramaniyam, A.; Srithayalan, S.; Selvakumar, P.M. An Insight into the antioxidant activity of the facial cream, solid soap and liquid soap made using the carotenoid extract of palmyrah (*Borassus flabellifer*) fruit pulp. *Ind. Crops Prod.* **2023**, *195*, 116413. [CrossRef]
3. Sasounian, R.; Martinez, R.M.; Lopes, A.M.; Giarolla, J.; Rosado, C.; Magalhães, W.V.; Velasco, M.V.R.; Baby, A.R. Innovative Approaches to an Eco-Friendly Cosmetic Industry: A Review of Sustainable Ingredients. *Clean Technol.* **2024**, *6*, 176–198. [CrossRef]
4. FAO and Non-Timber Forest Products-Exchange Programme. *Naturally Beautiful—Cosmetic and Beauty Products from Forests*; FAO and Non-Timber Forest Products-Exchange Programme: Bangkok, Thailand, 2020. [CrossRef]
5. Vandamme, E. Biocosmetics produced via microbial and enzymatic synthesis. *Agro Food Ind. Hi-Tech* **2001**, *12*, 11–18.
6. Goyal, N.; Jerold, F. Biocosmetics: Technological advances and future outlook. *Environ. Sci. Pollut. Res. Int.* **2023**, *30*, 25148–25169. [CrossRef] [PubMed]

Review

Medicinal Plants for Dermatological Diseases: Ethnopharmacological Significance of Botanicals from West Africa in Skin Care

Mokgadi Ursula Makgobole [1,*], Nomakhosi Mpofana [1] and Abdulwakeel Ayokun-nun Ajao [2]

[1] Department of Somatology, Durban University of Technology, 47 Steve Biko Road, Durban 4001, South Africa; nomakhosim@dut.ac.za

[2] Department of Botany and Plant Biotechnology, University of Johannesburg, P.O. Box 524, Auckland Park, Johannesburg 2006, South Africa; ajwak880@gmail.com

* Correspondence: mokgadim@dut.ac.za

Abstract: Skin disease is a severe health issue that affects a lot of people in Africa and is vastly underreported. Because of their availability, affordability, and safety, medicinal plants represent a major source of treatment for various skin diseases in West Africa. This review presents the medicinal plants used in treating skin diseases in West Africa and their available biological activities that have lent credence to their skin care usage. A total of 211 plant species from 56 families are implicated to be used in West Africa for several skin conditions such as aphthous ulcers, burns, eczema, scabies, sores, and wounds. Fabaceae is the most-implicated family (30 species) for the treatment of skin diseases, followed by Combretaceae (14 species) and Asteraceae (13 species). Most of the medicinal plants used are trees (93); leaves (107) were the most-used plant part, and decoction (73) was the preferred preparation method for the medicinal plants. The biological activities related to the pathology of skin diseases, such as antimicrobial and anti-inflammatory properties of 82 plants, have been evaluated. Based on their minimum inhibitory concentration, the most active antimicrobial plant is *Brillantaisia lamium*. Among the isolated phytochemicals, betulenic acid and lespedin were the most active, while plants such as *Kigelia africana* and *Strophanthus hispidus* showed significant wound-healing activities. This review highlights research gaps in the ethnobotanical studies of many West African countries, the biological activities of plants used to treat skin diseases, and the cosmetic potential of these plants.

Keywords: biological activities; cosmeceuticals; cosmetics; medicinal plants; skin diseases; West Africa

Citation: Makgobole, M.U.; Mpofana, N.; Ajao, A.A.-n. Medicinal Plants for Dermatological Diseases: Ethnopharmacological Significance of Botanicals from West Africa in Skin Care. *Cosmetics* **2023**, *10*, 167. https://doi.org/10.3390/cosmetics10060167

Academic Editor: Paulraj Mosae Selvakumar

Received: 14 November 2023
Revised: 2 December 2023
Accepted: 4 December 2023
Published: 7 December 2023

1. Introduction

The human body's largest and most vulnerable organ is the skin. It serves as a barrier and shield between the internal organs and direct microbial contamination and ultraviolet radiation [1]. It contains three layers: the epidermis, dermis, and subcutaneous. The epidermis, the outermost layer, protects the skin against infections caused by some microbes [2]. The dermis comprises follicles and glands and is essential for regulating body temperature, while the subcutaneous layer contains a network of connective tissues and fat [2]. Despite the dry and infertile nature of the skin, it is still home to millions of microbes, some of which are essential in preventing skin invasion by pathogens [3]. The nature of the skin significantly influences the type and abundance of microbes present, causing skin diseases or infections. For example, *Corynebacterium* and *Staphylococcus* species are abundant in the moist areas of the skin, while sebaceous areas are predominantly occupied by *Propionibacterium* species [4]. Also, different skin or systemic diseases may arise from broken skin or an imbalance between beneficial microbes and pathogens [5].

Skin disease is the fourth most common cause of illnesses in humans [6]. However, this may be inaccurate because reports have shown that a large percentage of people suffering from different skin conditions do not consult dermatologists [7]. Medicinal plants have long played a significant role in treating a wide range of illnesses, including skin diseases, which are not always reported but are self-treated [8]. In the last few decades, extensive efforts have focused on documenting medicinal plants to identify species that may be used in drug development [9]. Among these, some studies have reported or reviewed the plants used to treat skin diseases around the globe. For example, there is documentation of medicinal plants used for skin diseases in southern Africa, Pakistan, the south Balkan, and the eastern Mediterranean regions [10–12]. Many people in West Africa rely on traditional medicine/medicinal plants due to their accessibility and believe they are more potent and efficient in curing disease than conventional medicines or healthcare, which is inadequate in the region [13,14]. Also, many skin diseases tend to be so persistent and recurring that patients resort to many remedies, including herbal ones [1]. Therefore, quite a number of studies have aimed at documenting plants used to treat skin diseases in different West African countries. For example, the Akwa Ibom state of Nigeria [15], the northern region of the Republic of Benin [16], the northwest region of Cameroon [17], and many others.

Moreover, to validate the traditional uses of some of the documented plants, some other studies have tested the extracts from the plants for pharmacological activities related to skin diseases, such as antioxidant, anti-inflammatory, and antimicrobial activities, to prove their efficacy. For example, Rosamah et al. [1] investigated the pharmacological significance of *Macaranga* in dermatological diseases; Udegbunam et al. [18] confirmed the therapeutic activity of *Crinum jagus* in wound healing; and *Anacardium occidentale* extracts were tested on pathogenic microbes to verify its usefulness in skin care cosmetics by Gonçalves and Gobbo [19]. There are also many similar studies, but this information is scattered in the literature, which may account for the fact that none of the many cosmetic plants recognized by the European Pharmacopoeia are of West African origin [20]. This also underscores the significance of studies in this field. Therefore, this review aims to appraise all available information on the use of plants for treating skin diseases in West Africa to highlight research gaps and give a proper direction to future research on the dermatological significance of medicinal plants from West Africa.

2. Methodology

A comprehensive search of major journal websites and academic databases, such as African Journal Online, Google Scholar, ScienceDirect, Scopus, and Web of Science, was conducted. This aimed to retrieve all published ethnobotanical studies in the West African region up until September 2023. The keywords used in the search include the names of West African countries, ethnobotany, medicinal plants, indigenous plants, skin diseases, skin ailments, cosmetics, wounds, and West Africa. From the papers retrieved, plants used for skin diseases were extracted from the general ethnobotany. Plant names and their family were verified and updated using databases, namely: Tropicos (www.tropicos.org; accessed on 7 October 2023), World Flora Online (https://www.worldfloraonline.org/; accessed on 29 November 2023), and WFO Plant List website (https://wfoplantlist.org/plant-list; accessed on 7 October 2023). The selection of articles for this review was made on the basis of the following criteria:

1. It is published in English or translated to English.
2. At least one plant is listed for the treatment of skin diseases.
3. It studied the bioactivity of at least one of the plants in the list of plants documented.
4. If the ethnopharmacological study is carried out outside West Africa but examines the bioactivity of any documented plants used to treat skin diseases in the region. Figure 1 shows the PRISMA flowchart for the inclusion and exclusion procedure.

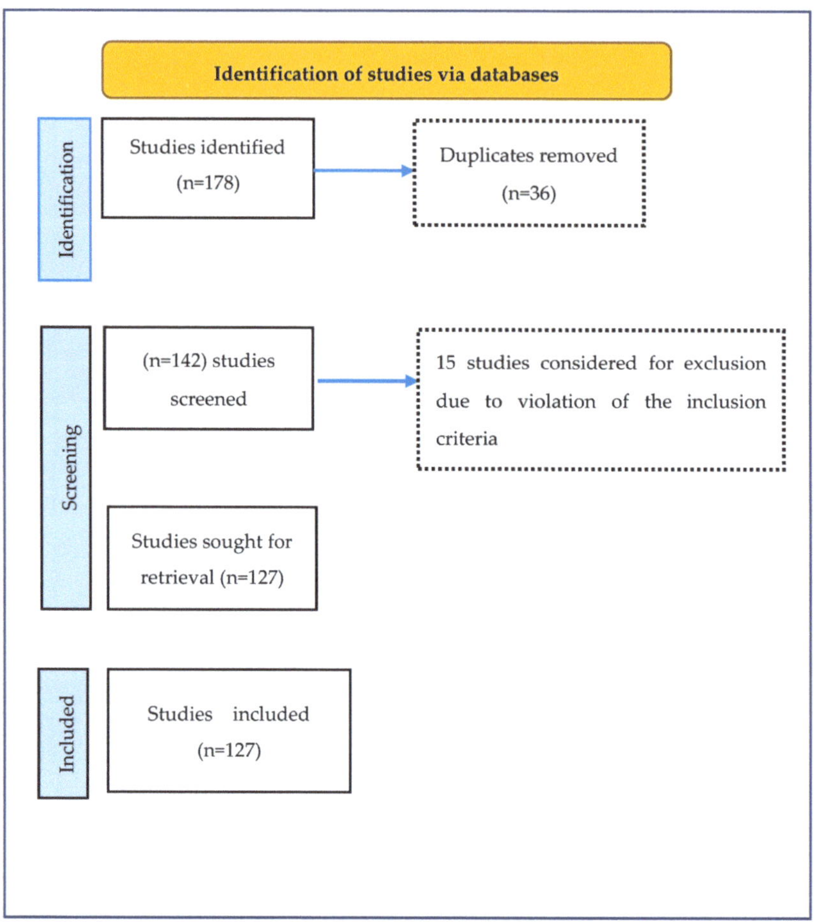

Figure 1. PRISMA flow chart showing total studies identified, removed, excluded, and included in the review.

3. Results and Discussion

3.1. Diversity of the Medicinal Plants Used for Skin Diseases in West Africa

The western part of Africa is a tropical region with high annual rainfall and temperature. Hence, it is known to have a wide flora diversity. A total of 211 taxa are reported in this review as used in the treatment of various skin diseases or infections in the western part of Africa (Table 1). In comparison, 100 plant species were reportedly used for skin diseases in southern Africa [10], 545 species in Pakistan [11], and 967 in the South Balkan and East Mediterranean region [12]. Over 90% of the plants were reported from Nigeria, the Republic of Benin, Togo, Ghana, Cameroon, and the Ivory Coast, implying a large research gap in the ethnobotanical studies of many West African countries. The reported taxa belong to 56 different families. Among these, plants of the Fabaceae family were overwhelmingly preferred for combating skin diseases in West Africa, as 30 species from 23 genera represent them. This is followed by the family Combretaceae, with 14 species, half of which are from the genus *Combretum*. Asteraceae and Euphorbiaceae are also well represented, with 13 and 12 species, respectively (Figure 2). Overall, nine families—Fabaceae (30), Combretaceae (14), Asteraceae (13), Euphorbiaceae (12), Phyllanthaceae (8), Rubiaceae (8), Moraceae (8), Malvaceae (7), and Apocynaceae (7)—represented over 50% of the recorded plant species, while 22 species, were the only members of their respective families. The legume family

(Fabaceae) is one of the most influential families of angiosperms in terms of medicinal uses across communities and regions worldwide [21]. Despite being the third largest family, its wide distribution worldwide could be a major reason for its widespread usage in traditional medicine [21]. The sustained use of the species of the Fabaceae family over time may also indicate their biological activities, and the phytochemistry of many species in this family has supported this hypothesis, showing that they contain critical active metabolites [22]. The biological activities and chemistry of the family Combretaceae have also revealed a wide range of useful phytochemicals [23]. However, other genera apart from the genera *Combretum* and *Terminalia* are rarely explored [24]. Asteraceae is the largest plant family and has always been highly represented in many ethnobotanical studies, and the family members are known to contain phytochemicals of medicinal importance [25].

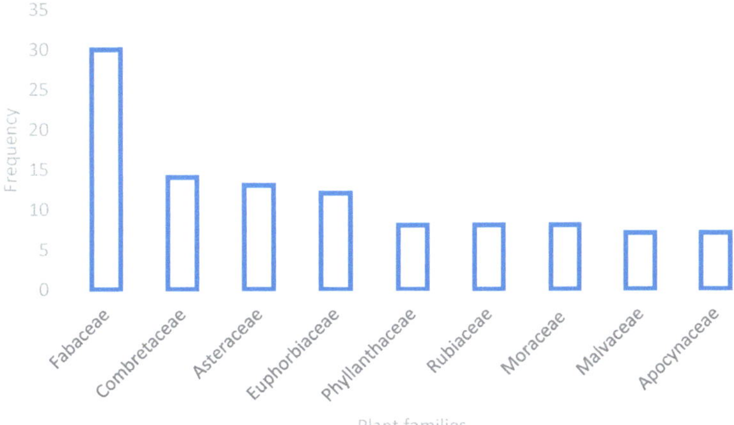

Figure 2. Most cited plant families.

3.2. Life Forms, Plant Parts Used, Mode of Preparation, and Conservation Statuses of the Plants

Over 60% of the plants implicated in this paper are woody species, comprising trees (93, 44%) and shrubs (40, 19%) (Figure 3). Other life forms are herbs (61, 29%) and climbers (17, 8%). Some studies have reported that herbs are the most-used medicinal plants due to their ease of harvesting [60,61]. Though this may be true for some individual studies, a holistic review of the ethnobotany of an area or an ailment category has revealed, in most cases, the prevalence of trees in traditional medicine [62,63]. Additionally, West Africa, being a tropical area, could be responsible for the abundance of trees in the region.

Table 1. Medicinal plants used to treat skin diseases in West Africa.

Family	Plant Species	Habit and Conservation Status	Country	Local Name	Plant Part(s) Used	Mode of Preparation	Ailment	References
Acanthaceae	*Acanthus montanus* (Nees) T. Anderson	Herb; LC	Nigeria	Mbara ekpe (Akwaibom)	Leaves	Poultice.	Abscess, boils, whitlow, and wounds	[15]
Acanthaceae	*Afrofittonia silvestris* Lindau	Herb; VU	Nigeria	Mmeme (Akwaibom)	Whole plant	Crushed and the juice is topically used for skin spots; Poultice is applied to whitlow.	Skin spots and whitlow	[15]
Acanthaceae	*Brillantaisia lamium* (Nees) Benth.	Herb; No Record	Cameroon	No record	Aerial parts	Decoction of aerial parts is used to bath.	Skin infections	[26]
Acanthaceae	*Justicia insularis* T. Anders	Herb; No Record	Nigeria	No record	Whole plant	Crushed and the juice applied; poultice.	Skin spots	[15]
Aiozoaceae	*Trianthema portulacastrum* L.	Herb; No Record	Nigeria	Ntia ntia ikon (Akwa Ibom)	Leaves	Decoction is used for bathing the affected area.	Wounds	[15]
Amaranthaceae	*Achyranthes aspera* L.	Herb; No Record	Nigeria	Udok mbiok	Leaves	Crushed and juice applied.	Skin ulcers	[15]
Amaranthaceae	*Alternanthera bettzickiana* (Regel) G. Nicholson	Herb; No Record	Nigeria	Nkpok isip essien	Leaves	Crushed and the juice applied.	Skin spots, measles	[15]
Amaranthaceae	*Amaranthus caudatus* L.	Herb; No Record	Nigeria	Inyan afia	Leaves	Leaf juice mixed with kaolin is applied.	Abscess, boil, eczema, and skin eruption	[15]
Amaranthaceae	*Celosia globosa* Schinz	Herb; No Record	Cameroon	NA	Leaves	Decoction of the leaves in Cameroon.	Athlete's foot	[26]
Amaranthaceae	*Cyathula prostrata* (L.) Blume	Herb; No Record	Cote d'Ivoire, Guinea and Nigeria	Nkibe ubuk	Leaves	Decoction is taken orally in Nigeria, and Cote d'Ivoire for leprosy; the juice from macerated leaves is applied to cuts and bruises in Guinea.	Leprosy, skin spots, scabies, sores, and rashes	[15,27,28]
Amaryllidaceae	*Allium cepa* L.	Herb; LC	Nigeria and West Africa	Alubosa (Yoruba)	Bulb	Poultice.	Scorpion sting and skin disease	[15,29]
Amaryllidaceae	*Allium sativum* L.	Herb; LC	Nigeria	Alubosa Ayu (Yoruba)	Clove	Poultice.	Skin spots, burns, ulcers, and scorpion sting	[15]

Table 1. *Cont.*

Family	Plant Species	Habit and Conservation Status	Country	Local Name	Plant Part(s) Used	Mode of Preparation	Ailment	References
Amaryllidaceae	*Crinum jagus* (Thompson) Dandy	Herb; LC	Nigeria	Ayim ekpo, ekop-eyen (Akwa ibom), Ogede odo (Yoruba)	Bulb	Poultice.	Whitlow	[15]
Anacardiaceae	*Anacardium occidentale* L.	Tree; LC	Nigeria	Kashu, Cashew	Leaves	Poultice	Ringworm and leprosy	[15]
Anacardiaceae	*Lannea acida* A. Rich.	Shrub; LC	Nigeria and Senegal	Ayara nsukakara	Leaves	Crushed and juice applied.	Burns and skin infections	[15,30,31]
Anacardiaceae	*Lannea microcarpa* A. Rich	Tree; LC	Republic of Benin	Sinman	Root bark	The root bark is ground to powder and applied topically.	Wounds	[16]
Anacardiaceae	*Mangifera indica* L.	Tree; DD	Nigeria	Mongoro	Leaves	Decoction for bathing and applied topically.	Skin spots	[15]
Anacardiaceae	*Ozoroa pulcherrima* (schweinf.) R. & A.	Shrub; No Record	Republic of Benin	Mukentétié	Root bark	The root bark is ground into powder and applied topically.	Wounds	[16]
Annonaceae	*Annickia chlorantha* (Oliv.) Setten & Maas	Tree; LC	Nigeria and West Africa	Osopa (Yoruba)	Leaves, stem bark	Crushed and juice applied.	Sores, ulcers, and wounds	[32]
Annonaceae	*Annona senegalensis* Pers.	Tree; LC	Ghana, Mali, Nigeria, and Togo	Sawa-sawa (Yoruba), Tchoutchourè (Togo)	Leaves and fruits	A poultice made from the leaves is used for leprosy, sores, and wounds in Ghana, Mali, and Nigeria. Decoction of the leaves and fruits is taken orally for aphthous ulcers in Togo.	Leprosy, sores, and wounds	[15,27,33, 34]
Annonaceae	*Monodora myristica* (Gaertn.) Dunal	Tree; LC	Nigeria	Enwun	Seeds	Seeds are ground to powder and applied externally.	Pediculosis and sores	[15]
Annonaceae	*Uvaria chamae* P. Beauv.	Shrub; LC	Nigeria and Senegal	Nkarika ekpo	Root	Sap from the crushed root is applied topically.	Snakebites and wounds	[15,35]
Annonaceae	*Xylopia aethiopica* (Dunal) A. Rich	Tree; No Record	Republic of Benin	Nadofacha	Seeds	Dried seeds are ground into powder and applied topically.	Wounds	[16]

Table 1. *Cont.*

Family	Plant Species	Habit and Conservation Status	Country	Local Name	Plant Part(s) Used	Mode of Preparation	Ailment	References
Apocynaceae	*Alstonia boonei* De Wild.	Tree; LC	Cameroon, Cote d'Ivoire; Nigeria, Senegal	Ahun (Yoruba)	Stem bark	Crushed and applied.	Snakebites	[15,36]
Apocynaceae	*Calotropis procera* (Aiton) W.T. Aiton	Herb; LC	Gambia, Nigeria, and Senegal	Bomubomu (Yoruba)	Leaves	The poultice made from the leaves.	Smallpox, skin eruption, snakebites, and wounds	[15,27,37]
Apocynaceae	*Funtumia elastica* (Preuss) Stapf	Tree; LC	Cameroon and Nigeria	Eto okpo	Leaves	Juice from the crushed leaves is applied topically.	Snakebites and wounds	[15,36]
Apocynaceae	*Holarrhena floribunda* (G.Don) T.Durand & Schinz	Tree; LC	Togo	Kororo (Togo)	Leaves	Decoction of the leaves is taken orally.	Aphthous ulcers	[34]
Apocynaceae	*Leptadenia hastata* (Pers.) Decne	Climber; No Record	Senegal	Mboom (wolof) Duto (mandingo)	Stem	Infusion of woody stems is taken orally.	Snakebites	[15]
Apocynaceae	*Rauvolfia vomitoria* Afzel	Tree; LC	Nigeria	kiko	Leaves	Crushed and applied.	Ringworm and itchy body	[15]
Apocynaceae	*Strophanthus hispidus* DC.	Shrub; LC	Cote d'Ivoire, Ghana, Guinea, Nigeria, and Senegal	Ibok idan	Root bark	Crushed and applied.	Snakebites, scorpion stings, cuts, skin ulcers, and sores	[15,28,38–40]
Araceae	*Anchomanes difformis* (Blume) Engl	Herb; LC	Cote d'Ivoire and Nigeria	Nkokot	Bulb	Crushed and sap applied.	Wounds	[15,41]
Araceae	*Colocasia esculenta* (L.) Schott	Herb; LC	Nigeria and Cameroon	Ikpon ekpo Ndai (Cameroon)	Whole plant, tubers	Crushed and applied to the sore. Paste from grated tubers is applied on the part affected by whitlow and tied with a band.	Insect bites, sores, and whitlow	[15,17]
Araceae	*Elaeis guineensis* Jacq.	Tree; LC	Cameroon, Ghana, and Nigeria	Eyop	Fruit pericarp	Peeled and applied.	Boil, scabies, and wounds	[15,36,40]
Araceae	*Pistia stratiotes* L.	Herb; LC	Nigeria	Amana mmon	Whole plant	Powdered dry plant is applied topically.	Wounds and sores	[15]
Araceae	*Xanthosoma sagittifolium* (L.) Schott	Herb; No Record	Nigeria	Ikpon mbakara	Leaves	Crushed and juice applied.	Smallpox, skin spots, and fungal skin infection	[15]
Asparagaceae	*Agave sisalana* Perrine	Herb; No Record	Togo	Kolgragou	Root	Decoction of the root is taken orally.	Aphthous ulcers	[34]

Table 1. *Cont.*

Family	Plant Species	Habit and Conservation Status	Country	Local Name	Plant Part(s) Used	Mode of Preparation	Ailment	References
Asparagaceae	*Dracaena arborea* (Willd.) Link	Tree; LC	Nigeria	Okno	Root bark	Poultice.	Boils and burns	[15]
Asparagaceae	*Sansevieria liberica* Gérôme & Labroy	Herb; No Record	Nigeria	Okono ekpe	Leaves, stem bark, Root	Decoction, poultice.	Eczema and snakebites	[15]
Asphodelaceae	*Aloe vera* (L.) Burm. f.	Herb; LC	Nigeria, Cameroon, Ghana, Togo, and Benin	Eti erin (Yoruba)	Leaves	Gel from the crushed leaves is applied topically.	Skin infections and wounds	[15,17]
Asteraceae	*Acanthospermum hispidum* DC	Herb; No Record	Togo	Kpangsoyè	Whole plant	Decoction of the whole plant is taken orally.	Aphthous ulcers	[34]
Asteraceae	*Ageratum conyzoides* L.	Herb; LC	Nigeria	Imi esu (Yoruba)	Whole plant	Crushed in water and applied topically. The same preparation is taken orally for general skin infections.	Rashes, skin ulcers, and wounds	[15,42]
Asteraceae	*Aspilia africana* (Pers.) C.D. Adams	Herb; No Record	Cameroon, Liberia, Nigeria, and Sierra Leone	Edemeron Wowoh (Cameroon)	Leaves	Decoction of the leaves in Cameroon. Leaves are crushed or squeezed on the wound in Nigeria and Cameroon.	Wounds	[15,17,20, 27]
Asteraceae	*Bidens pilosa* L.	Herb; No Record	Cameroon, Cote d'Ivoire, Nigeria	Ntafion ison Shoctesuc (Cameroon)	Leaves	Crushed and juice applied.	Insect bites and wounds	[15,17,41]
Asteraceae	*Chromolaena odorata* (L.) R.M. King & H. Rob.	Herb; LC	Cameroon, Ghana, Nigeria	Mbiet (Ghana) Awolowo (Yoruba) Twigi (Cameroon)	Leaves	Crushed and juice applied. A poultice made from the leaves is used to cover the wound.	Rashes, scorpion sting, snakebites, and Wounds	[15,17,40]
Asteraceae	*Crassocephalum biafrae* (Oliv. & Hiern) S. Moore	Herb; No Record	Nigeria	Mkpafit	Leaves	Dried leaves are ground into powder and applied to the wound.	Wounds	[15]
Asteraceae	*Crassocephalum crepidioides* (Benth.) S. Moore	Herb; No Record	Nigeria	Mkpafit	Leaves	Juice from the crushed leaves is applied.	Boil, burns, and wounds	[15]
Asteraceae	*Cyanthillium cinereum* (L.) H.Rob	Herb; No Record	Togo	Kogbèdiyè	Aerial part	Decoction is taken orally.	Aphthous ulcers	[34]

Table 1. *Cont.*

Family	Plant Species	Habit and Conservation Status	Country	Local Name	Plant Part(s) Used	Mode of Preparation	Ailment	References
Asteraceae	*Emilia coccinea* (Sims) G. Don	Herb; No Record	Nigeria, Cameroon	Utime nse Nsefouse (Cameroon) Femefouse (Cameroon)	Leaves	Juice from the crushed leaves is applied.	Measles, rashes, wounds	[15]
Asteraceae	*Emilia sonchifolia* (L.) DC	Herb; No Record	Nigeria	Utime nse, usio mmon	Leaves	Juice from the crushed leaves is applied.	Measles, rashes, and Wounds	[15]
Asteraceae	*Laggera decurrens* (Vahl) Hepper & J.R.I. Wood	Herb; No Record	Nigeria	Ewedorun (Yoruba)	Whole plants	Decoction of the whole plant is applied to the wound with cotton wool.	Wounds	[42]
Asteraceae	*Tridax procumbens* L.	Herb; LC	Nigeria	Ayara utimense (Akwa Ibom), imi esu or apasa funfun (Yoruba)	Leaves	Decoction of the leaves is taken orally.	Skin spots	[15]
Asteraceae	*Vernonia amygdalina* Delile	Shrub; LC	West Africa	Etidod (Akwa ibom) Ewuro (Yoruba) Ying (Cameroon)	Leaves	Juice from the crushed leaves is applied externally. The juice is mixed with palm oil in Yoruba culture.	Chickenpox, measles, ringworm skin spots, skin infections and wounds	[15,40]
Bignoniaceae	*Kigelia africana* (Lam.) Benth	Tree; LC	Cote d'Ivoire, Nigeria, and Senegal	Ntabinim	Stem bark	Dried stem bark is ground to powder and applied topically.	Leprosy, snakebites, sores, and wounds	[15,31,32]
Bignoniaceae	*Newbouldia laevis* (P. Beauv.) Seem	Tree; LC	Cote d'Ivoire and Nigeria	Tumo	Stem bark and root	Decoction of stem back and root is taken orally.	Boils and skin spots	[15]
Bignoniaceae	*Spathodea campanulata* P. Beauv.	Tree; LC	Nigeria	Esenim	Stem bark	Infusion of stem back is applied externally for burns, bruises, skin infections, ulcers and wounds.	Skin infections, ulcers and wounds	[15,41]
Bignoniaceae	*Stereospermum kunthianum* Cham.	Shrub; LC	Togo	Essogbalou	Leaves	Decoction is taken orally for herpes sores.	Herpes sores	[34]

Table 1. *Cont.*

Family	Plant Species	Habit and Conservation Status	Country	Local Name	Plant Part(s) Used	Mode of Preparation	Ailment	References
Boraginaceae	*Heliotropium indicum* L.	Herb; LC	Benin, Ghana, Nigeria, Senegal, and Togo	Ewe akuko (Yoruba); Soucondiè (Togo), Koklosou dinkpadja (Benin)	Leaves and whole plant	Decoction of the leaves is taken orally for boil in Nigeria; a poultice made from the leaves is applied to wounds and insect bites in Ghana and Senegal. Decoction of the whole plants is taken orally for aphthous ulcers in Togo.	Boil, insect bites, and ulcers	[15,31,32, 34,43]
Brassicaceae	*Brassica oleracea* L.	Herb; No Record	Nigeria	Efere mbakara	Leaves	Poultice.	Ringworm and skin ulcers	[15]
Burseraceae	*Commiphora africana* (A. Rich.) Engl.	Tree; LC	Nigeria	Eto komfi itiat	Stem bark	Decoction of the stem back is taken orally.	Rashes caused by measles	[15]
Burseraceae	*Dacryodes edulis* (G. Don) H.J. Lam	Tree; No Record	Cote d'Ivoire and Nigeria	Eben	Leaves	Decoction is applied externally.	Leprosy and skin spots	[15,27,44]
Burseraceae	*Dacryodes klaineana* (Pierre) H.J. Lam	Tree; LC	Nigeria	Eben ikot	Leaves and root	Decoction is of the leaves and root taken orally.	Skin spots	[15]
Cannabaceae	*Trema orientale* (L.) Blume	Tree; LC	Cameroon and Nigeria	No record	Whole plant	Decoction of the whole plant is used to bath or applied topically.	Abscesses and skin spots	[15,45]
Capparaceae	*Maerua angolensis* DC	Tree; LC	Republic of Benin	Fetounanfè	Root bark	The root bark is ground to powder and applied topically.	Wounds	[16]
Caricaceae	*Carica papaya* L.	Tree; LC	Cameroon	Pawpaw	Leaves	Leaf juice is applied on fresh wounds.	Wounds	[17]
Celastraceae	*Apodostigma pallens* (Planch. ex Oliv.) R.Wilczek	Climber; No Record	Republic of Benin	Mukentetie	Root	The chewed root is applied topically.	Wounds	[16]
Celastraceae	*Gymnosporia senegalensis* L. E. T. Loesener	Shrub; LC	Republic of Benin	Moukorou	Root bark	Root bark is ground to powder and applied topically.	Wounds	[16]
Celastraceae	*Maytenus senegalensis* (Lam.) Exell	Shrub; No Record	Togo	Liakpangsoyè (Togo)	Whole plant	Decoction of the leaves is taken orally.	Aphthous ulcers	[34]
Chrysobalanaceae	*Maranthes kerstingii* (Engl.) Prance	Tree; No Record	Togo	Poundoulayzay (Togo)	Leaves	Decoction of the leaves is taken orally.	Aphthous ulcers	[34]

Table 1. *Cont.*

Family	Plant Species	Habit and Conservation Status	Country	Local Name	Plant Part(s) Used	Mode of Preparation	Ailment	References
Chrysobalanaceae	*Parinari curatellifolia* Planch. ex Benth.	Tree; LC	Togo	Malay (Togo)	Leaves	Decoction of the leaves is taken orally.	Aphthous ulcers	[34]
Clusiaceae	*Allanblackia floribunda* Oliv	Tree; VU	Nigeria	Udiaebion, ekporo-enin	Leaves	Decoction of the leaves is used for bathing.	Skin spots	[15]
Clusiaceae	*Symphonia globulifera* L. f.	Tree; LC	Cameron and Nigeria	No record	Bark, roots, and resin	Boiled bark and roots are used as a wash to treat itch, and the resin is used to treat wounds and prevent skin infections in Cameroon. Leaves decoction for skin disease and skin spots in Nigeria.	Itching, skin infections, and wounds	[15,46,47]
Cochlospermaceae	*Cochlospermum planchonii* Hook. f.	Shrub; No Record	Togo	Kalantcheyah (Togo)	Leaves	Decoction of the leaves is taken orally.	Aphthous ulcers	[34]
Combretaceae	*Anogeissus leiocarpus* (DC.) Guill. & Perr.	Tree; No Record	Cote d'Ivoire and Nigeria	Kolou (Togo)	Leaves and Stem bark	Infusion of stem bark in water is mixed with honey for skin ulcers, sores, and wounds. Decoction of leaves for aphthous ulcer.	Wounds, skin ulcers, and sores	[15,34,41]
Combretaceae	*Combretum collinum* Fresen	Tree; LC	Republic of Benin	Gberukporo	Root bark	The root bark is ground to powder and applied topically.	Wounds	[16]
Combretaceae	*Combretum glutinosum* Perr. Ex DC	Tree; LC	Republic of Benin	Oudadaribou	Root bark	Powdered root bark is incinerated and applied topically.	Wounds	[16]
Combretaceae	*Combretum hypopilinum* Diels	Shrub; No Record	Gambia	Katanyangkungo	Leaves and root	Decoction of both leaves and root is used to bath.	Itchy body	[48]
Combretaceae	*Combretum micranthum* G. Don	Shrub; LC	Cote d'Ivoire, Nigeria, and West Africa	Asaka	Leaves	Infusion of the leaves is taken orally.	Leprosy, sores and skin spots	[27,31]
Combretaceae	*Combretum racemosum* P. Beauv.	Shrub; No Record	Nigeria	Uyai asaka	Leaves	Juice from the crushed leaves is taken orally.	Skin spots	[15]
Combretaceae	*Combretum sericeum* G. Don	Tree; No Record	Republic of Benin	Cocopourka	Root bark	The root bark is ground to powder and applied topically.	Wounds	[16]

Table 1. *Cont.*

Family	Plant Species	Habit and Conservation Status	Country	Local Name	Plant Part(s) Used	Mode of Preparation	Ailment	References
Combretaceae	*Combretum zeyheri* Engl. & Diels	Climber; LC	Nigeria	Ndia asaka	Leaves	Poultice.	Mump, skin eruption, and warts	[15]
Combretaceae	*Guarea thompsonii* Sprague & Hutch.	Tree; VU	Nigeria	Afia ikpok eto	Stem bark	Sap produced from the crushing of the stem bark is applied topically.	Skin diseases	[15]
Combretaceae	*Guiera senegalensis* J.F. Gmel.	Shrub; LC	Guinea, Senegal, and West Africa	No record	Leaves and twigs	Decoction of the leaves is taken for leprosy. The twigs are chewed for scorpion stings.	Leprosy and scorpion bites	[31,35,49]
Combretaceae	*Pteleopsis suberosa* Engl. et Diels	Tree; LC	Togo	Kézinzinang	Leaves and bark	Decoction is taken orally.	Aphthous ulcers	[34]
Combretaceae	*Terminalia avicennioides* Guill. & Perr. Fl. Seneg. Tent.	Tree; LC	Togo	Koyèkouloumryè	Aerial part	Decoction is taken orally	Aphthous ulcers	[34]
Combretaceae	*Terminalia ivorensis* A. Chev	Tree; VU	Cote d'Ivoire, Ghana, and Nigeria	Nkot ebene	Stem bark	Infusion of the stem bark is applied topically.	Sores and ulcers	[15,27,28]
Combretaceae	*Terminalia superba* Engl. & Diels	Tree; No Record	Nigeria	Afia eto	Leaves	Juice from the crushed leaves is applied externally.	Skin spots	[15]
Commelinaceae	*Commelina benghalensis* L.	Herb; LC	Cameroon	Wiwih	Latex	Latex is applied to the affected skin.	Ringworm	[17]
Commelinaceae	*Commelina diffusa* Burn. f.	Climber; LC	Nigeria	Ekpa ekpa ikpaha	Whole plant	Dried whole plant is ground to powder and applied externally.	Sores and burns	[15]
Convulvulaceae	*Ipomoea pileata* Roxb	Herb; No Record	Nigeria	Mkpafiafian	Leaves	Infusion of the leaves is applied topically.	Skin spots	[15]
Convulvulaceae	*Ipomoea quamoclit* L.	Herb; No Record	Nigeria	Ediam ikanikot	Leaves	Poultice.	Boil and wounds	[15]
Cucurbitaceae	*Citrullus colocynthis* (L.) Schrad.	Climber; No Record	Nigeria	Ikon	Seeds	Seeds are ground and applied topically.	Abscess and skin spots	[15]
Cucurbitaceae	*Cucurbita maxima* Duchesne	Climber; No Record	Ghana and Nigeria	Ikim	Leaves	Juice from the crushed leaves is applied topically.	Boil and skin spots	[15,27]
Cucurbitaceae	*Momordica balsamina* L.	Climber; No Record	Nigeria and Senegal	Mbiadon edon	Whole plant	Poultice.	Boil	[15]

Table 1. *Cont.*

Family	Plant Species	Habit and Conservation Status	Country	Local Name	Plant Part(s) Used	Mode of Preparation	Ailment	References
Cucurbitaceae	*Momordica charantia* L.	Climber; No Record	Ghana, Mali, Nigeria, and Senegal	Mbiadon edon Nyenyen (Ghana)	Fruit	Poultice; infusion of whole plants is taken orally in Ghana for snakebites.	Boil, burns snakebites, and ulcers	[15,40,50]
Dioscoraceae	*Dioscorea dumetorum* (Kunth) Pax	Climber; No Record	Nigeria	Enem (Akwa Ibom); Esuru (Yoruba)	Leaves	Decoction of the leaves is applied topically.	Skin spots	[15]
Dioscoraceae	*Dioscorea rotundata* Poir	Climber; No Record	Nigeria	Eko	Leaves	Infusion of the leaves is applied topically.	Burns and skin spots	[15]
Ebenaceae	*Diospyros canaliculata* De Wild.	Tree; LC	Cameroon	No record	Stem bark	No record.	Skin infections	[26]
Euphorbiaceae	*Acalypha fimbriata* Schumach. & Thonn	Herb; No Record	Nigeria	Okokho nyin	Leaves and twigs	Decoction of leaves and twigs is used topically to bath.	Skin spots and sores	[15]
Euphorbiaceae	*Acalypha hispida* Burm. f.	Shrub; No Record	Nigeria	Okokho nyin	Leaves	Decoction of leaves is used externally to bath.	Skin spots and sores	[15]
Euphorbiaceae	*Acalypha wilkesiana* Müll. Arg.	Shrub; No Record	Nigeria	Okokho nyin	Leaves	Decoction of leaves is used topically to bath.	Skin spots	[15]
Euphorbiaceae	*Alchornea cordifolia* (Schumach. & Thonn.) Müll. Arg.	Shrub; No Record	Ghana, Nigeria, and West Africa	Mbom	Leaves and fruits	Infusion of the leaves; juice from the crushed fruits is applied topically.	Skin spots and skin ulcers, skin spots, scorpion stings, and snakebites	[15,29,40]
Euphorbiaceae	*Alchornea laxiflora* (Benth.) Pax & K. Hoffm.	Shrub; LC	Nigeria	Nwariwa	Leaves	Infusion of the leaves is used for skin spots and skin ulcers.	Skin spots and ulcers	[15]
Euphorbiaceae	*Euphorbia hirta* L.	Herb; No Record	Nigeria	Etinkene ekpo	Leaves	The poultice made from the leaves is applied topically.	Snakebites, scorpion stings, and insect bites	[15]
Euphorbiaceae	*Jatropha curcas* L.	Shrub; LC	Togo	Essogbalou (Togo) Medjai (Cameroon)	Leaves and latex	Decoction of the leaves is taken for cancer sores. Latex from the cut stem is applied to the wounds.	Cancer sores and wounds	[17,34]
Euphorbiaceae	*Jatropha gossypiifolia* L.	Shrub; LC	Nigeria	Eto oko obio nsit	Leaves	Juice from the crushed leaves is applied topically.	Eczema, ringworm, and scabies	[15]

Table 1. *Cont.*

Family	Plant Species	Habit and Conservation Status	Country	Local Name	Plant Part(s) Used	Mode of Preparation	Ailment	References
Euphorbiaceae	*Macaranga barteri* Mull. Arg.	Shrub; LC	Ghana	Opam	Bark	Decoction of bark is taken orally.	Footrot	[50]
Euphorbiaceae	*Mallotus oppositifolius* (Geiseler) Müll. Arg.	Herb; LC	Cameroon, Ghana, and Nigeria	Uman nwariwa	Leaves or stem back	Decoction of the leaves is applied topically.	Skin spots	[15,40,51]
Euphorbiaceae	*Manniophyton fulvum* Müll. Arg.	Climber; No Record	Nigeria	Ekonikon	Leaves and stem bark	Infusion of the leaves and stem bark is applied topically.	Scabies, ringworm, and eczema	[15]
Euphorbiaceae	*Ricinus communis* L.	Shrub; LC	Nigeria	Eto kasto	Leaves and seeds	Infusion of the leaves; expression of the oil from the seeds.	Chickenpox smallpox, and skin spots	[15]
Fabaceae	*Abrus precatorius* L.	Climber; No Record	Nigeria	Nneminua (Akwa ibom); Oju ologbo (Yoruba)	Leaves	Juice from the crushed leaves is applied topically.	Skin spots	[15]
Fabaceae	*Afzelia africana* Sm.	Tree; VU	Nigeria	Eyin mbukpo	Stem bark	Sap produced from the crushed stem back is applied topically.	Leprosy, pimples, skin eruption, and wounds	[15]
Fabaceae	*Afzelia bella* Harms	Tree; LC	Nigeria	Enyin mbukpo	Leaves	Juice from the crushed leaves is applied topically.	Pimples	[15]
Fabaceae	*Aganope stuhlmannii* (Taub.) Adema	Tree; LC	Togo	Kpodougboou	Aerial part	Decoction is taken orally.	Aphthous ulcers	[34]
Fabaceae	*Albizia lebbeck* (L.) Benth.	Tree; LC	Nigeria	Ubam	Stem bark	Poultice.	Eczema and insect bites	[15]
Fabaceae	*Arachis hypogaea* L.	Herb; LC	Senegal	Gerte (wolof) Jamba katalig (mandingo)	Nut	Peanut oil mixed with powdered leaf of *A. digitata* is applied to wounds.	Burns	[52]
Fabaceae	*Baphia nitida* Lodd.	Tree; LC	Ghana and Nigeria	Afuo	Leaves	Juice from the crushed leaves is applied topically.	Boils, skin ulcers, and wounds	[15,40]
Fabaceae	*Burkea africana* Hook	Tree; LC	Togo	Tchangbali (Togo	Leaves	Decoction of the leaves is taken orally.	Aphthous ulcers	[34]

Table 1. *Cont.*

Family	Plant Species	Habit and Conservation Status	Country	Local Name	Plant Part(s) Used	Mode of Preparation	Ailment	References
Fabaceae	*Cajanus cajan* (L.) Huth	Herb; No Record	Nigeria and Togo	Nkoti (Akwa Ibom); Otili (Yoruba); Assongoyè (Togo	Seeds and whole plant	Seeds are ground into powder and applied topically for measles, smallpox, sores, skin ulcers, and skin spots in Nigeria. Decoction of the whole plants is taken orally for aphthous ulcers in Togo.	Measles, sores, skin ulcers, skin spots, and smallpox	[15,34]
Fabaceae	*Daniellia oliveri* (Rolfe) Hutch. & Dalziel	Tree; LC	Nigeria and Togo	Enan-eto (Akwa Ibom); Hemou (Togo)	Root, bark, and leaves	Sap from the crushed root bark is applied topically in Nigeria. Leaves and bark are macerated and taken orally for aphthous ulcers in Togo.	Aphthous ulcers and rashes	[15,34]
Fabaceae	*Detarium microcarpum* Guill. & Perr	Tree; LC	Nigeria and Togo	Kpayè (Togo)	Bark, leaves, and roots	Dried roots and leaves are ground into powder and applied externally for cuts, ulcers and wounds in Nigeria. Stem bark is macerated and taken orally for aphthous ulcers in Togo.	Aphthous ulcers and wounds	[15,34]
Fabaceae	*Distemonanthus benthamianus* Baill.	Tree; LC	Nigeria	Eto-afia	Root bark	Decoction of the root bark is taken orally.	Skin spot	[15]
Fabaceae	*Faidherbia albida* (Delile) A. Chev	Tree; LC	Gambia	Bubrick	Root	NA	Snakebites	[48]
Fabaceae	*Lonchocarpus cyanescens* (Schumach. & Thonn.) Benth.	Shrub; No Record	Nigeria	Awa	Leaves	Infusion of the leaves is applied topically.	Skin ulcers and skin spots	[15]
Fabaceae	*Lonchocarpus sericeus* (Poir.) Kunth	Tree; LC	Nigeria	Ipappo (Yoruba)	Leaves	Decoction of the leaves with *Vernonia macrocynus* O.Hoffm is taken orally.	Skin infections	[42]

Table 1. *Cont.*

Family	Plant Species	Habit and Conservation Status	Country	Local Name	Plant Part(s) Used	Mode of Preparation	Ailment	References
Fabaceae	*Parkia biglobosa* (Jacq.) R. Br. ex G. Don	Tree; LC	Nigeria and Togo	Ukon uyayak (Akwa Ibom); Igi iru (Yoruba); Soulou (Togo)	Stem bark	Dried stem bark is ground to powder for ringworms in Nigeria. Decoction of the stem bark is used externally for skin infection in Nigeria and taken orally for aphthous ulcers in Togo.	Aphthous ulcers, ringworm, and skin infection	[15,34]
Fabaceae	*Parkia clappertoniana* Keay	Tree; No Record	Nigeria and Ghana	Igba (Yoruba)	Leaves	Leaves are ground with *Loranthus* with potash and taken orally with pap (Nigeria); Extract from the husk is used for sores and wounds in Ghana.	Skin infections, sores, and wounds	[42,53]
Fabaceae	*Pentaclethra macrophylla* Benth	Tree; LC	Nigeria	Ukana	Stem bark	Decoction or infusion is used topically.	Skin spots	[15]
Fabaceae	*Piliostigma thonningii* (Schum.) Milne-Redh.	Tree; No Record	Republic of Benin, Nigeria, and Togo	Tilabaati (Benin); Pambakou (Togo) Abafe (Nigeria)	Root and root bark	Decoction of the root and fruit is taken orally for herpes sores and other skin diseases. The root bark is ground to powder and applied topically.	Herpes sores and skin disease	[16,34,42]
Fabaceae	*Pterocarpus erinaceus* Poir.	Tree; No Record	Nigeria and Togo	Ukpa (AkwaIbom, Nigeria); Tém (Togo)	Leaves and stem bark	Decoction of the leaves and stem bark is used externally for skin spots in Nigeria. Latex from the plant is applied topically for herpes sores and ringworm in Togo.	Herpes, ringworm sores, and skin spots	[15,34]
Fabaceae	*Pterocarpus santalinoides* L'Hér. ex DC.	Tree; No Record	Nigeria	Nkpa-inyan	Leaves	Decoction is used topically.	Skin spots	[15]
Fabaceae	*Senegalia ataxacantha* (DC.) Kyal. & Boatwr.	Tree; LC	Nigeria	Mbara okpok	Leaves	Juice from the crushed leaves is applied topically.	Burn and sores	[15]
Fabaceae	*Senna alata* L (Roxb	Shrub; LC	Nigeria	Asunwon (Yoruba); Akoria (Bennin)	Leaves and stem	Juice from the leaves and stem is applied externally.	Ringworms and skin spots	[42]

Table 1. *Cont.*

Family	Plant Species	Habit and Conservation Status	Country	Local Name	Plant Part(s) Used	Mode of Preparation	Ailment	References
Fabaceae	*Senna hirsuta* (L.) H.S.Irwin & Barneby	Herb; No Record	Cameroon	Tulushine	Leaves	Decoction of leaves is taken orally.	General skin diseases	[17]
Fabaceae	*Senna occidentalis* (L.) Link	Herb; LC	Nigeria	Flower uduk-ikot	Leaves	Juice from the crushed leaves is applied topically.	Abscess and chickenpox	[15]
Fabaceae	*Senna tora* (L.) Roxb.	Herb; No Record	Nigeria	Mfan udukikot	Leaves	Infusion of the leaves is applied topically.	Skin spots and sores	[15]
Fabaceae	*Tamarindus indica* L.	Tree; LC	Nigeria and Togo	Okukuk mbakara (Akwa Ibom, Nigeria); Nidié (Togo)	Leaves and root bark	Decoction of the root bark is used externally for bathing for skin spots in Nigeria. Decoction of the leaves is taken orally for aphthous ulcers in Togo.	Aphthous ulcers and skin spots	[15,34]
Fabaceae	*Tetrapleura tetraptera* (Schumach. & Thonn.) Taub.	Tree; LC	Nigeria	Uyayak (Akwa Ibom); Aidan (Yoruba)	Fruits	Oil from the expression of the fruit is used externally for skin spots.	Skin spots	[15]
Fabaceae	*Vachellia nilotica* (L.) P.J.H. Hurter & Mabb	Tree; LC	Senegal	Nep nep (wolof) Mbano (mandingo)	Root	Root infusion is applied topically.	Herpes	[52]
Fabaceae	*Zornia latifolia* Sm.	Herb; No Record	Nigeria	Ubok etikoriko	Leaves	Sap from the crushed leaves or dried leaves ground into powder is applied topically.	Snakebites and scorpion stings	[15]
Gentianaceae	*Anthocleista djalomensis* A. Chev.	Tree; LC	Nigeria	Ibu (Akwa Ibom); Sapo (Yoruba)	Stem bark	Sap from the crushed stem bark is used topically.	Skin spots, sores, ulcers, and wounds	[15]
Hypericaceae	*Harungana madagascariensis* (Lam.) ex Poir.	Tree; LC	Nigeria	Oton	Leaves, stem, root	Infusion of the leaves, stem, and root is used topically.	Skin spots	[15]
Hypericaceae	*Hypericum Lanceolatum* Lam.	Shrub; No Record	Cameroon	No record	Stem bark	No record.	Skin infections	[26]

Table 1. *Cont.*

Family	Plant Species	Habit and Conservation Status	Country	Local Name	Plant Part(s) Used	Mode of Preparation	Ailment	References
Hypericaceae	*Psorospermum febrifugum* Spach	Shrub; LC	Cameroon	No record	Stem bark and root	Decoction of the stem bark for skin sores in HIV/AIDS patients. Powdered root is used topically on parasitic skin diseases. It is used for pimples, eruptions, and wounds when ground up and mixed with oil.	Acne, leprosy, skin sores in HIV/AIDS patients, and skin infection.	[54,55]
Lamiaceae	*Clerodendrum splendens* G. Don	Climber; No Record	Mali and Nigeria	Mmon oyot adiaha ekiko	Leaves	Juice from the crushed leaves is applied topically.	Skin spots and snakebites	[15,29]
Lamiaceae	*Mesosphaerum suaveolens* (L.) Kuntze	Herb; No Record	Togo and Gambia	Pinbinè (Togo) Jammakarla (Gambia	Root	Root is macerated and taken orally for aphthous ulcers. Sap is applied to fresh cut.	Aphthous ulcers and fresh cuts	[34,48]
Lamiaceae	*Solenostemon monostachyus* (P. Beauv.) Briq	Herb; No Record	Nigeria	Ntorikwot	Leaves	Juice from the crushed leaves is added to water and applied topically.	Measles	[15]
Lamiaceae	*Vitex doniana* Sweet	Tree; LC	West Africa	Nkokoro	Root	Poultice.	Leprosy and wrinkles	[15,31,56]
Malvaceae	*Adansonia digitata* L.	Tree; No Record	Côte d'Ivoire, Gambia, and Nigeria and	Luru (Hausa) Buback (Gambia)	Leaves and fruits	Juice from the crushed leaves is applied topically. Fruit pulp is applied to body blisters.	Body blisters, scorpion stings, and snakebites	[15,48]
Malvaceae	*Bombax buonopozense* P. Beauv.	Tree; LC	Nigeria	Ukim	Stem bark	Infusion of the stem bark is applied externally.	Ringworm, rashes, and skin spots	[15]
Malvaceae	*Ceiba pentandra* (L.) Gaertn.	Tree; LC	Cote d'Ivoire and Nigeria	Akpu-ogwu (Igbo); Araba (Yoruba); Rimi (Hausa)	Stem bark	Decoction of the stem bark is used for bathing.	Leprosy, sores, and skin ulcers	[15,28]
Malvaceae	*Glyphaea brevis* (Spreng.) Monach	Shrub; LC	Nigeria	Ndodiro	Leaves	Poultice.	Burns and wounds	[15]
Malvaceae	*Gossypium hirsutum* L.	Shrub; VU	Nigeria	Eto-ofo	Leaves	Juice from the crushed leaves is applied topically.	Sores, skin eruption, and wounds	[15]

Table 1. *Cont.*

Family	Plant Species	Habit and Conservation Status	Country	Local Name	Plant Part(s) Used	Mode of Preparation	Ailment	References
Malvaceae	*Sterculia tragacantha* Lindl.	Tree; LC	Nigeria	Udot eto	Stem bark	Sap from the crushed stem bark is applied topically.	Boil, skin ulcers, and wounds	[15]
Malvaceae	*Triumfetta cordifolia* Guill., Perr. & A. Rich	Shrub; No Record	Nigeria	Nkibbe ubuk	Leaves	Juice from the crushed leaves is applied topically.	Skin spots	[15]
Marantaceae	*Thaumatococcus danielli* Benth	Herb; No Record	Nigeria	Ewe iran	Leaves	Powder from the dried leaves is mixed with oil and applied to the affected area.	Skin infections	[57]
Melastomataceae	*Heterotis rotundifolia* (Sm.) Jacq.-Fél.	Shrub; LC	Nigeria	Nyie ndan	Whole plant	Decoction of the whole plant is used externally for bathing.	Measles	[15]
Meliaceae	*Azadirachta indica* A. Juss.	Tree; LC	Ghana and Nigeria	Ibok utoenyin	Leaves, stem bark, root	Infusion is used topically.	Eczema, ringworm, skin spots, and scabies	[15,40]
Meliaceae	*Carapa procera* DC.	Tree; LC	Ghana, Guinea, and Nigeria	Mkpono ubom	Seeds	Oil from the crushed seed is used externally.	Burns, insect bites, and scabies	[40,58]
Meliaceae	*Khaya grandifoliola* A. Juss.	Tree; VU	Nigeria	Odala (Igbo), Oganwo (Yoruba)	Stem bark	Decoction is used topically.	Skin spots	[15]
Meliaceae	*Pseudocedrela kotschyii* (Schweinf.) Harms	Tree; LC	Togo	Helitétéwiyé	Root	Root maceration is taken orally.	Aphthous ulcers	[34]
Menispermaceae	*Chasmanthera dependens* Hochst	Climber; No Record	Republic of Benin	Boborou	Stem	Ground stem is applied topically or mixed with shea butter.	Wounds	[16]
Moraceae	*Afromorus mesozygia* (Stapf) E.M. Gardener	Tree; No Record	Cameroon	No record	Roots, stem, and leaves	No record.	Dermatitis	[26]
Moraceae	*Artocarpus altilis* (Parkinson) Fosberg	Tree; No Record	Cameroon	No record	Roots	No record.	Abscesses, boils, and skin infections	[26]
Moraceae	*Ficus sycomorus* L.	Tree; LC	Nigeria	Sikamo	Root	Poultice.	Snakebites	[15]

Table 1. *Cont.*

Family	Plant Species	Habit and Conservation Status	Country	Local Name	Plant Part(s) Used	Mode of Preparation	Ailment	References
Moraceae	*Ficus carica* L.	Tree; LC	Nigeria	Ukimo	Stem bark	Dried stem bark is ground to powder and applied to the wounds.	Wounds	[15]
Moraceae	*Ficus exasperata* Vahl	Tree; LC	Nigeria and Togo	Ukuok (Akwa ibom); Eepin (Yoruba); Laalayou (Togo)	Leaves and root	Sap from the crushed root is used externally for ringworms in Nigeria. Decoction of the leaves is taken orally for aphthous ulcers in Togo.	Ringworm	[15,34]
Moraceae	*Ficus ingens* (Miq.) Miq	Shrub; LC	Republic of Benin	Dekuru sanni	Root bark	The root bark is ground to powder and applied topically.	Wounds	[16]
Moraceae	*Ficus thonningii* Blume	Tree; LC	Benin republic	Kudoro	Roots	Adventitious roots of *F. thonningii* and bark of the root of *Newbouldia laevis* are ground into powder and applied topically.	Wounds	[16]
Moraceae	*Treculia obovoidea* N.E.Br	Tree; LC	Cameron	No record	Twigs	No record.	Skin disease	[26]
Myristicaceae	*Pycnanthus angolensis* (Welw.) Warb.	Tree; LC	Cameroon	No record	Stem bark	No record.	Fungal skin infection	[26]
Myrtaceae	*Eugenia uniflora* L.	Tree; LC	Nigeria	No record	Leaves	Decoction of the leaves is taken orally.	Skin spots	[15]
Ochnaceae	*Lophira lanceolata* Tiegh. ex Keay	Tree; LC	Nigeria and Togo	Tabsomang (Togo)	Leaves, root bark and stem	Decoction of the leaves and root back for chickenpox, fungal skin infection, and wounds in Nigeria. Stem is rubbed directly on the herpes sore in Togo.	Chicken pox, herpes sores, and skin infections	[15,34]
Ochnaceae	*Ochna rhizomatosa* (Tiegh.) Keay	Shrub; No Record	Republic of Benin	Yinkpenoka	Root bark	The root bark is ground to powder and applied topically for wounds.	Wounds	[16]
Ochnaceae	*Ochna schweinfurthii* F. Hoffm.	Shrub; LC	Republic of Benin	Yinkpenoka	Root bark	The root bark is ground to powder and applied topically for wounds.	Wounds	[16]

Table 1. *Cont.*

Family	Plant Species	Habit and Conservation Status	Country	Local Name	Plant Part(s) Used	Mode of Preparation	Ailment	References
Olacaceae	*Coula edulis* Baill.	Tree; LC	Cameron	No record	stem bark	No record.	Skin disease	[26]
Phyllanthaceae	*Bridelia ferruginea* Benth.	Tree; LC	Nigeria	Udia afua	Stem bark	Infusion is used topically.	Fungal skin infections and wounds	[15]
Phyllanthaceae	*Flueggea virosa*, (Roxb. ex Willd.) Royle	Shrub; LC	Republic of Benin	Opanko (Benin), Tchaakatchaka (Togo)	Root bark and aerial part	Root bark is incinerated and applied topically to wounds. Decoction is taken orally for aphthous ulcers.	Aphthous ulcers and wounds	[16,34]
Phyllanthaceae	*Hymenocardia acida* Tul.	Shrub; LC	Togo, Republic of Benin	KpaiKpai (Togo), Sinkakakou (Benin)	Leaves, Root bark	Decoction of the leaves is taken orally for aphthous ulcers. Root bark is ground to powder and applied topically to wounds.	Aphthous ulcers and wounds	[16,34]
Phyllanthaceae	*Maesobotrya barteri* (Baill.) Hutch	Tree; LC	Nigeria	Nnyanyatet	Root	Sap from the crushed root is applied externally.	Skin spots	[15]
Phyllanthaceae	*Maesobotrya dusenii* (Pax) Pax	Tree; No Record	Nigeria	Nnyanyatet	Root	Sap from the crushed root is applied externally.	Skin spots	[15]
Phyllanthaceae	*Phyllanthus amarus* Schumach. & Thonn.	Herb; No Record	Nigeria	Oyomokiso	Whole plant	Decoction of the whole plant is taken orally and for bathing.	Skin spots	[15]
Phyllanthaceae	*Phyllanthus pentandrus* Schum. And Thonn	Shrub; No Record	Nigeria	Ehin olobe	Leaves and fruit husks	The dried leaves are ground with *Vigna* plant, and the powder is then mixed with shea butter; the ointment is applied to boils.	Boil	[42]
Phyllanthaceae	*Uapaca togoensis* Pax	Tree; LC	Côte d'Ivoire	No record	Root and stem bark	Preparation from the root and stem bark.	Leprosy and skin diseases	[59]
Poaceae	*Andropogon gayanus* Kunth	Herb; No Record	Nigeria	Mbokko ekpo	Leaves	Dried leaves are ground to powder and used topically.	Wounds	[15]
Poaceae	*Imperata cylindrica* (L.) Raeusch.	Herb; LC	Nigeria	Ndan inwan	Rhizome	Rhuzome is crushed and applied topically.	Abscess and scorpion sting	[15]

Table 1. *Cont.*

Family	Plant Species	Habit and Conservation Status	Country	Local Name	Plant Part(s) Used	Mode of Preparation	Ailment	References
Poaceae	*Pennisetum polystachion* (L.) Schult.	Herb; LC	Nigeria	Nwan-mbakara	Shoot	Dried shoots are ground into powder and applied to the wounds.	Wounds	[15]
Poaceae	*Rottboellia cochinchinensis* (Lour.) Clayton	Herb; No Record	Nigeria	Mbokko enan ikot	Whole plant	Decoction of the whole plant is for bathing.	Measles	[15]
Polygalaceae	*Carpolobia lutea* G. Don	Shrub; LC	Nigeria	Ikpafun	Leaves	Decoction of the leaves is taken orally.	Skin spots	[15]
Polygalaceae	*Portulaca oleracea* L.	Herb; LC	Ghana and Nigeria	Uton ekpu	Whole plant	Decoction of the whole plant is used for bathing.	Dermatitis and skin spots	[15]
Rubiaceae	*Borreria verticillata* (L.) G. Mey.	Herb; No Record	Nigeria	No record	Leaves	Juice from the crushed leaves is applied topically.	Eczema and skin spots	[15]
Rubiaceae	*Chassalia kolly* (Schumach.) Hepper	Shrub; No Record	Togo	Tiyah (Togo)	Roots	Paste made from the roots is applied topically.	Ringworm	[34]
Rubiaceae	*Crossopteryx febrifuga* (Afzel. ex G.Don) Bents	Tree; LC	Republic of Benin	Otoupedou	Root bark	The root bark is ground to powder and applied topically.	Wounds	[16]
Rubiaceae	*Diodia sarmentosa* Sw.	Climber; No Record	Nigeria	No record	Leaves	Decoction of leaves is used topically.	Skin spots	[15]
Rubiaceae	*Gardenia ternifolia* schumach. & Thonn	Shrub; LC	Republic of Benin and Togo	Keyabouaka (Benin); Kaou (Togo)	Root and root bark	Root decoction is applied topically for ringworms in Togo. Root bark is incinerated and mixed with palm kernel oil and used topically for wounds in Benin.	Ringworm and wounds	[16,34]
Rubiaceae	*Morinda longiflora* G. Don	Climber; No Record	Nigeria	No record	Leaves	Infusion of the leaves in water and used externally.	Scabies	[15]
Rubiaceae	*Nauclea latifolia* Sm.	Tree; LC	Nigeria	No record	Leaves	Juice from the crushed leaves is applied externally.	Rashes	[15]
Rubiaceae	*Sarcocephalus latifolius* (Sm.) E.A.Bruce.	Tree; No Record	Togo	Kayou (Togo)	Root	Decoction of the root is taken orally.	Aphthous ulcers	[34]
Rutaceae	*Clausena anisata* (Willd.) Hook. f. ex Benth.	Shrub; LC	Nigeria	Mbiet ekpene	Stem bark	Decoction of the stem bark is taken orally.	Measles	[15]

Table 1. *Cont.*

Family	Plant Species	Habit and Conservation Status	Country	Local Name	Plant Part(s) Used	Mode of Preparation	Ailment	References
Rutaceae	*Zanthoxylum gilletii* (De Wild.) Waterman	Tree; LC	Ghana and Nigeria	Nkek	Root bark	Sap from the crushed root back is applied topically.	Boil	[15]
Rutaceae	*Zanthoxylum zanthoxyloides* (Lam.) Zepern. & Timler	Tree; LC	Togo	Kolgragu	Aerial part	Maceration is taken orally.	Aphthous ulcers	[34]
Salicaceae	*Homalium letestui* Pellegr.	Tree; No Record	Nigeria	Oton idim	Leaves	Decoction or infusion of the leaves is taken orally.	Skin spots	[15]
Sapindaceae	*Blighia sapida* K.D. Koenig	Tree; LC	Benin Republic, Ghana, Nigeria, and Togo	Ishin (Yoruba), Kpiziyè (Togo)	Leaves, fruit, and stem bark	Pounded bark is taken as an antidote to snake and scorpion bites. Decoctions of bark or fruit walls are applied to wounds. The ground-up leaves and salts are applied as a paste to treat yaws and ulcers. A calcinate made from the fruit or bark is applied externally for aphthous ulcers and herpes sores in Togo.	Aphthous ulcers, herpes sores, snakebites, scorpion bites, and wounds	[34,40,55]
Sapindaceae	*Paullinia pinnata* L.	Tree; No Record	Republic of Benin and Togo	Dikitinintibou (Benin); Adjandj kpouzou (Togo)	Aerial Part and root bark	Decoction of the aerial part is taken orally for aphthous ulcers in Togo. Root bark is ground to powder and applied topically for wounds.	Aphthous ulcers and wounds	[16,34]
Sapotaceae	*Vitellaria paradoxa* C.F. Gaertn	Tree; VU	Republic of Benin	Somou	Leaf	Leaves powder is mixed with butter and applied topically.	Wounds	[16]
Simaroubaceae	*Hannoa undulata* (Guill. & Perr.) Planch	Tree; No Record	Republic of Benin	Okoupopode	Root bark	The root bark is ground and applied topically.	Wounds	[16]
Smilacaceae	*Smilax anceps* Willd.	Climber; No Record	Nigeria	No record	Leaves, twigs	Decoction is applied topically.	Skin spot	[55]
Solanaceae	*Datura metel* L.	Herb; No Record	Nigeria	Nnya ekpo	Leaves	Juice from the crushed leaves is applied topically.	Insect bites and scorpion stings	[15]
Solanaceae	*Nicotiana rustica* L.	Herb; No Record	Nigeria	No record	Leaves	Poultice.	Skin ulcers, skin cancer, and wounds	[15]

Table 1. *Cont.*

Family	Plant Species	Habit and Conservation Status	Country	Local Name	Plant Part(s) Used	Mode of Preparation	Ailment	References
Solanaceae	*Nicotiana tabacum* L.	Herb; No Record	Ghana and Nigeria	Ewe taaba (Yoruba)	Leaves	Poultice.	Skin ulcers, skin cancer, and wounds	[15,40]
Ximeniaceae	*Ximenia americana* L.	Shrub; LC	Nigeria	No record	Root	Sap from the crushed root is applied topically.	Skin ulcers, rashes, and ringworm	[15]
Zingiberaceae	*Aframomum melegueta* (Roscoe) K.Schum	Herb; DD	Republic of Benin	Fètcharinanfè	Seed	The dried powder of the seed is applied topically.	Wounds	[16]

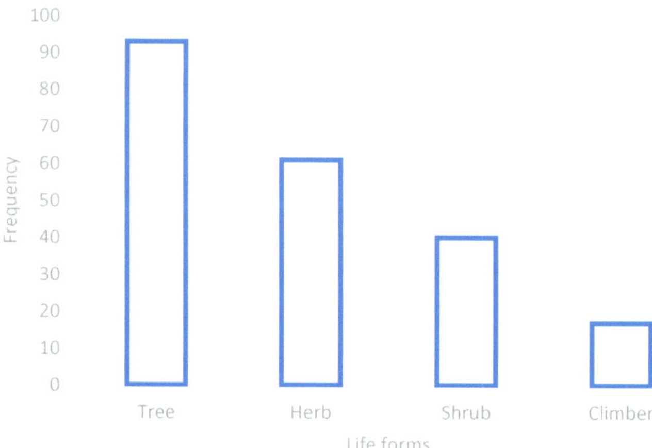

Figure 3. Life forms of the reported plants.

Regarding the plant parts or organs used in formulating or preparing various herbal medicines, leaves were more frequently used (107) than any other plant. Roots (55) and stems (39) were next, while all other plant parts had less than 20 use reports (Figure 4). The use of leaves in herbal remedies has always been highly reported, and the literature suggests it may be due to their ease of harvest [64]. It is also possible that leaves are especially well used for skin diseases because other reports on skin diseases [10,11] documented similar findings. However, the relatively high frequency of roots after the leaves could be related to the high prevalence of woody species implicated in this study. Also, the roots of plants are believed to contain a high quantity of phytochemicals because of their role in the absorption of nutrients [64]. Regrettably, harvesting medicinal plants' roots for medicinal purposes poses the greatest conservation risk to plants because they are difficult to regenerate. Considering the mode of preparation, decoction (73) was the most preferred method, followed by crushing/juicing (50), and powdering, grinding, or pounding (39) (Figure 5). The decoction method has been reported as the most common method of preparation of medicinal plants around the world [17]. However, there is no clear scientific reason why the method is widely used. The use of the method of crushing or juicing in this review may be attributed to the nature of the disease, as it is the easiest and quickest method to apply the extracts from the plant topically to the skin. Figure 6 shows the number of plants used to treat different skin diseases; wounds, being the highest, are treated with 65 plants. Skin ulcers are treated with 46 plants and skin spots with 58. The pain associated with wounds and ulcers may account for the observed figures. As for skin spots, it is general knowledge that many people love spotless skin.

An assessment of the recorded plants' conservation statuses revealed a lack of conservation status data for 43% of the recorded plants. Among the plants with conservation records, 113 have a conservation status of least concern (LC). Six plants (*Afrofittonia silvestris, Afzelia africana, Allanblackia floribunda, Gossypium hirsutum, Khaya grandifoliola,* and *Vitellaria paradoxa*) have a conservation status of vulnerable (VU), while two plants (*Aframomum melegueta* and *Mangifera indica*) have a conservation status of data deficient (DD). It was observed in this paper that most of the plants with no conservation assessments are herbs and climbers, implying that most conservation efforts have been centered on tree species. This highlights the need to intensify conservation studies on herbs and climbers that are used medicinally as they are also prone to extinction, like trees [65].

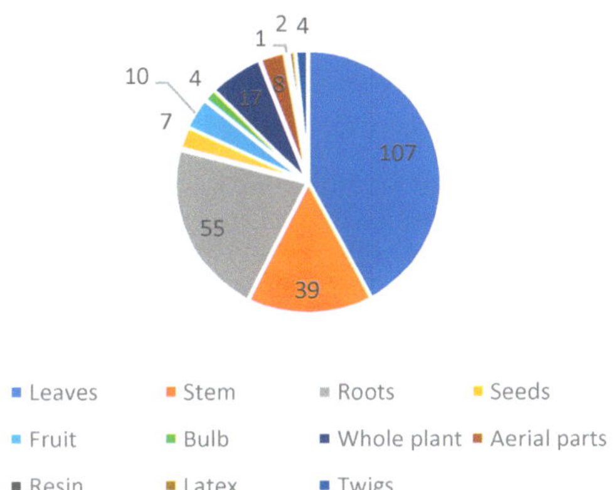

Figure 4. Parts of plants used to make herbal remedies.

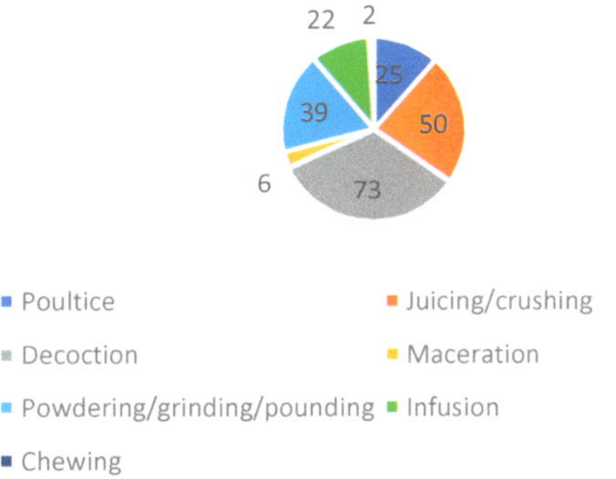

Figure 5. Mode of preparation.

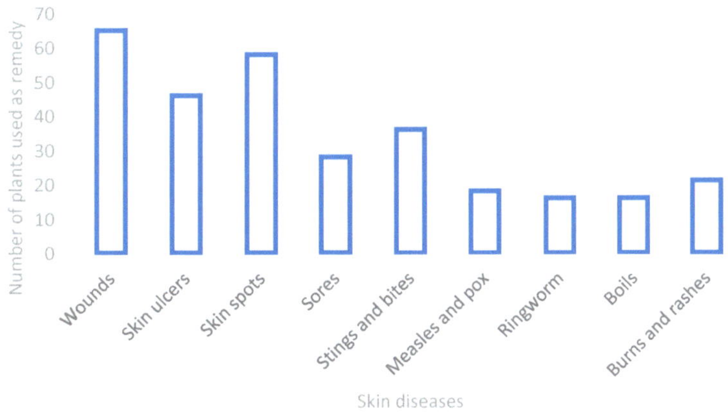

Figure 6. Number of plants used in treating different skin diseases.

3.3. Biological Activities of the Recorded Plants

Millions of beneficial microorganisms and pathogens inhabit the skin surface, and their imbalance or breaking of the skin could cause skin diseases [3,5]. Skin diseases recorded in this review and treated with medicinal plants include abscesses, athlete's foot, boils, measles, skin spots, as well as ulcers, whitlows, wounds, and many other skin infections. Due to the diversity and nature of these diseases, medicinal plants employed in treating skin disorders should possess pharmacological properties such as antibacterial, antifungal, antioxidant, anti-inflammatory, and wound-healing activities. Still, the review revealed that a large percentage of the studies have focused on the antibacterial and antifungal studies of the plants used in the treatment of skin diseases. In contrast, only a few studies explored other biological activities. However, it should be noted that researchers have addressed, to a large extent, the recommendations from a previous study [10] by studying the response of some neglected bacteria to botanicals used in skin diseases. Out of the 211 plant species recorded in this review, the biological activities of 82 plants have been assessed (Table 2), while over 60% of the plants are yet to be evaluated for any activities related to skin diseases, highlighting the wide gap in research into the biological activities of traditionally used medicinal plants.

Most of the studies in this review examined the antibacterial and antifungal activities of the plants using different types of dermatophytes. Some of the most commonly chosen skin pathogens include the bacteria *Staphylococcus aureus*, *Epidermophyton floccossum*, *Bacillus subtilis*, and *Pseudomonas aeruginosa*, while species of fungi in the genus *Candida*, especially *C. albicans*, were the most commonly tested fungi. Biological activities are a means of validating the efficacy of the traditional uses of plants. However, some of the activities recorded in Table 2 may not necessarily validate the use of the plants against skin diseases because the plant parts reportedly used in traditional medicine are different from those tested, and it is common knowledge that the type and quantity of the phytochemicals accumulated by different plant parts may be widely different. For example, the leaf juice of *Achyranthes aspera* is folklorically used to treat skin ulcers [15], but Gupta et al. [70] evaluated the antimicrobial activities of its root and stem. Likewise, the antimicrobial activities of the stem bark of *Lannea acida* were evaluated [133], but ethnobotanical records showed the use of its leaves in skin diseases [15,25]. Some of the most active antibacterial and antifungal plants based on their low MIC include the aerial part of *Brillantaisia lamium* (6.25 µg/mL), the essential oil of the seeds of *Monodora myristica* (8 µg/mL), the leaves of *Flueggea virosa* (8 µg/mL), the leaves of *Dacryodes edulis* (12.5 µg/mL), the essential oil from the aerial parts of *Ageratum conyzoides* (64 µg/mL), the aerial parts of *Clerodendrum splendens* (64 µg/mL), and the leaves of *Pistia stratiotes* (125 µg/mL). As regards the range of microorganisms susceptible to the extracts, alcohol extracts of *Aloe vera* leaf gel inhibited 115 skin pathogens [81]. Conversely, some plants were completely inactive, while others had very high MICs, thus making them inactive. For example, the aqueous extract from leaves of *Lannea microcarpa* was not active in all the skin pathogens tested [134], while the leaves of *Sansevieria liberica* and the roots of *Hannoa undulata* had a MIC value of 62.5 mg/mL each [127,151]. Some of these studies have confirmed or validated the ethnobotanical use of the plants, while others did not support their use. This may be due to the solvent extract used for the biological activities. It is also common practice to prepare herbal remedies using more than one plant for a single ailment for synergistic purposes or to treat the disease and symptoms [64]. Therefore, it is important to test the biological activities of medicinal plants following the traditional method of preparation and administration.

Table 2. Biological activities of the plants used against skin diseases in West Africa.

Plant Species	Plant Part(s) Used	Biological Activities/Extract or Isolated Compound Used	Results	Active Metabolites	References
Acalypha wilkesiana	Leaf	Aqueous	The extract inhibited *C. albicans*, *E. coli*, *P. aeruginosa*, *P. vulgaris*, *S. aureus* with zones of inhibition between 12 and 19 mm.	Alkaloids, glycoside, terpenes, 15-hydroxy pentadecanoic acid, 1,2,3-propanetriyl ester, and 9-octadecanoic acid	[66]
Acanthospermum Hispidum	Whole plant	Antifungal/Ethanol	The extract inhibited pathogenic fungi, namely, *C. albicans*, *M. luteus*, *M. roseus*, *P. aeruginosa*, and *S. aureus*, with zones of inhibitions between 12–18 mm.	Acanthospermal B	[67]
Acanthus montanus	Root Leaf	Anti-inflammatory/Aqueous Antifungal/Methanol	The extract significantly inhibited acute oedema in the mouse ear and the rat paw at $p < 0.05$. *Extract completely* inhibited dermatophytic fungi, namely, *Cladosporium* sp., *Fusarium* sp., *Trichophyton mentagrophyte*, *T. rubrum* and *T. soudanense* at 100 mg/mL.	Carbohydrates, alkaloids, tannins, glycosides, carbohydrates, flavonoids, and steroids.	[68,69]
Achyranthes aspera	Root and stem	Antifungal/Aqueous Methanol	Methanolic, aqueous root extract and methanolic extract of the stem were effective against *Epidermophyton floccossum*.	Alkaloids, carbohydrates, flavonoids, proteins, amino acids, tannins, phenols, steroids, glycosides and saponins	[70,71]
Aframomum melegueta	Seed	Antifungal/Ethanol	Ethanol extract was active against a range of dermatophytes.	Flavonoids, phenols, tannins, saponins, and terpenoids	[72]
Afzelia africana	Leaf	Antibacterial/Methanol	*Streptococcus pyogenes* was significantly susceptible to the extract with 25.0 mm zone of inhibition.	Alkaloids, anthraquinones, flavonoids, phenolics, glycosides, terpenoids, and steroids	[73,74]
Ageratum conyzoides	Aerial part	Antimicrobial/Essential oil	Essential oil was active against *E. coli*, *E. faecalis*, and *S. aureus* with zones of inhibition between 6.7 to 12.7 mm and MIC values between 64 to 256 µg/mL.	β-caryophyllene, β-copaene, α-calacorene, and 1,10-di-epi-cubenol	[75]

Table 2. *Cont.*

Plant Species	Plant Part(s) Used	Biological Activities/Extract or Isolated Compound Used	Results	Active Metabolites	References
Alchornea cordifolia	Leaf	Antimicrobial and wound healing/Aqueous ethanol	Both extracts demonstrated good activities against *C. albicans*, *B. subtilis*, *E. coli*, *P. aeruginosa* and *S. aureus*, with MIC values between 2.5–10.0 mg/mL. Both extracts also elicited significant wound-healing capacity on day 1 and day 9.	Anthraquinones, cardiac glycosides, flavonoids, saponins, sterols, and tannins	[76]
Allanblackia floribunda	Leaf, bark, stem bark, and root	Antibacterial/Butanol, chloroform, ethyl acetate, ethanol, and n-hexane	The extracts displaced activity against *A. flavus*, *B. subtilis*, *C. albicans*, *E. coli*, *S. aureus*, *P. vulgaris*, *P. aeruginosa* with zones of inhibition between 5 and 35 mm.	Alkaloids, anthraquinones, cardiac glycosides, flavonoids, saponins, tannins, and terpenoids	[77,78]
Allium sativum	Bulb	Antifungal/Aqueous	The extract had a significant antifungal effect on *T. rubrum*.	Allicin, alkaloids, and tannins	[79,80]
Aloe vera	Leaf and Gel	Antimicrobial/Ethanol	The gel extracts showed activity against 115 g-positive and gram-negative skin pathogens, while the leaf extracts showed no such activity.	Flavonoids, tannins, terpenoids, and saponins.	[81,82]
Alstonia boonei	Stem bark Root bark	Antibacterial/Methanol Wound healing/ Alkaloid extract	*Streptococcus pyogenes* was significantly susceptible to extract with zone of inhibition of 25.0 mm. The total alkaloid extract increased the wound contraction and decreased the epithelialization period.	Alkaloids, cyanogenetic glycosides, flavonoids, terpenoids, saponins, and steroids	[73,83]
Alternanthera bettzickiana	Aerial part	Antifungal/Aqueous and methanol	Aqueous and methanol extracts inhibited *Candida albicans* and *E. floccossum* inhibition zone of 9–11 mm.	Alkaloids, anthocyanin, diterpenes, leucoanthocyanin, saponins, phenols, diterpenes, tannin, terpenoids, steroids, and xanthoprotein,	[84]

Table 2. *Cont.*

Plant Species	Plant Part(s) Used	Biological Activities/Extract or Isolated Compound Used	Results	Active Metabolites	References
Amaranthus caudatus	Whole plant	Antifungal/Dichloromethane ethyl acetate, hexane, and methanol	The extracts were active against *C. albicans*.	Flavonoids, steroids, terpenoids, and cardiac glycosides	[85]
Anacardium occidentale	Fruit and stem bark	Antimicrobial/Water: Ethanol	Growth inhibition of *S. aureus* and *S. epidermidis* by the extracts.	Tannin and flavonoids	[19,86]
Anchomanes difformis	Leaf and Tuber	Antibacterial/Ethanol	Leaf extract was effective on *P. aeruginosa* with an inhibition zone of 21 mm; tuber extract inhibited *B. subtilis* and *P. aeruginosa* with zones of inhibitions of 35 and 29 mm, respectively.	Alkaloids, tannins, and saponins	[87]
Annickia chlorantha	Stem bark	Antifungal/Alkaloid fraction	The fraction and its ointment formulation demonstrated antifungal activities against *Candida* spp. and dermatophytes.	Alkaloids, flavonoids, glycosides, tannins, and saponins	[88,89]
Anogeisus leiocarpus	Leaf, stem bark, and root bark	Antibacterial/Ethanol	Individual extract and combination displayed strong activities against *E. coli S. aureus*, and *P. aeruginosa* with a zone of inhibition up to 15.80 mm.	Anthraquinones, alkaloids phenols, tannins, and ellagic acids	[90]
Anona senegalensis	Leaves	Antifungal/Aqueous and methanol	Aqueous and methanol were active against *Trichoderma* spp. with zones of inhibition of 14.5 mm and 8.3 mm respectively.	Alkaloids, flavonoids, and polyphenols	[91]
Anthocleista djalonensis	Root	Antibacterial/Methanol	The extract inhibited *E. coli, S. aureus*, and *Shigella* sp. with inhibition zones between 14 and 20 mm. The extracts also elicited significant wound healing in vivo.	Alkaloids, flavonoids, and volatile oils	[92]

Table 2. *Cont.*

Plant Species	Plant Part(s) Used	Biological Activities/Extract or Isolated Compound Used	Results	Active Metabolites	References
Aspilia africana	Leaf and root	Antimicrobial/Methanol	Both extracts were active against *B. substilis, E. coli, P. aeruginosa,* and *S. aureus,* with 3–6 mm inhibition zones.	Cardiac glycosides, flavonoids, saponins, tannins, and terpenoids	[93,94]
Azadirachta indica	Leaf	Antibacterial/Ethanol	The extract inhibited E. coli and S. aureus with a maximum inhibition zone of 16 mm.	β-sitosterol, flavonoid, lupeol, ferulic acid, and quercetin	[95]
Baphia nitida	Leaf and root	Antimicrobial/Ethanol	Extracts inhibited *C. albicans, E. coli, S. aureus, P. aeruginosa,* and *B. subtilis,* with a zone of inhibition up to 35 mm.	Flavonoids, glycosides, saponins, sterols, tannins	[96]
Bidens pilosa	Leaf	Antimicrobial/Ethanol	The ethanol extract was active against *E. coli, P. aeruginosa, S. pyogenes, S. aureus,* and *P. aeruginosa* inhibition zones between 2–5 mm.	Alkaloids, cardiac glycosides, flavonoids, and tannins	[97]
Brillantaisia Lamium	Aerial part	Antimicrobial activity/Dichloromethane and methanol (1:1)	Extract, fraction and lespedin, the isolated compound demonstrated antimicrobial activities against pathogenic bacterial and fungi, namely, *Candida tropicalis, Cryptococcus neoformans, Staphylococcus aureus* and *Enterococcus faecalis.*	β-sitosterol, sitosterol 3-O-β-D-glucopyranoside Aurantiamide acetate, campesterol, lespedin lupeol, and stigmasterol	[98]
Calotropis procera	Leaves	Antifungal/Ethanol	Extracts inhibited the growth of pathogenic fungi, namely, *Epidermophyton floccosum, M. gypreum, M. canis, T. mentagrophytes,* and *T. rubrum,* with MIC values between 250 and 1000 μg/mL.	Camphene, dodecanoic acid, linolenic acid, and thebaine	[99,100]

Table 2. *Cont.*

Plant Species	Plant Part(s) Used	Biological Activities/Extract or Isolated Compound Used	Results	Active Metabolites	References
Carapa procera	Leaf	Antibacterial and wound healing/Ethanol	The extract inhibited *E. coli* and *S. aureus* with MIC values between 2.5 and 5 mg/mL. The extract also demonstrated a significant wound-healing activity in rats.	Saponins, steroids, and tannins	[101]
Carica papaya	Flower	Antibacterial/Methanol	The flower extract inhibited *B. subtilis* and *E. coli* with the zones of inhibition between 10 to 40 mm.	Alkaloids, flavonoids, phenolics, saponins, and tannins	[102]
Chassalia kolly	Leaf	*Antioxidant and anti-inflammatory*/Ethanol	The extract elicited excellent antioxidant activity with IC_{50} = 0.05 µg/µL) and showed higher anti-inflammatory activities than aspirin.	Anthocyanins, flavonoids, and terpenes	[103]
Chromolaena odorata	Leaf, stem and root	Antimicrobial/Ethanol, hexane, and methanol	All the extracts inhibited the activities of pathogenic bacteria strains, such as *B. cereus*, *E. faecalis*, *S. epidermidis*, and *P. vulgaris*.	Alkaloids, aurone, chalcone, flavone flavonol, phytates, and tannins	[104,105]
Citrullus colocynthis	Fruit, seed and root	Antimicrobial/Ethanol	Extracts were active on pathogens, causing skin infections with zones of inhibition between 10 and 22 mm.	Alkaloids, flavonoids, glycosides, saponins, and tannins	[106,107]
Clerodendrum splendens	Aerial part	Antimicrobial and wound healing/Methanol	The extract was active against various bacteria and fungi with MIC values between 64 and 512 µg/mL. The extract also elicited wound-healing capacity by increasing wound epithelization, scar area and tensile strength.	D-glucopyranoside of (22E, 24S) and stigmasta-5,22,25-trien-3β-ol (3)	[108,109]
Colocasia esculenta	Leaf and tuber	Antibacterial/Methanol	Both extracts inhibited the growth of *E. coli*, *P. aeruginosa*, *P. mirabilis*, *S. aureus* with zones of inhibition between 4–30 mm.	10-fluoro trimethyl ester, 12,15-octadecatrienoic acid, decanoic acid, n-hexadecanoic acid, and pentadecanoic acid.	[110]

Table 2. *Cont.*

Plant Species	Plant Part(s) Used	Biological Activities/Extract or Isolated Compound Used	Results	Active Metabolites	References
Combretum collinum	Leaf	Antibacterial/Ethanol	The extract displayed activities against *S. epidermidis*, Methicillin-resistant *Staphylococcus aureus* and *S. aureus* with MIC values between 275.0 µg/mL and 385.5 µg/mL.	Myricetin-3-*O*-glucoside and myricetin-3-*O*-rhamnoside	[111]
Crinum jagus	Bulb	Antimicrobial/Methanol	The extract at 100 mg/mL inhibited *C. albicans*, *S. aureus* and *B. subtilis* with zones of inhibition of 14, 21, 25 mm, respectively.	Alkaloids, catechin, tannins, flavonoids, saponins, and triterpenes	[18,112]
Cyanthillium cinereum	Leaf	Antibacterial/Methanol	The extract inhibited the activity of *E. coli* and *S. aureus* with zones of inhibition of 21 mm and 19 mm, respectively.	Alkaloids, flavonoids, phenol, and terpenoids	[113,114]
Cyathula prostrata	Whole plant	Anti-inflammatory/Ethyl acetate	The extract inhibited carrageenan, arachidonic acid, and xylene-induced tests.	Flavonoids, phenols, cardiac glycosides, and terpenes	[115]
Dacryodes edulis	Leaf	Antibacterial/Ethanol	Extract elicited significant activities against *B. cereus*, *E. coli*, *S. aureus*, *P. aeruginosa*, with the zone of inhibition between 8 and 13 mm and MIC values between 12.5 and 250 µg/mL.	Ethylgallate and quercitrin	[116]
Daniellia oliveri	Leaf and stem bark	Antifungal/Aqueous and methanol	Extracts were active against several species of fungi, especially the *Candida* species.	Alkaloids, anthraquinones, flavonoids, and saponins	[117]
Dioscorea dumetorum	Tuber	Antifungal/Methanol	Extracts were active against *Aspergillus niger* and C. albicans.	Alkaloids, flavonoids, phenols, saponins, and tannin	[118]
Elaeis guineensis	Leaf	Antibacterial/Methanol	Extracts showed potent activities against *B. cereus*, *E. coli*, and *P. aeruginosa* with inhibition zones of 7.7–11.3 mm.	Alkaloids, coumarins, flavonoids, saponins, tannins, and terpenoids	[119]

Table 2. *Cont.*

Plant Species	Plant Part(s) Used	Biological Activities/Extract or Isolated Compound Used	Results	Active Metabolites	References
Emilia coccinea	Leaf	Antimicrobial/Methanol	The extract was active against *A. niger, C. albicans, E. coli, S. aureus, B. subtilis,* and *P. aeruginosa* with MIC value between 5–25 mg/mL and inhibition zone up to 22 mm.	Alkaloids, flavonoids, oxalate, tannins, phenols, and terpenoids	[120,121]
Euphorbia hirta	Leaf and stem	Antibacterial/ Methanol aqueous	Extracts inhibited *E. coli, P. mirabilis, P. aeruginosa* and *S. aureus.*	Anthraquinones, alkaloids, flavonoids, and terpenoids	[122]
Ficus thonningii	Stem bark	Antibacterial/Methanol	The extract inhibited a range of skin pathogens, such as *P. aeruginosa* and *S. pyogenes*, with inhibition zone up to 33.3 mm and MIC values between 1 and 1.25 mg/mL.	Alkaloids, anthraquinones, saponins, and tannins	[123]
Flueggea virosa	Leaf	Antibacterial/Ethanol, chloroform and petroleum ether	Extracts were active against *S. aureus* with MIC values between 8–16 µg/mL.	β-sitosterol,11-*O*-acetyl bergenin, bergenin, daucosterol, kaempferol, gallic acid, and virosecurinine	[124,125]
Funtumia elastica	Stem bark	Antifungal/Crude extract	The extract was active on *A. flavus, C. albicans, T. mentagrophytes* and *Trichosporon cutaneum* zones of inhibition between 11 to 17 mm.	Anthocyanins, butacyanin, flavonoids, and tannins	[126]
Hannoa undulata	Root	Antibacterial/Ethanol	The extract was active against *Cutibacterium acnes, K. pneumonia, P. aeruginosa,* and *S. aureus* with zonesof inhibition between 12 and 15 mm and MIC value of 62.5 mg/mL.	Alkaloids, flavonoids, saponins, and triterpenes	[127]
Heliotropium indicum	Whole plant	Ethanol	The extract showed activities against selected skin pathogens with zones of inhibition between 12 and 25 mm.	Phenols, saponins, terpenoids, and cardiac glycosides	[128]

Table 2. *Cont.*

Plant Species	Plant Part(s) Used	Biological Activities/Extract or Isolated Compound Used	Results	Active Metabolites	References
Jatropha curcas	Leaf	Antibacterial/Methanol	Extract inhibited *E. coli* and *S. aureus* with zones of inhibition of 26 and 18 mm, respectively, as well as MIC of 0.125 mg/mL.	Alkaloids, Flavonoids, saponins, tannins, and saponins	[129]
Khaya grandifoliola	Leaf, Stem bark, and root	Antibacterial/Methanol ethyl acetate	Extracts were active against *S. aureus* and *S. pyogenes* with MIC value of 0.25 mg/mL.	Alkaloids, phenols, flavonoids, and terpenoids	[130]
Kigelia africana	Leaf and stem bark	Antimicrobial and wound healing/Methanol	Both extracts demonstrated healing properties by increasing wound contraction significantly by 72% at 7 days. Extracts from the plant were also active against *E. coli*, *P. aeruginosa*, *S. aureus*, *B. subtilis*, and *C. albicans* with MIC values between 2.5–7.5 mg/mL.	Flavonoids, iridoids, coumarins, naphthoquinones, terpenes, and terpenoids	[131,132]
Lannea acida	Stem bark	Antifungal/Crude	The extract inhibited the growth of fungi, namely *Aspergillus favus*, *C. albicans*, *C. tropicalis*, *Fusarium solani*, *Rhyzopus stolonifera*, with zones of inhibition ranging from 28.33 to 9.66 mm.	Flavonoids, phenols, terpenoids	[133]
Lannea microcarpa	Leaves	Antimicrobial/Aqueous	The extract was not effective against *E. coli*, *Pseudomonas aeruginosa* and *S. aureus*.	Flavonoids, phenolics, polyphenols, and terpenoids	[134]
Macaranga barteri	Bark	Anti-inflammatory and wound healing/Ethanol	Extract reduced inflammation and skin hyperalgesia in rats with carrageenan-induced edema.	Gallic acid, macabarterin, 3-*O*-methylellagic acid, 4-*O*-b-D-xylopyranoside, and 3-*O*-methylellagic acid	[1]

Table 2. *Cont.*

Plant Species	Plant Part(s) Used	Biological Activities/Extract or Isolated Compound Used	Results	Active Metabolites	References
Mallotus oppositifolius	Leaf	Antifungal/Crude extract, fractions, and isolated compounds	The extract, fractions and isolated compounds were active against dermatophytes (*Microsporum langeroinii, M. audouin, Trichophyton rubrum, T. soudanense*) with MIC values between 1.86 and 25,000 µg/mL.	Betulinic acid, quercetin, and quercitrin	[135]
Mangifera indica	Seed	Antimicrobial/Essential oil	The extract demonstrated antimicrobial activities against tested organisms, namely, *E. coli, C. albicans, S. aureus,* and *Mycobacterium smegmatis,* with a zone of inhibition between 10–18 mm.	Carbohydrates, polyphenols, terpenoids, sterols, carotenoids, fatty acids, and amino acids	[136,137]
Momordica charantia	Leaf and fruit	Antimicrobial/Methanol	Extracts from the leaf and fruit inhibited *C. albicans, E. coli, P. aeruginosa,* and *S. aureus.*	Anthocyanin, coumarin, cardiac glycosides, and tannins	[138,139]
Monodora myristica	Seed	Antimicrobial/Essential oil	The essential oil was active against different strains of *S. aureus* and *E. coli* with MIC values between 8 to 512 µg/mL.	Monoterpenes and sesquiterpenes	[140]
Nauclea latifolia	Stem bark	Antimicrobial/Methanol	Extracts inhibited *B. subtilis, C. albicans P. aeruginosa,* zones of inhibition between 13 and 18 mm and MIC values between 0.5 and 4 mg/mL.	Alkaloids, flavonoids, saponins, phytates, and tannins	[141,142]
Newbouldia laevis	Leaf	Antimicrobial/Methanol	The extract inhibited the growth *C. albicans, E. coli, P. aeruginosa,* and *S. aureus.*	Flavonoids, cardiac glycosides, and tannins, terpenes,	[143]
Piliostigma thonningii	Leaf	Antifungal/Methanol	The extract was active against dermatophytes with MIC values ranging from 13 to 24 mg/mL	Alkaloids, flavonoids, and terpenoids	[144]

Table 2. *Cont.*

Plant Species	Plant Part(s) Used	Biological Activities/Extract or Isolated Compound Used	Results	Active Metabolites	References
Pistia stratiotes	Leaf	Antifungal/Methanol	The extract was active against pathogenic fungi, namely, *E. floccosum*, *M. gypseum* and *M. nanum T. rubrum*, and *T. mentagrophytes* with MIC values between 125 µg/mL to 250 µg/mL.	Alkaloids, flavonoids, glycosides, and phytosterols	[145]
Psorospermum febrifugum	Stem bark	Antibacterial/Ethanol, Methanol, and aqueous	Extracts were active against *E. coli*, *S. aureus*, *P. aeruginosa* and *S. pyogenes* with inhibition zones between 11 and 19 mm and MIC value of 6.25 mg/mL.	Alkaloids, anthraquinones, flavonoids, steroids, tannins, terpenes, and xanthones	[146]
Pycnanthus angolensis	Stem bark	Antimicrobial/Aqueous and ethanol	Extracts were active against *C. albicans P. mirabilis*, *P. aeruginosa* and *S. aureus*.	Alkaloids, essential oils, glycosides, flavonoids, saponins, and tannins	[147]
Rauvolfia vomitoria	Stem bark	Antifungal/Dichloromethane	The extract was active against *A. niger* and *C. albicans* with a zone of inhibition of up to 19 mm.	Alkaloids, saponins, tannins, carbohydrates, and reducing sugars	[148,149]
Ricinus communis	Leaf	Antimicrobial/Aqueous and ethanol	Extracts exhibited strong inhibitory activities against *E. coli*, *K. pneumonia*, *S. aureus*, and *P. aeruginosa* with zones of inhibition up to 35 mm.	Tannins, saponins, terpenoids, flavonoids, and reducing sugar	[150]
Sansevieria liberica	Leaf	Antimicrobial/Methanol	The extract was active against *B. cereus* and *S. aureus* with MIC value of 62.5 mg/mL.	Carbohydrates, flavonoids, and triterpenes	[151]
Senna alata	Leaf	Antifungal/Methanol	The extract inhibited *C. albicans* and *S. pyogenes* with 25.0 mm zone of inhibition.	Alkaloids, flavonoids, tannins, saponins, and terpenoids	[73]
Spathodea campanulata	Leaf and flower	Antibacterial/Ethanol	Both extracts inhibited the growth of *B. subtilis*, *E. coli*, *S. aureus*, and *P. vulgaris*, with zones of inhibition between 6–11 mm.	Alkaloids, flavonoids, glycosides, phenolics, saponin, steroids, tannin, and terpenoids	[152]

Table 2. *Cont.*

Plant Species	Plant Part(s) Used	Biological Activities/Extract or Isolated Compound Used	Results	Active Metabolites	References
Stereospermum kunthianum	Leaf	Antibacterial/crude extract	Extract inhibited *E. coli*, *S. aureus*, and *P. aeruginosa* with a zone of inhibition of up to 35 mm. The MIC values of the extract on the organism also ranged between 2.09 mg/mL and 4.17 mg/mL.	Coumarins, fatty acids, and sterols	[153]
Strophanthus hispidus	Leaf and root	Antimicrobial and wound healing/Methanol	Both extracts improved wound contraction at day 11. The extracts were also active against *E. coli*, *P. aeruginosa*, *S. aureus*, *B. subtilis*, and *C. albicans* with MIC values between 2.5–7.5 mg/mL.	Alkaloids, flavonoids, cardiac and cyanogenic glycosides, and saponins	[131,154]
Symphonia globulifera	Leaf and stem bark	Antimicrobial/Aqueous	Leaf extract was active against *C. albicans*, *E. coli*, *S. aureus*, and *P. aeruginosa* with zones of inhibition between 13 and 21 mm.	Alkaloids, anthraquinones, flavonoids, tannins, and quinones	[155]
Terminalia avicennioides	Leaf	Wound healing and antioxidant/Methanol	The extract inhibited pathogens associated with wound infection and increased the concentration of superoxide dismutase and catalase from the healed skin tissues.	Ellagic acids, flavonoids, phenols, and tannins	[90,156]
Tetrapleura tetraptera	Leaf	Antibacterial/Methanol	The extract was active on *S. aureus* and *S. pyogenes* with a zone of inhibition of between 21.5 to 25 mm	Phenols, flavonoids, saponins, and alkaloids	[73]
Trianthema portulacastrum	Leaf	Wound healing/n-Butanol fraction of hydroethanol	Extract ointment (5% and 10% *w/w*) significantly promotes wound healing.	Caffeic acid, ferulic acid Chlorogenic acid, and protocatechuic acid	[157]

Table 2. *Cont.*

Plant Species	Plant Part(s) Used	Biological Activities/Extract or Isolated Compound Used	Results	Active Metabolites	References
Tridax procumbens	Leaf	Antimicrobial/Ethanol	The extract was active against bacterial, namely, *E. coli*, *S. pyogenes*, *S. aureus*, *B. subtilis* and *P. aeruginosa*, with inhibition zones between 4–9 mm.	Alkaloids, flavonoids, cardiac glycosides, and tannins	[97]
Uvaria chamae	Stem bark	Antimicrobial/Methanol	The extract was active against *E. coli*, *B. subtilis* and *S. aureus* with zones of inhibition between 18 and 28 mm.	Flavonoids, glycosides, saponins, tannins,	[158]
Vernonia amygdalina	Leaf	Antimicrobial/Aqueous	The extract displayed significant activities against different strains of multi-drug resistant *S. aureus*.	Alkaloids, flavonoids, phenol, tannins and terpenoids	[159]
Vitellaria paradoxa	Nuts, leaf, stem and root	Antimicrobial/Ethanol	Extracts showed activity against pathogens with zones of inhibition between 11 and 30 mm and MIC values between 60 and 70 mg/mL.	Alkaloids, carbohydrates, saponins, steroids, and tannins	[160,161]
Xanthosoma sagittifolium	Leaf, stalk, and root	Antifungal/Ethanol	The extract inhibited the pathogenic fungi, *T. rubrum*, with an inhibition diameter of 18 mm, close to the standard drug, amphotericin (20 mm).	Cardiac glycoside, flavonoids, tannins, phenols, and terpenoids	[162,163]
Xylopia aethiopica	Fruit	Antimicrobial/Aqueous. methanol and essential oil	The extract was active against the pathogenic fungi, namely *Microsporum canis*, *M. equinum*, and *T. mentagrophyte* with inhibition zones of 2.84 to 3.5 mm. The oil extract also showed significant activity against *B. subtilis*, *Enterobacter aerogenes*, *E. coli*, *S. pyogenes*, *S. aureus*, and *Serratia marcescens*. *Candida albicans* was highly susceptible to the methanol extract zone of inhibition of 25.0 mm.	α-pinene, β-linalool, α-terpineol, pinocarveol, terpinene-4-acetate, α-thugene, β-phellandrene, β-caryophyllene, γ-terpinene, 1,8-cineole, acetyleugenol, benzylbenzoate, eugenol, *cis*-ocimene, and sabinene	[73,164,165]

Plant secondary metabolites are the ingredients that confer therapeutic effects on medicinal plants [72]. An assessment of the antimicrobial activities of the phytochemicals of the plants in Table 2 was carried out. Though many of the plants have been evaluated for their phytochemistry, only a few phytochemicals have their antimicrobial activities carried out against skin pathogens. The alkaloidal fraction of *Annickia chlorantha* stem back showed significant antifungal activities both in vitro and in vivo [88]. Tamokou et al. [98] isolated seven compounds from the aerial part of *B. lamium*, among which lespedin (MIC = 6.25 µg/mL) and aurantiamide acetate (MIC = 50 µg/mL) significantly inhibited the growth of the tested skin pathogens. Betulenic acid isolated from the leaves of *Mallotus oppositifolius* also displayed a noteworthy activity against a dermatophyte, *Microsporum langeronii* with MIC value of 1.86 µg/mL [135]. Betulenic acid elicited the best antimicrobial activity with the lowest MIC value in this review.

Further research is required on the metabolite to ascertain its biotoxicity and synergistic effect with other antimicrobial metabolites to further explore its potency in developing a commercially available antimicrobial agent. Regarding the antioxidant and anti-inflammatory activities of the plants, ethanol extracts of the leaves of *Chassalia kolly* showed excellent activities with an IC_{50} of 0.05 µg/µL and also showed higher anti-inflammatory activities than aspirin [103]. This is traditionally used in the treatment of ringworm [34]. Similarly, using the xylene and chorioallantoic membrane (CAM) anti-inflammatory model, *Cyathula prostrata* showed moderate activity [115], despite its traditional use for sores and rashes [28]. It is recommended that further studies into the activities of *C. kolly* be carried out to assess its complete activities, isolate active metabolites, and harness its potential in treating skin ailments.

Wounds are one of the most common skin problems that may occur in any part of the skin due to breaking/puncturing the skin or rupturing other body tissues. For the wound-healing activities of the plants, the n-butanol fraction of the hydroethanolic leaf extract of *Trianthema portulacastrum* accelerated wound healing in rats by increasing the contraction and epithelialization of the wound and decreasing the level of inflammatory markers [157]. This result validated the use of the plant in the management of wounds in Nigeria [15]. The alkaloidal extract from the stem bark of *Alstonia boonei* also significantly increases the rate of contraction of the wound and reduced the epithelialization period in vivo [83]; the plant is traditionally used for snakebites [36]. Similarly, in an in vivo study, the methanol extracts of the leaves and roots of *Strophanthus hispidus* and the roots and stem bark of *Kigelia africana* significantly increased wound contraction at days 11 and 7, respectively [131]. Agyare et al. [76] revealed that the aqueous and methanol extracts of *Alchornea cordifolia* displayed wound-healing capacity at days 1 ($p < 0.05$) and 9 ($p < 0.001$). Other plants implicated in the study that have demonstrated wound-healing capacity include *Anthocleista djalonensis* [92], *Clerodendrum splendens* [108], and *Carapa procera* [101].

4. Conclusions

This review compiled the list of plants used traditionally in Western Africa for combating various skin ailments and the available scientific studies that have been carried out on the plants. A large percentage of the data was reported from a handful of countries, showing a large research gap in many West African countries on the traditional use of plants for skin ailments. The family Fabaceae is by far the most used, while Combretaceae, Asteraceae, and Euphorbiaceae were also well used. The most common habit of the plants was tree; leaf was the most-used plant part; and decoction was the most-preferred method of preparation. The biological activities of 82 out of the 211 plant species have been carried out, which means many plants still need to be investigated for biological activities related to skin diseases in West Africa. Plants such as *Brillantaisia lamium*, *Kigelia africana*, and *Strophanthus hispidus* that have demonstrated strong biological activities related to skin diseases are recommended for further research to identify the active metabolites and their mode of action.

Author Contributions: Conceptualization, A.A.-n.A., M.U.M. and N.M.; methodology, data curation, A.A.-n.A., M.U.M. and N.M.; writing—original draft preparation, A.A.-n.A., M.U.M. and N.M.; writing—review and editing, A.A.-n.A., M.U.M. and N.M.; funding acquisition, M.U.M. and N.M. All authors have read and agreed to the published version of the manuscript.

Funding: This research received no external funding.

Institutional Review Board Statement: Not applicable.

Informed Consent Statement: Not applicable.

Data Availability Statement: Not applicable.

Acknowledgments: Many thanks to Mukaila Yusuf Ola for proofreading the manuscript.

Conflicts of Interest: The authors declare no conflict of interest.

References

1. Rosamah, E.; Haqiqi, M.T.; Putri, A.S.; Kuspradini, H.; Kusuma, I.W.; Amirta, R.; Arung, E.T. The potential of *Macaranga* plants as skincare cosmetic ingredients: A review. *J. Appl. Pharm. Sci.* **2023**, *13*, 001–012. [CrossRef]
2. McGrath, J.A.; Eady, R.A.J.; Pope, F.M. Anatomy and organization of human skin. *Rook's Textb. Dermatol.* **2004**, *1*, 2–3.
3. Paulino, L.C.; Tseng, C.H.; Strober, B.E.; Blaser, M.J. Molecular analysis of fungal microbiota in samples from healthy human skin and psoriatic lesions. *J. Clin. Microbiol.* **2006**, *44*, 2933–2941. [CrossRef] [PubMed]
4. Byrd, A.L.; Belkaid, Y.; Segre, J.A. The human skin microbiome. *Nat. Rev. Microbiol.* **2018**, *16*, 143–155. [CrossRef] [PubMed]
5. Belkaid, Y.; Segre, J.A. Dialogue between skin microbiota and immunity. *Science* **2014**, *346*, 954–959. [CrossRef] [PubMed]
6. Hay, R.J.; Johns, N.E.; Williams, H.C.; Bolliger, I.W.; Dellavalle, R.P.; Margolis, D.J.; Naghavi, M. The global burden of skin disease in 2010: An analysis of the prevalence and impact of skin conditions. *J. Investig. Dermatol.* **2014**, *134*, 1527–1534. [CrossRef] [PubMed]
7. Basra, M.K.; Shahrukh, M. Burden of skin diseases. *Expert. Rev. Pharmacoecon. Outcomes Res.* **2009**, *9*, 271–283. [CrossRef] [PubMed]
8. Tabassum, N.; Hamdani, M. Plants used to treat skin diseases. *Pharmacogn. Rev.* **2014**, *8*, 52–55. [CrossRef]
9. Mukaila, Y.O.; Oladipo, O.T.; Ogunlowo, I.; Ajao, A.A.N.; Sabiu, S. Which plants for what ailments: A quantitative analysis of medicinal ethnobotany of Ile-Ife, Osun State, Southwestern Nigeria. *Evid.-Based Complement. Altern. Med.* **2021**, *2021*, 5711547. [CrossRef]
10. Mabona, U.; Van Vuuren, S.F. Southern African medicinal plants used to treat skin diseases. *S. Afr. J. Bot.* **2013**, *87*, 175–193. [CrossRef]
11. Alamgeer, S.A.; Asif, H.; Younis, W.; Riaz, H.; Bukhari, I.A.; Assiri, A.M. Indigenous medicinal plants of Pakistan used to treat skin diseases: A review. *Chin. Med.* **2018**, *13*, 52. [CrossRef]
12. Tsioutsiou, E.E.; Amountzias, V.; Vontzalidou, A.; Dina, E.; Stevanović, Z.D.; Cheilari, A.; Aligiannis, N. Medicinal plants used traditionally for skin related problems in the south balkan and east mediterranean region—A review. *Front. Pharmacol.* **2022**, *13*, 936047. [CrossRef] [PubMed]
13. Kankara, S.S.; Ibrahim, M.H.; Mustafa, M.; Go, R. Ethnobotanical survey of medicinal plants used for traditional maternal healthcare in Katsina state, Nigeria. *S. Afr. J. Bot.* **2015**, *97*, 165–175. [CrossRef]
14. Zizka, A.; Thiombiano, A.; Dressler, S.; Nacoulma, B.M.; Ouédraogo, A.; Ouédraogo, I.; Ouédraogo, O.; Zizka, G.; Hahn, K.; Schmidt, M. Traditional plant use in Burkina Faso (West Africa): A national-scale analysis with focus on traditional medicine. *J. Ethnobiol. Ethnomed.* **2015**, *11*, 9. [CrossRef] [PubMed]
15. Ajibesin, K.K. Ethnobotanical survey of plants used for skin diseases and related ailments in Akwa Ibom State, Nigeria. *Ethnobot. Res. Appl.* **2012**, *10*, 463–522.
16. Codo Toafode, N.M.; Oppong Bekoe, E.; Vissiennon, Z.; Ahyi, V.; Vissiennon, C.; Fester, K. Ethnomedicinal information on plants used for the treatment of bone fractures, wounds, and sprains in the northern region of the Republic of Benin. *Evid.-Based Complement. Altern. Med.* **2022**, *2022*, 8619330. [CrossRef] [PubMed]
17. Simbo, D.J. An ethnobotanical survey of medicinal plants in Babungo, Northwest Region, Cameroon. *J. Ethnobiol. Ethnomed.* **2010**, *6*, 8. [CrossRef] [PubMed]
18. Udegbunam, S.O.; Udegbunam, R.I.; Nnaji, T.O.; Anyanwu, M.U.; Kene, R.O.C.; Anika, S.M. Antimicrobial and antioxidant effect of methanolic *Crinum jagus* bulb extract in wound healing. *J. Intercult. Ethnopharmacol.* **2015**, *4*, 239. [CrossRef]
19. Gonçalves, G.M.S.; Gobbo, J. Antimicrobial effect of Anacardium occidentale extract and cosmetic formulation development. *Braz. Arch. Biol. Technol.* **2012**, *55*, 843–850. [CrossRef]
20. González-Minero, F.J.; Bravo-Díaz, L.; Moreno-Toral, E. Pharmacy and fragrances: Traditional and current use of plants and their extracts. *Cosmetics* **2023**, *10*, 157. [CrossRef]
21. Maroyi, A. Medicinal uses of the Fabaceae family in Zimbabwe: A Review. *Plants* **2023**, *12*, 1255. [CrossRef]
22. Ajao, A.A.; Sibiya, N.P.; Moteetee, A.N. Sexual prowess from nature: A systematic review 315 of medicinal plants used as aphrodisiacs and sexual dysfunction in sub-Saharan Africa. *S. Afr. J. Bot.* **2019**, *122*, 342–359. [CrossRef]
23. Rogers, C.B.; Verotta, L. Chemistry and biological properties of the African Combretaceae. In *Chemistry, Biological and Pharmacological Properties of African Medicinal Plants*; Hostettman, K., Chinyanganga, F., Maillard, M., Wolfender, J.-L., Eds.; University of Zimbabwe Publications: Harare, Zimbabwe, 1996.

24. Eloff, J.N.; Katerere, D.R.; McGaw, L.J. The biological activity and chemistry of the southern African Combretaceae. *J. Ethnophar-macol.* **2008**, *119*, 686–699. [CrossRef] [PubMed]

25. Ccana-Ccapatinta, G.V.; Monge, M.; Ferreira, P.L.; Da Costa, F.B. Chemistry and medicinal uses of the subfamily Barnadesioideae (Asteraceae). *Phytochem. Rev.* **2018**, *17*, 471–489. [CrossRef]

26. Ntie-Kang, F.; Lifongo, L.L.; Mbaze, L.M.A.; Ekwelle, N.; Owono Owono, L.C.; Megnassan, E.; Efange, S.M. Cameroonian medicinal plants: A bioactivity versus ethnobotanical survey and chemotaxonomic classification. *BMC Complement. Altern. Med.* **2013**, *13*, 147. [CrossRef] [PubMed]

27. Dalziel, J.M. The Useful Plants of West Tropical Afica. In *Being an Appendix to "the Flora of West Tropical Africa"*; Crown Agents for the Colonies: London, UK, 1937.

28. Bouquet, A.; Debray, M. Plantes médicinales de la Côte d'Ivoire. *Trav. Doc. L'office Rech. Sci. Tech. Outre-Mer.* **1974**, *32*, 5–229.

29. Lewis, W.H.; Elvin-Lewis, M.P. *Medical Botany: Plants Affecting Human Health*; John Wiley & Sons: Hoboken, NJ, USA, 2003.

30. Kerharo, J.; Adam, J.G. Deuxième inventaire des plantes médicinales et toxiques de la Casamance (Sénégal). *Ann. Pharm. Françaises* **1963**, *21*, 773–792.

31. Kerharo, J.; Adam, J.G. *La Pharmacopée Sénégalaise Traditionnelle: Plantes Medicinales et Toxiques*; Vigot: Paris, France, 1974.

32. Irvine, F.R. *Plants of the Gold Coast*; Oxford University Press: London, UK, 1930.

33. Chevalier, A.; Laffitte, M. Une enquête sur les plantes médicinales de l'Afrique occidentale. *Rev. Bot. Appl. Agric. Trop.* **1937**, *27*, 165–175. [CrossRef]

34. Alfa, T.; Anani, K.; Adjrah, Y.; Batawila, K.; Ameyapoh, Y. Ethnobotanical survey of medicinal plants used against fungal infections in prefecture of sotouboua central region, Togo. *Eur. Sci. J.* **2018**, *14*, 342. [CrossRef]

35. Burkill, H.M. *The Useful Plants of West Tropical Africa*, 2nd ed.; Family A–D; Royal Botanic Gardens, Kew: London, UK, 1985; Volume 1.

36. Betti, J.L. An ethnobotanical study of medicinal plants among the Baka pygmies in the Dja biosphere reserve, Cameroon. *Afr. Stud. Monogr.* **2004**, *25*, 1–27.

37. Chopra, R.N.; Nayar, S.L.; Chopra, I.C. *Glossary of Indian Medicinal Plants*; CSIR: New Delhi, India, 1956.

38. Pobéguin, H. *Plantes Médicinales de la Guinée*; Challamel: Paris, France, 1912.

39. Kerharo, J.; Adam, J.G. Premier inventaire des plantes médicinales et toxiques de la Casamance (Senegal). *Ann. Pharm. Françaises* **1962**, *20*, 76–84.

40. Agbovie, T.; Amponsah, K.; Crentsil, O.R.; Dennis, F.; Odamtten, G.T.; Ofusohene-Djan, W. Conservation and sustainable use of medicinal plants in Ghana. *Ethnobot. Surv.* **2002**, *15*, 32–37.

41. Kerharo, J.; Bouquet, A. Indigenous Conceptions of Leprosy and its Treatment in the Ivory Coast and Haute-Volta. *Bull. Société Pathol. Exot.* **1950**, *43*, 56–65.

42. Bhat, R.B.; Etejere, E.O.; Oladipo, V.T. Ethnobotanical studies from central Nigeria. *Econ. Bot.* **1990**, *44*, 382–390. [CrossRef]

43. Dossou-Yovo, H.O.; Vodouhè, F.G.; Kaplan, A.; Sinsin, B. Application of ethnobotanical indices in the utilization of five medicinal herbaceous plant species in Benin, West Africa. *Diversity* **2022**, *14*, 612. [CrossRef]

44. Bouquet, A. Féticheurs et médicines traditionalles du Congo (Brazzaville). *Mémoires Off. Rech. Sci. Technol. Outre-Mer.* **1969**, *36*, 4–304.

45. Chifundera, K. Contribution to the inventory of medicinal plants from the Bushi area, South Kivu Province, Democratic Republic of Congo. *Fitoterapia* **2001**, *72*, 351–368. [CrossRef]

46. Oyen, L.P.A. Symphonia globulifera L.f. In *PROTA (Plant Resources of Tropical Africa/Ressources Végétales de l'Afrique Tropicale)*; Louppe, D., Oteng-Amoako, A.A., Brink, M., Eds.; Wageningen University: Wageningen, The Netherlands, 2005.

47. Fromentin, Y.; Cottet, K.; Kritsanida, M.; Michel, S.; Gaboriaud-Kolar, N.; Lallemand, M.C. *Symphonia globulifera*, a widespread source of complex metabolites with potent biological activities. *Planta Medica* **2015**, *81*, 95–107. [CrossRef]

48. Madge, C. Therapeutic landscapes of the Jola, The Gambia, West Africa. *Health Place* **1998**, *4*, 293–311. [CrossRef]

49. Aubréville, A. *Flore Forestière Soudano-Guinéene: A.O.F., Cameroun, A.E.F*; Société d'Éditions Géographiques Maritimes et Coloniales: Paris, France, 1950.

50. Boadu, A.A.; Asase, A. Documentation of herbal medicines used for the treatment and management of human diseases by some communities in southern Ghana. *Evid.-Based Complement. Altern. Med.* **2017**, *2017*, 3043061. [CrossRef]

51. Kabran, F.A.; Maciuk, A.; Okpekon, T.A.; Leblanc, K.; Seon-Meniel, B.; Bories, C.; Champy, P.; Djakouré, L.A.; Figadère, B. Phytochemical and biological analysis of *Mallotus oppositifolius* (Euphorbiaceae). *Planta Med.* **2012**, *78*, 1381. [CrossRef]

52. Ellena, R.; Quave, C.L.; Pieroni, A. Comparative medical ethnobotany of the Senegalese community living in Turin (Northwestern Italy) and in Adeane (Southern Senegal). *Evid.-Based Complement. Altern. Med.* **2012**, *2012*, 604363. [CrossRef] [PubMed]

53. Kuma, D.N.; Boye, A.; Kwakye-Nuako, G.; Boakye, Y.D.; Addo, J.K.; Asiamah, E.A.; Atsu Barku, V.Y. Wound healing properties and antimicrobial effects of *Parkia clappertoniana* keay fruit husk extract in a rat excisional wound model. *BioMed Res. Intern.* **2022**, *2022*, 9709365. [CrossRef] [PubMed]

54. Komakech, R. Herbs & Plants. Psorospermum febrifugum Spach. A Medicinal Plant for Skin Diseases. Available online: southworld.net/herbs-plants-psorospermum-febrifugum-spach-a-medicinal-plant-for-skin-diseases (accessed on 1 April 2023).

55. Tropical lants Database, Ken Fern. Available online: tropical.theferns.info/viewtropical.php?id=Psorospermum+febrifugum (accessed on 1 April 2023).

56. Rukangira, E. The African herbal industry: Constraints and challenges. Proceedings of the Natural Products and Cosmeceuticals, August 2001. *Erbor. Domani* **2001**, *1*, 1–15.

57. Adediwura, F.J.; Ajibesin, K.K.; Odeyemi, T.; Ogundokun, G. Ethnobotanical studies of folklore phytocosmetics of South West Nigeria. *Pharm. Biol.* **2015**, *53*, 313–318.
58. DeFilipps, R.A.; Maina, S.L.; Crepin, J. Medicinal Plants of the Guianas (Guyana, Surinam, French Guiana). In *Medicinal Plants of the Guianas (Guyana Surinam Fr. Guiana)*; Department of Botany, National Museum of Natural History, Smithsonian Institution: Washington, DC, USA, 2004.
59. Kitambala, M.M. Uapaca guineensis Müll.Arg. In *PROTA (Plant Resources of Tropical Africa/Ressources Végétales de l'Afrique Tropicale)*; Schmelzer, G.H., Gurib-Fakim, A., Eds.; Wageningen University: Wageningen, The Netherlands, 2023.
60. Kokwaro, J.O. *Medicinal Plants of East Africa*; University of Nairobi Press: Nairobi, Kenya, 2009.
61. Haq, I. Safety of medicinal plants. *Pak. J. Med. Res.* **2004**, *43*, 203–210.
62. Mlilo, S.; Sibanda, S. An ethnobotanical survey of the medicinal plants used in the treatment of cancer in some parts of Matebeleland, Zimbabwe. *S. Afr. J. Bot.* **2022**, *146*, 401–408. [CrossRef]
63. Makgobole, M.U.; Onwubu, S.C.; Nxumal, C.T.; Mpofana, N.; Ajao, A.A.N. In search of oral cosmetics from nature: A review of medicinal plants for dental care in West Africa. *S. Afr. J. Bot.* **2023**, *162*, 644–657. [CrossRef]
64. Petran, M.; Dragos, D.; Gilca, M. Historical ethnobotanical review of medicinal plants used to treat children diseases in Romania (1860s–1970s). *J. Ethnobiol. Ethnomed.* **2020**, *16*, 15. [CrossRef]
65. Howes, M.J.R.; Quave, C.L.; Collemare, J.; Tatsis, E.C.; Twilley, D.; Lulekal, E.; Nic-Lughadha, E. Molecules from nature: Reconciling biodiversity conservation and global healthcare imperatives for sustainable use of medicinal plants and fungi. *Plants People Planet* **2020**, *2*, 463–481. [CrossRef]
66. Iyekowa, O.; Oviawe, A.P.; Ndiribe, J.O. Antimicrobial activities of *Acalypha wilkesiana* (Red Acalypha) extracts in some selected skin pathogens. *Zimb. J. Sci. Technol.* **2016**, *11*, 48–57.
67. Ram, A.J.; Bhakshu, L.M.; Raju, R.V. In vitro antimicrobial activity of certain medicinal plants from Eastern Ghats, India, used for skin diseases. *J. Ethnopharmacol.* **2004**, *90*, 353–357.
68. Okoli, C.O.; Akah, P.A.; Onuoha, N.J.; Okoye, T.C.; Nwoye, A.C.; Nworu, C.S. *Acanthus montanus*: An experimental evaluation of the antimicrobial, anti-inflammatory and immunological properties of a traditional remedy for furuncles. *BMC Complement. Altern. Med.* **2008**, *8*, 27. [CrossRef] [PubMed]
69. Ndam, P.; Onyemelukwe, N.; Nwakile, C.D.; Ogboi, S.J.; Maduakor, U. Antifungal properties of methanolic extracts of some medical plants in Enugu, south east Nigeria. *Afr. J. Cli. Experiment. Microbiol.* **2018**, *19*, 141–148. [CrossRef]
70. Gupta, S.; Toppo, K.I.; Karkun, D.; Kumar, A.; Mishra, N.; Dadsena, R.; Thakur, T. Antimicrobial activity of *Achyranthes aspera* against some human pathogenic bacteria and fungi. *Intern. J. Pharmacol. Biol. Sci.* **2013**, *7*, 43–54.
71. Abhaykumar, K. Phytochemical studies on *Achyranthes aspera*. *World Sci. News* **2018**, *100*, 16–34.
72. Olatunji, B.P.; Idris, O.O.; Ogunmefun, O.T.; Abuka, D.U. Assessment of *Andrographis paniculata* and *Aframomum melegueta* on bacteria isolated from wounds. *J. Chem. Pharm. Sci.* **2019**, *12*, 2349–8552. [CrossRef]
73. Gbadamosi, I.T.; Oyedele, T.O. The efficacy of seven ethnobotanicals in the treatment of skin infections in Ibadan, Nigeria. *Afr. J. Biotechnol.* **2012**, *11*, 3928–3934. [CrossRef]
74. Vigbedor, B.Y.; Osei-Owusu, J.; Kwakye, R.; Neglo, D. Bioassay-guided fractionation, ESI-MS scan, phytochemical screening, and antiplasmodial activity of *Afzelia africana*. *Biochem. Res. Int.* **2022**, *2022*, 6895560. [CrossRef]
75. Kouame, B.K.F.P.; Toure, D.; Kablan, L.; Bedi, G.; Tea, I.; Robins, R.; Tonzibo, F. Chemical constituents and antibacterial activity of essential oils from flowers and stems of *Ageratum c onyzoides* from Ivory Coast. *Rec. Nat. Prod.* **2018**, *12*, 160–168. [CrossRef]
76. Agyare, C.; Ansah, A.O.; Ossei, P.P.S.; Apenteng, J.A.; Boakye, Y.D. Wound healing and anti-infective properties of *Myrianthus arboreus* and *Alchornea cordifolia*. *Med. Chem.* **2014**, *4*, 533–539. [CrossRef]
77. Ajibesin, K.K.; Rene, N.; Bala, D.N.; Essiett, U.A. Antimicrobial activities of the extracts and fractions of *Allanblackia floribunda*. *Biotechnol.* **2008**, *7*, 129–133. [CrossRef]
78. Ayoola, G.A.; Ipav, S.S.; Sofidiya, M.O.; Adepoju-Bello, A.A.; Coker, H.A.; Odugbemi, T.O. Phytochemical screening and free radical scavenging activities of the fruits and leaves of *Allanblackia floribunda* Oliv (Guttiferae). *Intern. J. Health Res.* **2008**, *1*, 87–93. [CrossRef]
79. Samuel, J.K.; Andrews, B.; Jebashree, H.S. In vitro evaluation of the antifungal activity of *Allium sativum* bulb extract against Trichophyton rubrum, a human skin pathogen. *World J. Microbiol. Biotechnol.* **2000**, *16*, 617–620. [CrossRef]
80. Yusuf, A.; Fagbuaro, S.S.; Fajemilehin, S.O.K. Chemical composition, phytochemical and mineral profile of garlic (*Allium sativum*). *J. Biosci. Biotechnol. Discov.* **2018**, *3*, 105–109. [CrossRef]
81. Bashir, A.; Saeed, B.; Mujahid, T.Y.; Jehan, N. Comparative study of antimicrobial activities of *Aloe vera* extracts and antibiotics against isolates from skin infections. *Afr. J. Biotechnol.* **2011**, *10*, 3835–3840.
82. Arunkumar, S.; Muthuselvam, M. Analysis of phytochemical constituents and antimicrobial activities of *Aloe vera* L. against clinical pathogens. *World J. Agric. Sci.* **2009**, *5*, 572–576.
83. Fetse, J.P.; Kyekyeku, J.O.; Dueve, E.; Mensah, K.B. Wound healing activity of total alkaloidal extract of the root bark of *Alstonia boonei* (Apocynacea). *Br. J. Pharm. Res.* **2014**, *4*, 26–42. [CrossRef]
84. Pamila, U.A.; Karpagam, S. Antimicrobial activity of *Alternanthera bettzickiana* (regel) g. Nicholson and its phytochemical contents. *Intern. J. Pharm. Sci. Res.* **2017**, *8*, 2594–2599.
85. Maiyo, Z.C. Chemical Compositions and Antimicrobial Activity of *Amaranthus hybridus, Amaranthus caudatus, Amaranthus spinosus* and *Corriandrum sativum*. Ph.D. Dissertation, Egerton University, Egerton, Kenya, 2008.

86. Silva, J.G.; Souza, I.A.; Higino, J.S.; Siqueira-Junior, J.P.; Pereira, J.V.; Pereira, M.S.V. Atividade antimicrobiana do extrato de *Anacardium occidentale* Linn. em amostras multiresistentes de *Staphylococcus aureus*. *Rev. Bras. Farm.* **2007**, *17*, 572–577. [CrossRef]
87. Oyetayo, V.O. Comparative Studies of the Phytochemical and Antimicrobial Properties of the Leaf, Stem and Tuber of *Anchomanes difformis*. *J. Pharmacol. Toxicol.* **2007**, *2*, 407–410. [CrossRef]
88. Nyong, E.E.; Odeniyi, M.A.; Moody, J.O. In vitro and in vivo antimicrobial evaluation of alkaloidal extracts of *Enantia chlorantha* stem bark and their formulated ointments. *Acta Pharma* **2015**, *72*, 14–52.
89. Odoh, U.; Okwor, I.; Ezejiofor, M. Phytochemical, trypanocidal and antimicrobial studies of *Enantia chlorantha* (Annonaceae) root. *J. Pharm. Allied Sci.* **2011**, *7*, 4.
90. Mann, A.; Yahaya, Y.; Banso, A.; John, F. Phytochemical and antimicrobial activity of *Terminalia avicennioides* extracts against some bacteria pathogens associated with patients suffering from complicated respiratory tract diseases. *J. Med. Plants Res.* **2008**, *2*, 9–97.
91. Tukur, A.; Musa, N.M.; Bello, H.A.; Sani, N.A. Determination of the phytochemical constituents and antifungal properties of *Annona senegalensis* leaves (African custard apple). *ChemSearch J.* **2020**, *11*, 16–24.
92. Chah, K.F.; Eze, C.A.; Emuelosi, C.E.; Esimone, C.O. Antibacterial and wound healing properties of methanolic extracts of some Nigerian medicinal plants. *J. Ethnopharmacol.* **2006**, *104*, 164–167. [CrossRef] [PubMed]
93. Obioma, A.; Chikanka, A.T.; Dumo, I. Antimicrobial activity of leave extracts of *Bryophyllum pinnatum* and *Aspilia africana* on pathogenic wound isolates recovered from patients admitted in University of Port Harcourt Teaching Hospital, Nigeria. *Ann. Clin. Lab. Res.* **2017**, *5*, 185–189. [CrossRef]
94. Johnson, E.C.; Eseyin, O.A.; Udobre, A.E.; Ike, P. Antibacterial Effect of Methanolic Extract of the Root of *Aspilia africana*. *Niger. J. Pharm. Appl. Sci. Res.* **2011**, *1*, 44–50.
95. Pandey, G.; Verma, K.K.; Singh, M. Evaluation of phytochemical, antibacterial and free radical scavenging properties of *Azadirachta indica* (neem) leaves. *Int. J. Pharm. Pharm. Sci.* **2014**, *6*, 444–447.
96. Agyare, C.; Oguejiofor, S.; Adu-Amoah, L.; Boakye, Y.D. Anti-inflammatory and anti-infective properties of ethanol leaf and root extracts of *Baphia nitida*. *Br. Microbiol. Res. J.* **2016**, *11*, 1–11. [CrossRef]
97. Owoyemi, O.O.; Oladunmoye, M.K. Phytochemical screening and antibacterial activities of *Bidens pilosa* L. and *Tridax procumbens* L. on skin pathogens. *Intern. J. Mod. Biol. Med.* **2017**, *8*, 24–46.
98. Tamokou, J.D.D.; Kuiate, J.R.; Tene, M.; Nwemeguela, T.J.K.; Tane, P. The antimicrobial activities of extract and compounds isolated from *Brillantaisia lamium*. *Iran. J. Med. Sci.* **2011**, *36*, 24. [PubMed]
99. Goyal, S.; Kumar, S.; Rawat, P.; Dhaliwal, N. Antifungal activity of *Calotropis procera* towards dermatoplaytes. *Intern. J. Adv. Pharm. Biol. Chem.* **2013**, *2*, 2277–4688.
100. Naser, E.H.; Kashmer, A.M.; Abed, S.A. Antibacterial activity and phytochemical investigation of leaves of *Calotropis procera* plant in Iraq by GC-MS. *IJPSR* **2019**, *10*, 1988–1994.
101. Udoumoh, A.F.; Eze, C.A.; Chah, K.F.; Etuk, E.U. Antibacterial and surgical wound healing properties of ethanolic leaf extracts of *Swietenia mahogoni* and *Carapa procera*. *Asian J. Trad. Med.* **2011**, *6*, 272–277.
102. Dwivedi, M.K.; Sonter, S.; Mishra, S.; Patel, D.K.; Singh, P.K. Antioxidant, antibacterial activity, and phytochemical characterization of *Carica papaya* flowers. *Beni-Suef Univ. J. Basic Appl. Sci.* **2020**, *9*, 1–11. [CrossRef]
103. Alain, K.Y.; Morand, A.J.; Andreea, B.D.; Théophile, O.; Pascal, A.D.C.; Alain, A.G.; Dominique, S.C.K. Phytochemical analysis, antioxidant and anti-inflammatory activities of *Chassalia kolly* leaves extract, a plant used in Benin to treat skin illness. *GSC Biol. Pharm. Sci.* **2021**, *15*, 063–072. [CrossRef]
104. Thophon, S.H.S.; Waranusantigul, P.; Kangwanrangsan, N.; Krajangsang, S. Antimicrobial activity of *Chromolaena odorata* extracts against bacterial human skin infections. *Mod. Appl. Sci.* **2016**, *10*, 159–171.
105. Ngozi, I.M.; Jude, I.C.; Catherine, I.C. Chemical profile of *Chromolaena odorata* L. (King and Robinson) leaves. *Pak. J. Nutr.* **2009**, *8*, 521–524. [CrossRef]
106. Najafi, S.; Sanadgol, N.; Nejad, B.S.; Beiragi, M.A.; Sanadgol, E. Phytochemical screening and antibacterial activity of *Citrullus colocynthis* (Linn.) Schrad against Staphylococcus Aureus. *J. Med. Plants Res.* **2010**, *4*, 2321–2325.
107. Hameed, B.; Ali, Q.; Hafeez, M.M.; Malik, A. Antibacterial and antifungal activity of fruit, seed and root extracts of *Citrullus colocynthis* plant. *Biol. Clin. Sci. Res. J.* **2020**, *2020*. [CrossRef]
108. Gbedema, S.Y.; Emelia, K.; Francis, A.; Kofi, A.; Eric, W. Wound healing properties and kill kinetics of *Clerodendron splendens* G. Don, a Ghanaian wound healing plant. *Pharmacogn. Res.* **2010**, *2*, 63. [CrossRef] [PubMed]
109. Oscar, N.D.Y.; Joel, T.N.S.; Ange, A.A.N.G.; Desire, S.; Brice, S.N.F.; Barthelemy, N. Chemical constituents of *Clerodendrum splendens* (Lamiaceae) and their antioxidant activities. *J. Dis. Med. Plants* **2018**, *4*, 120–127.
110. Chakraborty, P.; Deb, P.; Chakraborty, S.; Chatterjee, B.; Abraham, J. Cytotoxicity and antimicrobial activity of *Colocasia esculenta*. *J. Chem. Pharm. Res.* **2015**, *7*, 627–635.
111. Marquardt, P.; Seide, R.; Vissiennon, C.; Schubert, A.; Birkemeyer, C.; Ahyi, V.; Fester, K. Phytochemical characterization and in vitro anti-inflammatory, antioxidant and antimicrobial activity of *Combretum collinum* Fresen leaves extracts from Benin. *Molecules* **2020**, *25*, 288. [CrossRef] [PubMed]
112. Mvongo, C.; Noubissi, P.A.; Kamgang, R.; Minka, C.S.M.; Mfopa, A.; Oyono, J.L.E. Phytochemical studies and in vitro antioxidant potential of two different extracts of *Crinum jagus*. *Int. J. Pharm. Sci. Res.* **2015**, *6*, 2354–2358.
113. Suja, S.; Varkey, I.C. Medicinal and Pharmacological Values of *Cyanthillium Cinereum* (Poovamkurunilla) Extracts: Investigating the Antibacterial and Anti-Cancer Activity in Mcf-7 Breast Cancer Cell Lines. *Int. J. Res. Anal. Rev.* **2019**, *6*, 412–415.

114. Bharti, S.; Yadav, S.; Panday, J. Evaluation of anti-anxiety activity of the leaves of *Cyanthillium cinereum*. *NeuroQuantol.* **2023**, *21*, 733–746.
115. Ibrahim, B.; Sowemimo, A.; van Rooyen, A.; Van de Venter, M. Anti-inflammatory, analgesic and antioxidant activities of *Cyathula prostrata* (Linn.) Blume (Amaranthaceae). *J. Ethnopharmacol.* **2012**, *141*, 282–289. [CrossRef]
116. Ajibesin, K.K.; Essien, E.E.; Adesanya, S.A. Antibacterial constituents of the leaves of *Dacryodes edulis*. *Afr. J. Pharm. Pharmacol.* **2011**, *5*, 1782–1786. [CrossRef]
117. Coker, M.E.; Ogundele, O.S. Evaluation of the antifungal properties of extracts of *Daniella oliveri*. *Afr. J. Biomed. Res.* **2016**, *19*, 55–60.
118. Ifediba, C.J.; Okezie, U.M.; Onyegbule, F.A.; Gugu, T.H.; Egbuim, T.C.; Ugwu, M.C. Antifungal activity of the methanol tuber extract of *Dioscorea Dumetorum* (Pax). *World Wide J. Multidiscip. Res. Dev.* **2017**, *3*, 376–380.
119. Yin, N.S.; Abdullah, S.Y.; Phin, C.K. Phytochemical constituents from leaves of *Elaeis guineensis* and their antioxidant and antimicrobial activities. *Int. J. Pharm. Pharm. Sci.* **2013**, *5*, 137–140.
120. Erhabor, J.O.; Oshomoh, E.O.; Timothy, O.; Osazuwa, E.S.; Idu, M. Antimicrobial activity of the methanol and aqueous leaf extracts of *Emilia coccinea* (Sims) G. Don. *Niger. J. Biotechnol.* **2013**, *25*, 37–45.
121. Unegbu, C.C.; Ajah, O.; Amaralam, E.C.; Anyanwu, O.O. Evaluation of phytochemical contents of *Emilia coccinea* leaves. *J. Med. Bot.* **2017**, *1*, 47–50.
122. Edrees, W.H.A. The inhibitory effect of *Euphorbia hirta* extracts against some wound bacteria isolated from Yemeni patients. *Chron. Pharm. Sci.* **2019**, *3*, 780–786.
123. Usman, H.; Abdulrahman, F.I.; Usman, A. Qualitative phytochemical screening and in vitro antimicrobial effects of methanol stem bark extract of *Ficus thonningii* (Moraceae). *Afr. J. Tradit. Complement. Altern. Med.* **2009**, *6*, 289–295. [CrossRef] [PubMed]
124. Dickson, R.A.; Houghton, P.J.; Hylands, P.J.; Gibbons, S. Antimicrobial, resistance-modifying effects, antioxidant and free radical scavenging activities of *Mezoneuron benthamianum* Baill., *Securinega virosa* Roxb. &Wlld. and *Microglossa pyrifolia* Lam. *Phytother. Res. Intern. J. Devot. Pharmacol. Toxicol. Evaluat. Nat. Prod. Deriv.* **2006**, *20*, 41–45.
125. Guo-Cai, W.A.N.G.; Liang, J.P.; Ying, W.A.N.G.; Qian, L.I.; Wen-Cai, Y.E. Chemical constituents from *Flueggea virosa*. *Chin. J. Nat. Med.* **2008**, *6*, 251–253.
126. Adekunle, A.A.; Ikumapayi, A.M. Antifungal property and phytochemical screening of the crude extracts of *Funtumia elastica* and *Mallotus oppositifolius*. *West. Indian Med. J.* **2006**, *55*, 219. [CrossRef]
127. Kombate, B.; Metowogo, K.; Kantati, Y.T.; Afanyibo, Y.G.; Fankibe, N.; Halatoko, A.W.; Aklikokou, K.A. Phytochemical screening, antimicrobial and antioxidant activities of *Aloe buettneri, Mitracarpus scaber* and *Hannoa undulata* used in Togolese Cosmetopoeia. *J. Drug Deliv. Ther.* **2022**, *12*, 19–24. [CrossRef]
128. Tolulope, O. Phytochemical Screening, Antibacterial Activity and Fatty Acids from *Heliotropium indicum*. *Pharmacogn. J.* **2023**, *15*, 350–352.
129. Okorondu, S.I.; Akujobi, C.O.; Okorondu, J.N.; Anyado-Nwadike, S.O. Antimicrobial activity of the leaf extracts of *Moringa oleifera* and *Jatropha curcas* on pathogenic bacteria. *Intern. J. Biol. Chem. Sci.* **2013**, *7*, 195–202.
130. Agbo, I.A.; HlangothI, B.; Didloff, J.; Hattingh, A.C.; Venables, L.; Govender, S.; van de Venter, M. Comparative evaluation of the phytochemical contents, antioxidant and some biological activities of *Khaya grandifoliola* methanol and ethyl acetate stem bark, root and leaf extracts. *Trop. J. Nat. Prod. Res.* **2023**, *7*, 2829–2836.
131. Agyare, C.; Dwobeng, A.S.; Agyepong, N.; Boakye, Y.D.; Mensah, K.B.; Ayande, P.G.; Adarkwa-Yiadom, M. Antimicrobial, antioxidant, and wound healing properties of *Kigelia africana* (Lam.) Beneth. and *Strophanthus hispidus* DC. *Adv. Pharmacol. Pharm. Sci.* **2013**, *2013*, 692613.
132. Bello, I.; Shehu, M.W.; Musa, M.; Asmawi, M.Z.; Mahmud, R. *Kigelia africana* (Lam.) Benth.(Sausage tree): Phytochemistry and pharmacological review of a quintessential African traditional medicinal plant. *J. Ethnopharmacol.* **2016**, *189*, 253–276. [CrossRef] [PubMed]
133. Isaiah, O.O.; Olusegun, O.A.; Adesola, O.C.; Samson, A.O. Anti-infective Properties and time-killing assay of *Lannea acida* extracts and its constituents. *Biosci. Bioeng.* **2021**, *6*, 100.
134. Ouattara, M.B.; Bationo, J.H.; Kiendrebeogo, M.; Nacoulma, O.G. Evaluation of Acute Toxicity, Antioxidant and Antibacterial Potential of Leaves Extracts from Two Anacardiaceae's Species: *Lannea microcarpa* Engl. & K. Krause and *Mangifera indica* L. *J. Biosci. Med.* **2022**, *10*, 125–134.
135. Ngouana, V.; Fokou, P.V.; Menkem, E.Z.; Donkeng, V.F.; Fotso, G.W.; Ngadjui, B.T.; Boyom, F.F. Phytochemical analysis and antifungal property of *Mallotus oppositifolius* (Geiseler) Müll. Arg.(Euphorbiaceae). *Int. J. Biol. Chem.* **2021**, *15*, 414–426.
136. El-Gied, A.A.A.; Abdelkareem, A.M.; Hamedelniel, E.I. Investigation of cream and ointment on antimicrobial activity of *Mangifera indica* extract. *J. Adv. Pharm. Technol. Res.* **2015**, *6*, 53. [CrossRef]
137. Kumar, M.; Saurabh, V.; Tomar, M.; Hasan, M.; Changan, S.; Sasi, M.; Mekhemar, M. Mango (*Mangifera indica* L.) leaves: Nutritional composition, phytochemical profile, and health-promoting bioactivities. *Antioxidants* **2021**, *10*, 299. [CrossRef] [PubMed]
138. Daniel, P.; Supe, U.; Roymon, M.G. A review on phytochemical analysis of *Momordica charantia*. *Int. J. Adv. Pharm. Biol. Chem.* **2014**, *3*, 214–220.
139. Mwambete, K.D. The in vitro antimicrobial activity of fruit and leaf crude extracts of *Momordica charantia*: A Tanzania medicinal plant. *Afr. Health Sci.* **2009**, *9*, 34–39. [PubMed]
140. Dongmo, S.C.M.; Njateng, G.S.S.; Tane, P.; Kuiate, J.R. Chemical composition and antimicrobial activity of essential oils from *Aframomum citratum, Aframomum daniellii, Piper capense* and *Monodora myristica*. *J. Med. Plants Res.* **2019**, *13*, 173–187.
141. Azubuike, C.P.; Obiakor, C.V.; Igbokwe, N.H.; Usman, A.R. Antimicrobial and Physical Properties of Herbal Ointments Formulated with Methanolic extracts of *Persea americana* seed and *Nauclea latifolia* stem bark. *J. Pharm. Sci. Pharm. Pract.* **2014**, *10*, 166–172.

142. Eze, S.O.; Ernest, O. Phytochemical and nutrient evaluation of the leaves and fruits of *Nauclea latifolia* (Uvuru-ilu). *Communicat. Appl. Sci.* **2014**, *2*, 1.

143. Usman, H.; Osuji, J.C. Phytochemical and in vitro antimicrobial assay of the leaf extract of *Newbouldia laevis*. *Afr. J. Tradit. Complement. Altern. Med.* **2007**, *4*, 476–480. [CrossRef]

144. Chukwunonye, U.C.E.; Ebele, O.P.; Kenne, T.M.; Gaza, A.S.P. Phytochemical screening and antimicrobial activity of methanol extract and fractions of the leaf of *Piliostigma thonningii* Schum (Caesalpiniaceae). *World Appl. Sci. J.* **2017**, *35*, 621–625.

145. Premkumar, V.G.; Shyamsundar, D. Antidermatophytic activity of *Pistia stratiotes*. *Indian J. Pharmacol.* **2005**, *37*, 127.

146. Namukobe, J.; Sekandi, P.; Byamukama, R.; Murungi, M.; Nambooze, J.; Ekyibetenga, Y.; Asiimwe, S. Antibacterial, antioxidant, and sun protection potential of selected ethno medicinal plants used for skin infections in Uganda. *Trop. Med. Health* **2021**, *49*, 49. [CrossRef]

147. Chukwudozie, I.K.; Ezeonu, I.M. Antimicrobial properties and acute toxicity evaluation of *Pycnanthus angolensis* stem bark. *Sci. Afr.* **2022**, *16*, e01185. [CrossRef]

148. Kisangau, D.P.; Hosea, K.M.; Lyaruu, H.V.; Joseph, C.C.; Mbwambo, Z.H.; Masimba, P.J.; Sewald, N. Screening of traditionally used Tanzanian medicinal plants for antifungal activity. *Pharm. Boil.* **2009**, *47*, 708–716. [CrossRef]

149. Olatokunboh, A.O.; Kayode, Y.O.; Adeola, O.K. Anticonvulsant activity of *Rauvolfia vomitoria* (Afzel). *Afr. J. Pharm. Pharmacol.* **2009**, *3*, 319–322.

150. Donkor, A.M.; Mosobil, R.; Suurbaar, J. In vitro bacteriostatic and bactericidal activities of *Senna alata*, *Ricinus communis* and *Lannea barteri* extracts against wound and skin disease causing bacteria. *J. Anal. Pharm. Res.* **2016**, *3*, 00046.

151. Adelanwa, E.B.; Habibu, I. Phytochemical screening and antimicrobial activities of the methanolic leaf extract of *Jacaranda mimosifolia* D. DON and *Sansevieria liberica* THUNB. *J. Trop. Biosci.* **2015**, *10*, 1–6.

152. Kowti, R.; Harsha, R.; Ahmed, M.G.; Hareesh, A.R.; Thammanna Gowda, S.S.; Dinesha, R.; Satish Kumar, B.P.; Irfan Ali, M. Antimicrobial activity of ethanol extract of leaf and flower of *Spathodea campanulata* P. Beauv. *Res. J. Pharm. Biol. Chem. Sci.* **2010**, *1*, 691–698.

153. Aliyu, M.S.; Hanwa, U.A.; Tijjani, M.B.; Aliyu, A.B.; Ya'u, B. Phytochemical and antibacterial properties of leaf extract of *Stereospermum kunthianum* (Bignoniaceae). *Niger. J. Basic Appl. Sci.* **2009**, *17*, 235–239. [CrossRef]

154. Ojiako, O.A.; Igwe, C.U. A time-trend hypoglycemic study of ethanol and chloroform extracts of *Strophanthus hispidus*. *J. Herbs Spices Med. Plants* **2009**, *15*, 1–8. [CrossRef]

155. Lukubye, B.; Ajayi, C.O.; Wangalwa, R.; Kagoro-Rugunda, G. Phytochemical profile and antimicrobial activity of the leaves and stem bark of *Symphonia globulifera* Lf and *Allophylus abyssinicus* (Hochst.) Radlk. *BMC Complement. Med. Ther.* **2022**, *22*, 223. [CrossRef]

156. Adewumi, S.S.; Akinpelu, B.A.; Akinpelu, D.A.; Aiyegoro, O.A.; Alayande, K.A.; Agunbiade, M.O. Studies on wound healing potentials of the leaf extract of *Terminalia avicennioides* (Guill. & parr.) on wistar rats. *S. Afr. J. Bot.* **2020**, *133*, 285–297.

157. Yadav, E.; Singh, D.; Yadav, P.; Verma, A. Attenuation of dermal wounds via downregulating oxidative stress and inflammatory markers by protocatechuic acid rich n-butanol fraction of *Trianthema portulacastrum* Linn. in wistar albino rats. *Biomed. Pharmacother.* **2017**, *96*, 86–97. [CrossRef] [PubMed]

158. Ebi, G.C.; Ifeanacho, C.J.; Kamalu, T.N. Antimicrobial properties of *Uvaria chamae* stem bark. *Fitoterapia* **1999**, *70*, 621–624. [CrossRef]

159. Akinduti, P.A.; Emoh-Robinson, V.; Obamoh-Triumphant, H.F.; Obafemi, Y.D.; Banjo, T.T. Antibacterial activities of plant leaf extracts against multi-antibiotic resistant *Staphylococcus aureus* associated with skin and soft tissue infections. *BMC Complement. Med. Ther.* **2022**, *22*, 47. [CrossRef] [PubMed]

160. Kouakou, A.B.; Megnanou, R.M. Potential treatment of mycosic dermatoses by shea (*Vitellaria paradoxa*) nuts hulls and press cakes: In vitro efficacy of their methanolic extracts. *Arab. J. Med. Aromat. Plants* **2021**, *7*, 342–351.

161. Ndukwe, I.G.; Amupitan, J.O.; Isah, Y.; Adegoke, K.S. Phytochemical and antimicrobial screening of the crude extracts from the root, stem bark and leaves of *Vitellaria paradoxa* (GAERTN. F). *Afr. J. Biotechnol.* **2007**, *6*, 1905–1909. [CrossRef]

162. Schmourlo, G.; Mendonça-Filho, R.R.; Alviano, C.S.; Costa, S.S. Screening of antifungal agents using ethanol precipitation and bioautography of medicinal and food plants. *J. Ethnopharmacol.* **2005**, *96*, 563–568. [CrossRef]

163. Ashalata, N.; Swarnalata, N.; Laitonjam, W.S. Phytochemical Constituents, Total Flavonoid and Phenolic Content of *Xanthosoma sagittifolium* Stem Extracts. *J. Acad. Indust. Res.* **2021**, *10*, 1–4.

164. Onuora, C.C.; Florence, C.C. Antimycotic Efficacy of Aqueous Extract from *Xylopia aethiopica* Against Some Zoophilic Dermatophytes. *GSJ* **2020**, *8*, 4595–4604.

165. Usman, L.A.; Akolade, J.O.; Odebisi, B.O.; Olanipekun, B. Chemical Composition and antibacterial activity of fruit essential oil of *Xylopia aethiopica* D. grown in Nigeria. *J. Ess. Oil Bear. Plants* **2016**, *19*, 648–655. [CrossRef]

Review

An Insight into the Cosmetic and Dermatologic Applications of the Molecules of Palmyra Palm

Sunehra Sayanhika [1,2] and Paulraj Mosae Selvakumar [2,*]

[1] Bioinformatics and Biotechnology Program, Asian University for Women, Chittagong 4000, Bangladesh; sunehra.sayanhika@auw.edu.bd

[2] Environmental Science Program, Asian University for Women, Chittagong 4000, Bangladesh

* Correspondence: p.selvakumar@auw.edu.bd

Abstract: Palmyra palm is a resourceful plant species that can be put to use in superabundance. Its extensive range of use stretches in all directions—making it useful for consumption in countless forms, and the cosmetic industry is not a deviation from this. In accordance with the beneficial molecules generated in it, such as polyphenols, flavonoids, carotenoids and others, the addition of Palmyra raises incentives like reduction and microbe prevention when developed into therapeutic products. Therefore, the virtues of Palmyra fruit, sap, leaves, pulp, bark, haustoria and other parts are being diversely exploited in the beauty and health industry at present. To summarise the compass of Palmyra palm and its products in cosmetology and dermatology, an overview is drafted exploring the extant literature on the topic. Following the description of the available molecules, their adoption into skincare products and in vivo effects was analysed in this study. Aiming to highlight the prospects of Palmyra in skin and personal care formulations, this article discusses the span of its potential in light of its physicochemical attributes.

Keywords: Palmyra palm; cosmetic products; antioxidant; anti-microbe; cosmetology; dermatology

Citation: Sayanhika, S.; Selvakumar, P.M. An Insight into the Cosmetic and Dermatologic Applications of the Molecules of Palmyra Palm. *Cosmetics* **2024**, *11*, 196. https://doi.org/10.3390/cosmetics11060196

Academic Editor: Maria Manconi

Received: 18 September 2024
Revised: 6 November 2024
Accepted: 12 November 2024
Published: 20 November 2024

1. Introduction

Borassus flabellifer, commonly known as Palmyra palm, is a versatile species that can be wielded in multifarious ways (Figure 1). The culture of nurturing Palmyra palms has been passed down for generations in certain Asian and African countries where it grows easily [1]. Besides being largely popular for their contributions to the timbre, paper and handicraft industries, the nutraceutical and functional benefits from the flavonoids, carotenoids, phenol groups, hemicellulose, lignin, pectin, flabelliferins, tannins, sugars and vitamins extracted from the different parts of the tree are utilised to a great extent. At the same time, value-added products like toddy, palm jaggery and palm sugar have a range of uses in various forms [2]. With the knowledge of its molecular components and the correct technique, tapping into the advantages of the tree is beyond rewarding [3]. Traditionally, Palmyra has been manoeuvred for soothing skin irritations like dermatitis, redness, inflammations, prickly heats and boils [2]. For example, Palmyra tappers in Southern India have long been tending to their cuts and injuries from tree climbing and handling inflorescence with scaly extracts from the Palmyra leaves' basal parts [4]. Noting how Palmyra's impressive mineral content and anti-parasitic characteristics bears it to an even wider spectrum of exploitation for beautification and medicinal purposes [5], this review assesses the extent of its molecular application in the fields of cosmetology and dermatology.

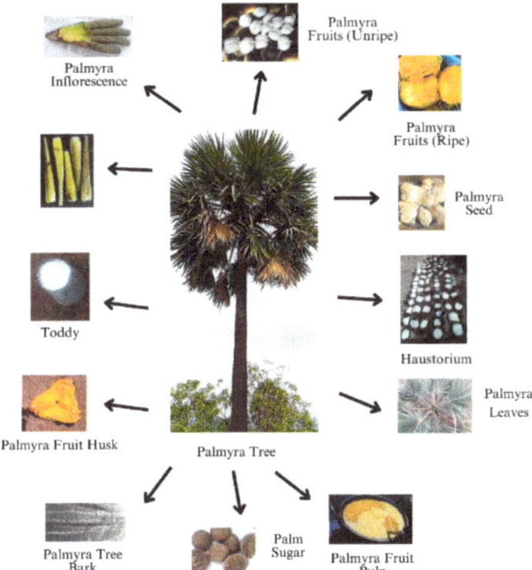

Figure 1. Parts of Palmyra palm.

2. Molecules of Palmyra Palm

2.1. Phenols: Flavonoids and Tannins

The biomolecules found in Palmyra palm have shown noteworthy potential as cosmetic ingredients (Figure 2). Palmyra fruit has 36.3–99.34 µg/100 mg of total phenolic content and 98.48–222 µg/100 mg of total flavonoid content [6]. Phenolic compounds and flavonoids found in the sap, tender fruit and seeds give rise to antioxidant and anti-inflammatory features [7]. Flavonoids also possess the ability to counteract oxidant, inflammatory, mutagenic, and carcinogenic activity and thus have a positive influence in terms of being administered in medical and cosmetic formulations [8]. Phenolics such as gallic acid, chlorogenic acid, rutin and caffeic acid are present in the plant and enable free radical scavenging properties [9]. While examining the secretions of the young palm plants, an oxidising reaction confirmed the presence of tannin, a phenolic compound, due to a colour change from white to a tannin colour. Subsequently, its correlation with antimicrobial activity during the early growth stage of shoots was suggested [4].

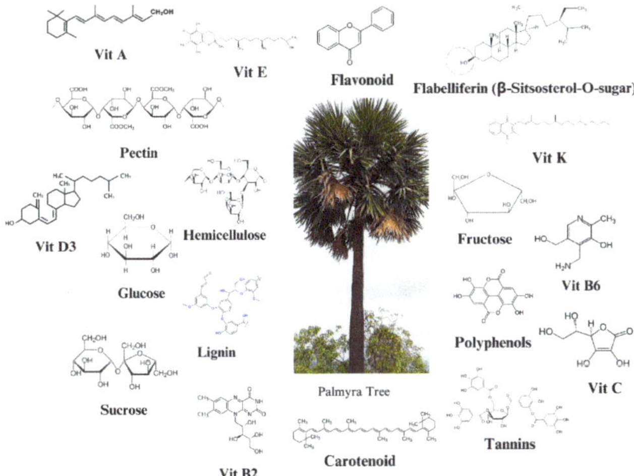

Figure 2. Molecules of Palmyra palm.

The different types of phenolic compounds available in the species create the possibility of choosing the target molecule from the total content for desired effects in products. In fact, achieving the aforementioned benefits from directing the phenolics from the Palmyra into therapeutic and cosmetic products is dependent on the successful retrieval of the components from the plant. By breaking down the powder of grinded young Palmyra palm fruits in petroleum ether, active ingredients can be extracted through soxhlation with ethanol. Followed by a qualitative phytochemical screening, the separation of phenolic and flavonoid contents can be achieved with assays [10].

2.2. Carotenoids

Research on Palmyra pulp has revealed its carotenoid alongside polyphenolic components, confirming its potential as a natural skin cleanser and moisturiser. The antioxidants available from the carotenoids are able to nourish the skin and safeguard it by feeding on the free radicals from exposure to the UV rays of sunlight [11]. Estimates of beta-carotenoid in the fruit pulp revealed an amount of 3 ppm/g [12]. Atmospheric liquid extraction using methanol and dichloromethane (DCM) is able to extract carotenoids from the Palmyra palm fruit pulp that can be sufficiently utilised for scaled-up procedures [13].

2.3. Saponins

Steroidal saponins, called flabelliferins, of four types are attributed to the bitterness of the Palmyra tubers and ripe fruits [14]. Using flash chromatography, flabelliferins can be isolated from the fruit pulp of Palmyra palm [15]. A total saponin content of 0.63 g/100 g was detected in Palmyra fruit [6]. In total, the flabelliferins detected in the fruit, tuber and inflorescence of Palmyra are of 14 types. Among them, Flabelliferin FB acts against microbial attack [16]. The category of metabolites has a long history of being adopted in folk skincare and remedies [17]. For cosmetic or therapeutic product development, the isolation of each class of flabelliferins from the total content has to be screened for their effects on skin-related benefits or cures.

2.4. Vitamins

Scrutiny of the composition of Palmyra unearthed vitamins and minerals that are healthy for human biological systems [7]. In research centring Palmyra fruit pulp, peaks for Vitamins D_3 (0.970 µg/100 mg), C (16.9 mg/100 g), E (2 µg/100 mg) and K (0.4 µg/100 mg) were found among the tested vitamins during an HPLC analysis, suggesting that their presence in high quantities [12]. An experiment based on spectrophotometric measurement explored the quantity of Vitamins A (8.89 ± 0.03 µg/100 g), B2 (0.03 ± 0.003 mg/100 g) and B6 (3.50 ± 0.20 mg/100 g) in Palmyra fruit [18]. On the other hand, the presence of Vitamin E was confirmed in young roots [9]. The fresh pulp of Palmyra is also enriched with Vitamins A and C. Processing Palmyra palm syrup in a vacuum state or membrane filter or both, the vitamin profile can be checked with a thermal process and ultrafiltration process before further differentiation [19]. While developing skincare and medicinal products, all available vitamins might not be used together as their outcomes and target areas differ. Since a good range of vitamins is detected from the assessment, their separation in sufficient amounts has to be conducted following the detection of the presence of vitamin types.

2.5. Fibres and Reducing Sugars

Furthermore, the biochemical reactions in Palmyra palm produce a range of polymeric compounds, like pectin, hemicellulose and lignin, whose incorporation in cosmetic products has proven to refine skin health upon usage [20]. The chemical composition of the raw fruit fibres showed 2.58% pectin, 25.48% hemicellulose and 16.19% lignin content. The fibre polymers of Palmyra can be separated through chemical screening after implementing wet processing techniques (i.e., scouring, bleaching and softening) on manually dry-processed fibres obtained from the fruit [21]. The non-centrifugal sugar (or jaggery) prepared from

Palmyra sap shows reducing sugars like glucose (5.0–18.5 g/L), sucrose (92.9–174.4 g/L) and fructose (5.0–18.1 g/L) [22].

Since the success and safety regarding the application of these molecules essentially depend on the biological effects they produce, isomerisation must be taken into consideration during the extraction as well as incorporation into products.

3. Cosmetics Developed from Palmyra Palm

Referring to its commercial value, it remains one of the three most economically important species of one of the most broadly distributed genera of Palmae—numerous contemporary uses of Palmyra have been documented to date. The translucent heart of ice apple leaves a natural soothing effect on the skin during scorching summer days. Its phytonutrients and phytochemicals are effective in treating pimples from prickly heats [23]. A study on the value-added products of Palmyra palm notes the anti-inflammatory impacts of sugar palm fruits [24]. If applied in a thin coat and washed off upon drying, a face pack of freshly ground Palmyra palm fruit and powdered pure sandalwood with a coconut water base can soothe the skin from sunburn [25]. The beneficial effects from applying certain parts of the species resulted from the molecules present in them, indicating their potential as raw materials for cosmetics and skin remedies.

Palmyra has been credited as one of the savanna plants comprising a variety of cosmetic ingredients. Anti-aging products developed from such sources are equipped with anti-elastase and anti-collagenase properties. Anti-wrinkle cosmetics from the plant were said to be viable because of the antioxidants. In contrast, its antibacterial cosmetics clear acne and assist in sound dental and dermal health [26]. The bark of this celestial tree was documented as being used in mouthwash and dentifrice production [27].

To assess the antioxidant effects of the Palmyra palm fruit pulp, facial cream, solid soap and liquid soap made out of its carotenoid extract was examined (Figure 3). All three cosmetic products containing carotenoid oil exhibited a considerably higher inhibition in comparison to commercial creams. The facial cream created from all-natural ingredients had a moisturising effect similar to solid lotion bars besides being cost-effective, non-toxic and kid-friendly. Carotenoids present in the products break down in contact with heat and sunlight, thus the authors warned against application as a day cream. Both organic soaps held no synthetic chemicals or harmful salts and had satisfactory levels of antioxidant and antimicrobial properties. More importantly, they neither irritated skin nor clogged washroom sinks owing to the trace amount of free alkali content. The soapberry extracts in the liquid soap made it comparatively more beneficial. Storage in a dark and cool environment was notably stressed for all of these cosmetics [28].

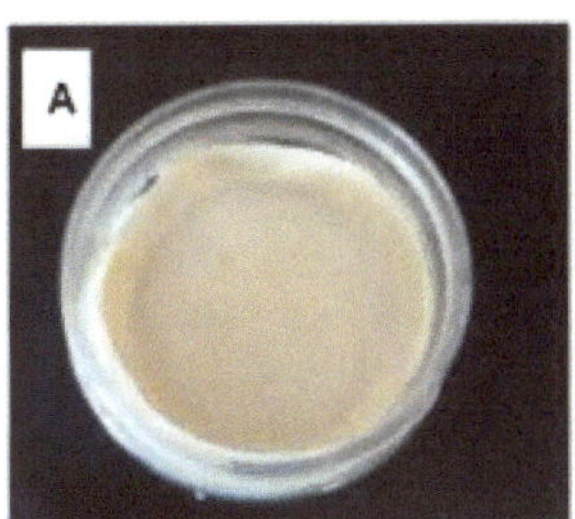
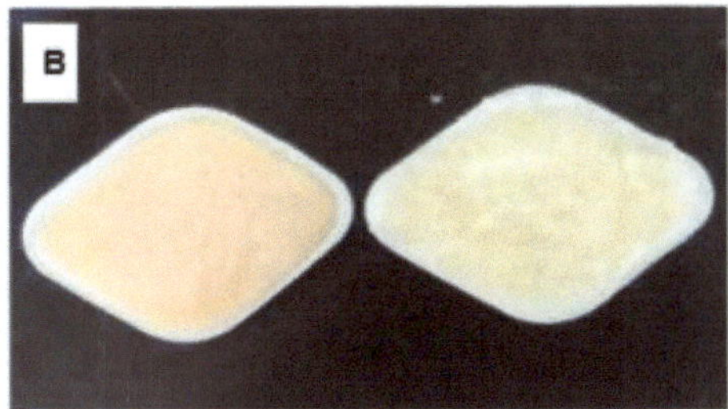
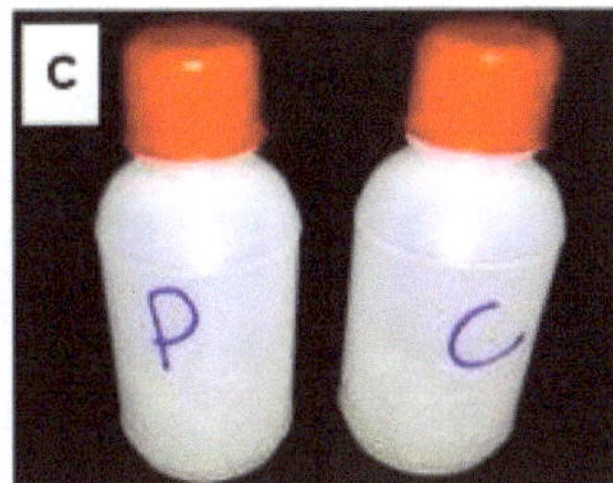

Figure 3. Images of facial cream (**A**), solid soap (**B**) and liquid soap (**C**) prepared from Palmyra palm (Source: [28], Copyright obtained from Elsevier).

Combining Palmyra fruit pulp with aloe vera gel, a soothing face wash with worthwhile antioxidants and antimicrobial properties was produced. Keeping the pulp to aloe vera ratio as 2:3, the best results for physicochemical and sensory analyses were retrieved among multiple test samples [29]. To avoid the generic drying effect of soaps, a group of researchers created a natural liquid soap from Palmyra fruit pulp by creatively pairing it with rock salt and cold-pressed virgin coconut oil. The use of soapberry ensured the free alkali count was low [30].

The Indian brand "Reju Ayur" launched an "Under Eye Gel with Ice Apple & Cucumber" that combats dark circles, wrinkles and signs of tiredness in the eye area. Created in the union of saffron oil extract and aloe vera extract with ice apple and cucumber, the formula supposedly modulates the skin microbiome and accelerates collagen synthesis for bright and moisturised skin. Gentle application on a clean face is recommended for the eye cream, and it is usable as both a day and night cream. Cautiousness has to be maintained during its use so as to not irritate or damage the eyes [31]. The interaction between Palmyra pulp and the other natural constituents, like soapberry, aloe vera gel, rock salt, virgin coconut oil and cucumber, has to be investigated to determine the extent of each of their contributions towards the benefits. Nonetheless, Palmyra pulp clearly qualifies as a safe and instrumental component of known merits that has been successfully employed in cosmetic development.

Research aiming to manufacture an eco-friendly Palmyra pulp toothpaste was conducted successfully by surveying 12 product versions with variable amounts of Carboxymethyl Cellulose, turmeric powder and citric acid [32]. In 2020, Balinese pharmacists managed to create hand sanitizers with 96% alcohol percent, altering fermented palm wine to tackle the shortage of disinfectant supply [33]. The repurposing of a local produce-based ethanol into a germicide is exemplary in terms of the versatility of plant-based secondary products. Research and development on the transformed attributes of Palmyra molecules upon their conversion into secondary products can guide the manufacture of creative formulations.

For example, a body polish, made by processing Palmyra palm sugar crystals with Vitamin B5 and macadamia oil, has been distributed in the market by a brand called "Cocoon". Inspired by the specialty of the An Giang province of Vietnam, a grainy body exfoliation formula was developed from the concept of fragrant palm sugar cakes. The product claims to remove dead skin and dirt from the body surface, nourishing skin to be soft, smooth and supple. It is noteworthy that its usage on open wounds is forbidden [34].

Nano-fibrillated cellulose from palm leaf and stalk was used for the green synthesis of carbon nanodots (CD) in a reliable, efficient and cost-effective way. As CD can penetrate cells well, the scope of their application in cosmetics is wide [35]. In fact, CD derived from plant extracts amplifies the ability of cosmetics with respect to brightening, antioxidant, anti-inflammatory and UV-absorption activity [36]. Given the widespread use of nanoparticles in the field of cosmetics at present, the molecules of Palmyra palm discussed in this review can be screened to investigate their abilities as nanoparticles in biological systems. In most of the existing Palmyra-based cosmetic formulations, a group of molecules from plant parts was utilised to garner a number of positive effects. Exploiting the capacity of individual molecules simplifies the selection process according to the purpose of the cosmetic or therapeutic product.

Biocosmetics has emerged as a growing industry in recent times to promote safe alternatives to fossil-dependent sources. Given the growing demand for natural cosmetics, recent technologies are focussing on viable extraction methods of bioactive ingredients so that smart formulations can be produced for the best results in dermal and transdermal administrations [37]. Biocosmetics are bio-based products that are completely naturally sourced from ingredients such as plants, animals, microorganisms, enzymes and insects. They are developed by focussing on skin and hair care as well as oral hygiene. Contrary to hazardous and non-biodegradable components present in inorganic beauty products, biocosmetics offer eco-friendly alternatives that are preserved without chemical pesticides while underscoring ecological concerns in the manufacturing process [38]. The increasing interest in sustainable and organic consumerism is driving the market demand for bio-derived products. People are more conscious about the environmental impacts and ethical sourcing of their purchases. Consequentially, the use of the likes of plant extracts, natural antioxidants and essential oil has peaked in contrast to synthetic options. For personal care and beautification, people are inclined toward such options due to their dependability in terms of health and safety [39]. Currently, the industry for biocosmetics is looking hopeful as the switch from fossil-based to bio-based cosmetics is being backed by the majority of producers and brands [38]. Since the ingredients of biocosmetics, like antioxidants, are obtained by physical and/or chemical methods, biotechnology plays an essential role in its production. It is mandatory to highlight the targeted property of the material while pursuing the bioactivation of molecules, such as antioxidant or phenolic traits [38]. Palmyra palm molecules have a wide spectrum of properties of interest that can be extensively manoeuvred in cosmetic and therapeutic formulations by themselves or in combinations. Therefore, with substantial investment in research and development, cosmetic products exhibiting the beneficial qualities of Palmyra palm have immense possibilities to attract business opportunities in the field.

In-Vivo Effects of Palmyra-Based Formulations

Treating dry, patchy and wrinkly skin, the facial cream from Palmyra palm was able to enhance the radiance and smoothness of skin in subjects. None of the users reported skin irritation. As it contained high oil content, the cream was pronounced unsuitable for oily skin [28]. Checking all parameters, two final versions of the eco-friendly Palmyra pulp toothpaste were chosen through sensory evaluation [32]. The Balinese hand sanitizers from fermented palm wine met WHO standards and thus were greatly availed by the population during a price hike crisis in the COVID-19 period [33]. The final face wash created with aloe vera and Palmyra fruit pulp extract spreads, foams and washes off properly alongside having a pleasant colour, texture and appearance [29]. The Palmyra fruit pulp liquid soap with rock salt and cold-pressed virgin coconut oil was reportedly gentle and hydrating upon usage [30]. Except for the sanitiser and marketed cosmetics, all Palmyra-based experimental cosmetics included in this section were pronounced safe owing to sensory tests. For these prototypes to be commercialised, additional tests involving the microbiological and side-effect parameters will be required to meet the requirements of the Food and Drug Administration (FDA). For now, the positive reviews from sensory evaluations can be considered as a foundation for further inquiry.

4. Scopes of Palmyra Palm Biomolecules in Cosmetology

A number of molecules accessible from the *B. flabellifer* are commonly used in skincare. According to Jamkhande et al., the phytochemicals found in the leaf extracts of the tree show antioxidant and antibacterial activity [40]. A study by Paschapur et al. reported the anti-inflammatory properties of its male inflorescences [41]. In addition to hydrating the body, Palmyra fruits are effective in healing wounds and immunity modulation given their abilities to minimise bacterial and inflammation effects. Secondary Palmyra products like toddy resist fungal activity while palm sugar by-products aid the skin barrier by preventing prickly heat, boils and redness from inflammation [42]. The popularity of the parts of Palmyra as home remedies for skin problems provides a reason to probe into the molecular basis of the claims of prophylactic effect.

Quantifying the efficiency of the antioxidants present in the Palmyra pulp using a DPPH assay, the IC50 i.e. the concentration needed to inhibit free radical activity by 50 percent was found to be 0.85–0.82 mg/mL [11]. The endogenous antioxidant system of skin can be boosted by the available polyphenols employing their role in transcending the epidermal layers in order to target certain receptor molecules. As a result, their use in treating skin diseases led to repairing skin damage [43]. The in-vitro evaluation of flavonoids and predominantly found phenolics from crop by-products, such as chlorogenic acid, caffeic acid and gallic acid, showed efficacy in the treatment of critical dermal diseases as well as minor skin issues [44]. Beta-carotene is used in personal care formulations for the pigment's ability to impart moisturisation to dry skin conditions and flakiness [45]. Since vitamins add to the antioxidant system across the layers of human skin, their topical use can be availed as an alternative to balance the gradient of keratinocytes [46]. The implementation of the molecule groups derived from Palmyra can potentially be multidirectional based on the chosen choice or combination. In any case, the plethora of options are worth examining for skin improvement.

Acquiring Vitamins C and E from parts of the Palmyra palm can contribute to plenty of pharmacological benefits. The collagen and elastic fibres that form the dermal layer connective tissues are responsible for maintaining skin elasticity [47]. Vitamin C or L-ascorbic acid plays an essential role in boosting collagen production, keratinocyte differentiation and skin barrier formation. At the same time, it decreases melanocyte biosynthesis and therefore counteracts pigmentation [48]. It also helps with masking signs of aging by reducing puffiness and wrinkle lines, cracks and crevices in the periorbital area [49]. The efficiency of Vitamin C is low for transdermal penetration, and its lipidic derivatives are more popular for topical applications. More study is required to improve the stability of Vitamin C in cosmetic products [48].

A study incubated cultured skin substitutes in Vitamin C-rich media to check the anatomic and physiological effects upon exposure to UV-B. The findings revealed better cellular viability as well as the notable development of the epidermal barrier and basement membrane. Cutaneous levels were especially elevated, and the extent of damage from erythema and formation of sunburn cells was lower [50]. The examination of a conjugate of Vitamin C in human whole skin explants showed favourable production of collagen III resulting in improved skin thickness [49]. This suggests that Vitamin C extracted from Palmyra might be effectual for both skin condition improvement as well as post-operational scars.

Clinical reviews have noted significant refinement in terms of brightness, skin tone, elasticity, compactness and general appearance of the skin after the use of Vitamin E. Moreover, fine lines around the eyes and roughness in texture subsided [46]. The isoforms or derivatives of Vitamin E are used in dermo-cosmetic topical applications for antioxidant, anti-pollution and photoprotective effects [51]. The intensity of oedema and sensitivity due to UV-B-caused damage to the skin is alleviated [46]. It minimises the impact of UV radiations on the skin utilising the ability to lessen lipid peroxidation, inflammation and erythema. In turn, the endogenous antioxidants present on site are protected. Since it aids the photostabilisation of chemical filters and reduces the degenerative effects of oxidative stress upon penetration into the skin, its addition as oil-in-water emulsion forms in sunblock improves the Sun Protection Factor (SPF). As a free radical scavenger, it smoothens the skin by working on expression lines and wrinkles. Its application stunts skin aging by boosting collagen production and enables the stratum corneum to have better control over conducting epithelialization and retaining humidity. Nanotechnology-based systems harness the properties to moisturise and repair the damage. The topical application of Vitamin E in combination with antioxidants is known to show better results. Among these, Vitamin C is a common option given its potential to rejuvenate oxidised Vitamin E [51]. In comparison to oral consumption, developing topical products comprising Vitamin E has a wide range of options to make it multifunctional as well as safe. The evaluation process of such therapy is also easier comparatively [46]. Acquiring the abundance of natural Vitamin E from the Palmyra palm, there is ample scope to produce a therapeutic dermo-cosmetic line of fine quality.

Through immunomodulation, active Vitamin D_3 metabolites regulate the proliferation and differentiation of keratinocytes to form a protective epidermal layer for skin homeostasis. In the process, the premature signs of aging, inflammation and other UV-associated damage are repaired due to anti-oxidative signals. It assists the growth of hair follicles [52]. Its analogs balance cell turnover and prevent the formation of psoriasis plaques by inhibiting the accumulation of dead cells on the skin's surface [53]. Although UV-B predisposes humans to non-melanoma skin cancer, depending on the dose and temperature, it is required to meet more than 90% of the body's Vitamin D_3 requirement. Again, as the skin's capacity to synthesize Vitamin D_3 lowers with age, the prescribed topical administration of cosmeceuticals made of isolated Vitamin D_3 precursors can protect the skin from environmental stressors [52]. So, developing cosmetic products with Palmyra palm-derived Vitamin D_3 will bear the chance of supplementing the skin's requirement for it in adults and resisting the proneness to non-melanoma skin cancer at the same time.

Vitamin K is introduced in creams for its superior healing ability. By advancing collagen and fibroblast buildup and resisting oxidative stress, it paces up repair after injury. When paired with calming agents like chamomile extract, it boosts blood circulation in narrowed-down target sites such as post-surgical bruises and rosacea-caused redness. Its nature of supporting vascular walls during blood clotting is put to work in this mechanism. The investigation indicates that Vitamin K can improve blood circulation in spots like bluish-dark circles that are caused by lymphatic drainage upon the disruption of blood flow. After blood pigments pile up underneath the dark circles, their purplish colour starts fading gradually, restoring the skin to normal condition [54]. When exposed to contaminants and photodamage, Vitamin K protects the skin. Furthermore, it protects collagen and helps with wrinkles and stretch marks. Importantly, it is safe for all skin types. Usually, Vitamin K is used in combination with other cosmetic ingredients in skincare for optimised results [55]. The difference in effects upon combining Vitamin K with each of the other vitamins found in Palmyra can be recorded to establish a basis for producing biocosmetic products from the species.

Among others, the role of the vitamins found in Palmyra palm, namely A, B2, B6, C and E, in cosmetics was uncovered in terms of their influence on retention, penetration and absorption capacity [56]. Hemicellulose is used in cosmetics for its moisturising, whitening, exfoliating and anti-aging properties [57]. In addition to its function as an antioxidant, anti-inflammation, anti-aging and antibacterial agent, lignin acts as a natural UV blocker and a tyrosinase inhibitor and is thus a skin-whitener. The cytotoxicity level of lignin is low as well, making it a user-friendly ingredient option for cosmetics. Its pairing with inorganic constituents achieves amplified positive effects in combination, making it a key component of bio-based cosmetic lines [58]. Pectin is a well-known ingredient in cosmetic products and also a demulcent [59]. On the other hand, glucose [60], sucrose [61] and fructose [62] have shown potential as humectants. Thus, several molecules found in Palmyra Palm have immense possibilities in the cosmetic industry.

5. Dermatologic Prospects of Palmyra Palm

As per the Siddha system of medicine practiced in ancient Southern India, a decoction called Panai Karpam could be prepared by mixing small bits of Palmyra roots with two parts of water and drying the solution under the sun. The concentrated mixture was allegedly able to treat alopecia when applied thrice a day [5]. There are literary records of ethnic communities using mixtures of Palmyra palm extracts with other herbs to treat scabies, sores, burns, secondary syphilis and leprosy [27]. In Ayurveda, the tree was cited as the cure for skin allergies and pruritus [63].

Phenolic metabolites or polyphenols from Palmyra are effectual in promoting health benefits when consumed by means of diet or topical application. These natural components can help with skin disorders and injuries while also posing low toxicity. Apart from skin conditions like wrinkles and ageing, further progression of acne can be resolved by their use in cosmetics [64]. The fruit husk of Palmyra can be made into an extract containing about 15 bioactive molecules that prevent microbial activity [9]. Flavonoids, as discussed to be found in Palmyra, have been helpful with chronic skin diseases like vitiligo, psoriasis, acne and atopic dermatitis [8]. Moreover, research on the functions and uses of the Palmyra haustorium has revealed phytochemical and bioactive compounds with pharmaceutical and ethnobotanical relevance [65]. Using the plant for curing skin ailments since ancient times renders a proposition for its credibility in anti-microbial formulations. The impact of its molecules on controlling or resolving certain skin complications and disorders has to be credited for their application in therapeutic products.

The extraction of Vitamin C available in Palmyra palm can be used in the development of revolutionary therapeutic formulations. The topical use of Vitamin C is more functional for wound healing as its oral administration is not sufficient in terms of bioavailability [66]. For the healing of surgical wounds and tissue repair, the local application of Vitamin C is beneficial. In several skin diseases, Vitamin C has reaped impressive results when applied as a therapeutic adjuvant. For example, treating acne scars with Vitamin C helps restore the tension and texture of skin alongside lessening post-inflammatory discolouration. It assists in turning back the keratinocytes to the normal state in psoriatic skin conditions and preventing the relapse of Progressive Pigmented Purpuric Dermatosis (PPPD) by strengthening the vascular system and blood vessel collagen. With its supplementary application, hyperpigmentation in the pigment-lost spots and patches was reported among vitiligo patients [48]. As an antioxidant that dissolves in water, a clinical study conducted on 30 s degree burn patients suggested healing effects from a solution of Vitamin C. In burn sites, it was seen to play a part in expediting collagen production in tissues and simultaneously removing free radicals [67].

Vitamin E, obtainable from Palmyra, can have great implementation as an ancillary to maximise the output from certain treatment combinations. To treat lesions from a keratinization disorder called Darier disease (DD) or keratosis follicularis, a study adopted a topical application of Vitamin E, retinyl palmitate and urea on the affected regions of a patient for two months. The keratotic papules were seen to improve with no additional complications. The genetic disorder is usually treated with analogs of topical aromatic vitamin A, but the side effects from such a procedure discouraged patients from pursuing the entire course of medication. However, with Vitamin E, not only did the mentioned combination achieve the desired results but also the bioavailability of vitamin A escalated, making drug delivery easier [68]. Furthermore, Vitamin E addresses conditions related to dermatitis, acne and Melasma [51].

Besides maintaining an optimum functional state of the skin, active Vitamin D forms oversee cutaneous immunity [52]. Compromised immunity from its deficiency leads to drying out of the skin bearing it to infections. As the skin barrier becomes fragile, events like acne, eczema and rosacea might worsen due to the increased severity of inflammation. Such imbalance may deteriorate acne conditions and collagen might prematurely stiffen and age [53]. In addition to treating skin problems, the extent of Palmyra's molecular constituents in boosting skin immunomodulation can be focussed on for research to list this quality as one of the potential benefits of medicinal products from the species.

Wound contraction due to the use of palm leaf secretion extracts was inspected in another study. Treating with 10% (*w*/*w*) extract with dimethyl sulfoxide (DMSO) resulted in noticeable wound healing just after the fourth day of treatment, followed by complete healing by day 12. Based on the outcome, a wound dressing was produced from the secretions of scaly palm leaves. In the same research, the leaf secretion extracts of Palmyra displayed an inhibition zone a diameter larger than that of standard antibiotics against *Vibrio mimicus* [4]. The specimen of the said pathogen was isolated from wound infections [69]. Complementing the advantage of DMSO as an integrant, wound healing was successfully achieved by the dressing harnessing the anti-microbial property of Palmyra leaf extracts.

Probing into the development of a hydrogel from the *B. flabellifer* L. male flower extract, the available bioactive compounds, such as gallic acid, coumarin and quercetin, were found to show antioxidant, anti-inflammatory and antibacterial attributes against *Cutibacterium acnes*. The Minimum Inhibitory Concentration (MIC) against the strain was concluded as 250 mg/mL. Measured as IC50, the ferric ion-reducing power of the male flower extract was high, indicating good reductive antioxidant action. The hydrogel effectively worked on the activity of acne vulgaris caused by *C. acnes*, a Gram-positive bacteria produced in sebaceous follicles, by treating the compromised skin condition through deploying antioxidants to balance out cellular redox reactions in the system [70]. Developing topical medicine for *C. acnes* that also serves as a sunblock by exploiting the antioxidant feature targeting both criteria can be a viable option.

Testing the probiotic Lactic Acid Bacteria (LAB) isolated from Palmyra palm sugar has shown resistance against methicillin-resistant *Staphylococcus aureus* (MRSA) [71]. Sites of *S. aureus* infections, including MRSA, are often identified in the form of scrapes or cuts on the skin. The affected area might appear to have a red and swollen bump that is painful and warm to the touch [72]. Streaking against *S. aureus*, the aqueous extract of Palmyra palm (L.) showed zones of inhibition of 30 µg/mL on Mueller Hinton agar [73].

Palmyra palm fruit pulp has been identified for the isolate *Acetobacter ghanensis*, used to produce acetic acid [74]. Bacterial and fungal infections are commonly treated with acetic acid due to the presence of antimicrobial and antioxidant functions. The same goes for its ascendency in treating skin conditions like pruritus and striae gravidarum and removing head lice [75].

A report on the isolation of 2,3,4-trihydroxy-5-methylacetophenone from Palmyra palm syrup revealed a broad range of antibacterial inhibition against *Escherichia coli*, *Mycobacterium smegmatis* and *Staphylococcus simulans* [76]. *E. coli*, although dominantly found in guts, was proven to cause skin infections like cellulitis or even type 1 necrotizing soft tissue infections (NSTIs) on some occasions dwelling on damaged tissues from burns, surgeries or eczema. In such cases, the bacterial infection affects the dermis, subcutaneous tissue, superficial fascia or muscle layers [77]. Due to *M. smegmatis* contamination, cases of patients suffering chronic skin and soft tissue-associated infections on injection or surgical areas post-cosmetic procedures have been recorded [78]. On the other hand, skin and soft tissue infections from *S. simulans* have been reported among people operating in close proximity with animals—e.g., veterinarians and butchers [79]. Therefore, it is apparent that the research and development of topical medication from Palmyra palm are promising in terms of fungal and bacterial skin contagions.

Overall, the molecules of Palmyra palm have excellent properties that can be diversely explored and implemented in the cosmetics and therapeutics sectors in the coming days (Table 1). However, their adoption into cosmetic and dermatologic products will have to be optimised with respect to in vitro and in vivo experiments so as to safely work on targeted skin issues.

Table 1. Scope of Palmyra palm in the cosmetic and dermatologic field.

Molecules of Palmyra Palm	Effects	Types of Potential Cosmetic or Therapeutic Products
Polyphenols	Antioxidant, anti-inflammation [43]	Sunscreen, anti-aging cream, demulcent [43], topical treatment for acne, skin disorders, injury, cut [64]
Carotenoids	Antioxidant [11]	Sunscreen, anti-aging cream [11]
Flavonoids	Antioxidant [7], anti-inflammation [8]	Sunscreen, anti-aging cream, demulcent [7], topical treatment for vitiligo, psoriasis, acne, atopic dermatitis [8]
Flabelliferins	Anti-microbe [16]	Topical treatment
Tannins	Anti-microbe [40]	Topical treatment
Hemicellulose	Antioxidant, anti-tyrosinase [57]	Moisturiser, exfoliator, anti-aging cream, whitening and brightening cream [57]
Lignin	Antioxidant, anti-inflammation, anti-tyrosinase, anti-microbe [58]	Sunscreen, anti-aging cream, whitening and brightening cream, topical treatment [58]
Pectin	Anti-inflammation [59]	Demulcent [59]
Vitamin C	Increase in collagen production, skin barrier repair, anti-UV-B, reduction of hyperpigmentation, wound healing [48]	Therapeutic adjuvant for acne scars, psoriatic skin conditions psoriatic, vitiligo [48] and burn wounds [67], Topical treatment of surgical wounds and tissue repair [48]
Vitamin K	Wound healing, collagen protection [55]	Topical treatment for dark circles, wrinkles and stretch marks [55], therapeutic adjuvant for post-surgical bruise and rosacea-caused redness [54]
Vitamin D_3	Antioxidant, anti-inflammation, immunomodulation [52]	Hair care product, anti-aging cream [52], cosmetics for overall skin health improvement [53].
Vitamin E	Antioxidant, anti-pollution, anti-inflammation [51], anti-UV-B [46], increase in collagen production [51]	Multifunctional dermocosmetic product for Darier disease [68], dermatitis, acne and Melasma [51], anti-aging cream, topical treatment of erythema [51], oedema [46]
Glucose, fructose, sucrose	Moisturisation [60–62]	Humectant [60–62]
Biomolecules from Palmyra palm fruit pulp	Anti-microbe [75]	Topical treatment for pruritus, striae gravidarum, head lice [75]
Biomolecules from Palmyra palm syrup	Anti-microbe [76]	Topical treatment for *Escherichia coli*, *Mycobacterium smegmatis*, *Staphylococcus simulans* [76]
Biomolecules from Palmyra palm leaf secretion	Anti-microbe [4]	Topical treatment for *Vibrio mimicus* [4]
Biomolecules from male flower extract	Anti-microbe [70]	Topical treatment for *Cutibacterium acnes* [70]
Biomolecules from Palmyra palm sugar	Anti-inflammation [42], anti-microbe [71]	Topical treatment for methicillin-resistant *Staphylococcus aureus* [71], prickly heat, boils, redness [42]
Biomolecules from toddy	Anti-fungal [42]	Topical treatment
Biomolecules from Palmyra palm extracts	Anti-inflammation, wound healing, anti-microbe	Topical treatment for *Staphylococcus aureus* [73], scabies, sores, burns, secondary syphilis, leprosy [27]

6. Conclusions

Alongside being a plant that is exceptional in the sense that it can be adapted in multiple dimensions while providing a reliable source of stable sustenance for many populations across the world, Palmyra palm evidently carries a bunch of molecules that promote healthy skin. Thus, the trend of undertaking investigations to explore the expanse of exercising its constituents has been fruitful. In the meantime, the advent of cosmetics based on it has achieved success. Through incorporation into cosmetic products, significant antioxidant and antimicrobial effects can be acquired by the biomolecules of the plant, such as flavonoids, carotenoids, polyphenols, hemicellulose, lignin, pectin, flabelliferins, tannins, sugars and vitamins, dissected in the article. It is important to optimise the plausible counter effects of the molecules while introducing them into biocosmetics to ensure the safety of users. Depending on the combination of beneficial molecules available, several cosmetics created from utilising parts of Palmyra have already been marketed. However, guiding product development primarily based on the properties of individual molecule groups can also be a possibility. Thus, there is ample room for exploration in the research and development of Palmyra in both cosmetology and dermatology. Moving forward, more attention to the optimization of these components in the conceptualization of Palmyra cosmetics and topical medications will be essential for the finest possible merchandise.

Funding: This research received no external funding.

Institutional Review Board Statement: Not applicable.

Informed Consent Statement: Not applicable.

Data Availability Statement: Not applicable.

Conflicts of Interest: The authors declare no conflict of interest.

References

1. Jana, H.; Jana, S. Palmyra Palm: Importance in Indian Agriculture. *Rashtriya Krishi* **2017**, *12*, 35–40. Available online: https://www.researchgate.net/publication/356323854_Palmyra_palm_Importance_in_Indian_agriculture (accessed on 21 October 2024).
2. Murugesan, R.; Vathsala, V. Palmyra palm (*Borassus flabellifer*): A multifaceted plant with diverse uses and cultivation practices. *Agric. Mag.* **2023**, *2*, 440–444. Available online: https://www.researchgate.net/publication/376649351_Palmyra_Palm_Borassus_flabellifer_A_Multifaceted_Plant_with_Diverse_Uses_and_Cultivation_Practices (accessed on 21 October 2024).
3. Davis, T.A.; Johnson, D.V. Current utilization and further development of the Palmyra palm (*Borassus flabellifer* L.; Arecaceae) in Tamil Nadu state, India. *Econ. Bot.* **1987**, *41*, 247–266. [CrossRef]
4. Mariselvam, R.; Ignacimuthu, S.; Ranjitsingh, A.; Mosae, S.P. An insight into leaf secretions of Asian Palmyra palm: A wound healing material from nature. *Mater. Today Proc.* **2021**, *47*, 733–738. [CrossRef]
5. Basava Prasad, A.; Vignesh, S.; Elumalai, A.; Anandharaj, A.; Chidanand, D.; Baskaran, N. Nutritional and pharmacological properties of Palmyra palm. *Food Humanit.* **2023**, *1*, 817–825. [CrossRef]
6. Behera, S. Phytochemical constituents and nutritional potential of Palmyra palm: A review. *Rev. Contemp. Sci. Acad. Stud.* **2022**, *2*. [CrossRef]
7. Vijayanchali, S.S. *Proximate Composition and Health Benefits of Palmyra (Borassus flabellifer) an Overview*; Writers Row Publications: Bengaluru, India, 2024.
8. Čižmárová, B.; Hubková, B.; Tomečková, V.; Birková, A. Flavonoids as promising natural compounds in the prevention and treatment of selected skin diseases. *Int. J. Mol. Sci.* **2023**, *24*, 6324. [CrossRef] [PubMed]
9. Basava Prasad, A.; Arunkumar, A.; Vignesh, S.; Chidanand, D.; Baskaran, N. Exploring the nutritional profiling and health benefits of Palmyra palm haustorium. *S. Afr. J. Bot.* **2022**, *151*, 228–237. [CrossRef]
10. Renuka, K.; Devi, V.R.; Subramanian, S.P. Phytochemical screening and evaluation of in vitro antioxidant potential of immature Palmyra palm (*Borassus flabellifer* Linn.) fruits. *Int. J. Pharm. Pharm. Sci.* **2018**, *10*, 77. [CrossRef]
11. Debenthini, S.; Jeganathan, B.; Sarananda, K.H. Antioxidant properties of Palmyrah (*Borassus flabellifer* L.) fruit pulp and effect of heat treatment on bitterness. In Proceedings of the 1st Faculty of Agriculture Undergraduate Research Symposium, Kandy, Sri Lanka, 23 December 2014. Available online: https://www.researchgate.net/publication/273717971_Antioxidant_Properties_of_Palmyrah_Borassus_flabellifer_L_Fruit_Pulp_and_Effect_of_Heat_Treatment_on_Bitterness (accessed on 21 October 2024).
12. Vijayakumari, B.; Kiranmayi, P.; Vengaiah, P.C. Estimation of vitamins, minerals and amino acids in Palmyra palm (*Borossus flabellifer* L.) fruit pulp. *Int. Res. J. Pharm.* **2017**, *7*, 70–73.

13. Srithuvaragan, R.; Anuluxshy, B. Extracting Pigments from Palmyrah Fruit Pulp (*Borassus flabellifer* L.) for the Production of Natural Colorants for Food. *Int. J. Sci. Basic Appl. Res.* **2019**, *48*, 145–151. Available online: https://www.gssrr.org/index.php/JournalOfBasicAndApplied/article/view/10402 (accessed on 21 October 2024).

14. Nikawala, J.K.; Jansz, E.; Baeckstrom, P.; Abeysekera, A.; Wijeyaratne, S.C. Flabelliferins of Naringinase Debittered Palmyrah Fruit Pulp. 2000. Available online: https://dr.lib.sjp.ac.lk/handle/123456789/978 (accessed on 21 October 2024).

15. Nikawela, J.; Abeysekara, A.; Jansz, E. Flabelliferins-steroidal saponins from palmyrah (*Borasus flabellifer* L.) fruit pulp 1. isolation by flash chromatography, quantification and saponin related activity. *J. Natl. Sci. Found. Sri Lanka* **1998**, *26*, 9. [CrossRef]

16. Alexis Thayaparan, C.T.; Kuldeep, S.A.; Dewanjee, S.; Kazi, E.; Tharmalingam, S.; Dhanaraj, P.; Paulraj, M.S. An insight into the structure and functions of flabelliferins and borassosides; nutraceuticals in Asian palm (*Borassus flabellifer*). *Food Nutr. Chem.* **2023**, *1*, 31. [CrossRef]

17. Lautenschlaeger, H. (n.d.). Saponins in Skin Care. Hautschutz—Hautpflege—Dermatika—Willkommen. Available online: https://dermaviduals.de/english/publications/special-actives/saponins-in-skin-care.html (accessed on 21 October 2024).

18. Danbature, W.L.; Fai, F.Y.; Khadijat, M.M. Quantitative determination of vitamins A, B1, B2 and B6 in Palmyra fruits marketed in Gombe metropolis. *Bima J. Sci. Technol.* **2023**, *1*, 22–27. [CrossRef]

19. Thi Le, D.H.; Chiu, C.; Chan, Y.; Wang, C.R.; Liang, Z.; Hsieh, C.; Lu, W.; Mulio, A.T.; Wang, Y.; Li, P. Bioactive and physicochemical characteristics of natural food: Palmyra palm (*Borassus flabellifer* Linn.) syrup. *Biology* **2021**, *10*, 1028. [CrossRef] [PubMed]

20. Selvakumar, P.M.; Thanapaul, R.J. An insight into the polymeric structures in Asian Palmyra palm (*Borassus flabellifer* Linn). *Org. Polym. Mater. Res.* **2021**, *2*, 16–21. [CrossRef]

21. Sindhu, G.; Kulloli, S.D.; KJ, S. Extraction and characterization of Palmyra palm (*Borassus flabellifer*) fruit fiber. *Int. J. Adv. Biochem. Res.* **2024**, *8*, 196–198. [CrossRef]

22. Rayappa, M.K. Palmyrah palm (*Borassus flabellifer*) non-centrifugal sugar—Current production practices as a natural sugar and a promising functional food/additive. *J. Agric. Food Res.* **2023**, *14*, 100829. [CrossRef]

23. Bafna, J. Ice Apple or tadgola: The fruit of Palmyra palm tree. *Health Vision*. 25 March 2023. Available online: https://healthvision.in/ice-apple-or-tadgola-the-fruit-of-palmyra-palm-tree/ (accessed on 21 October 2024).

24. Savalapurapu, B.; Sanyasi, G.; Bhavana, M.D.; Deverapalli, S. Development and formulation of value-added product from Palmyrah palm. *J. Chem. Health Risks* **2023**, *13*, 833–838. Available online: https://jchr.org/index.php/JCHR/article/view/927 (accessed on 21 October 2024).

25. Banerjee, P.S. Health benefits of Palmyra fruit. *Medindia*. 10 August 2015. Available online: https://www.medindia.net/health/diet-and-nutrition/top-4-health-benefits-of-palmyra-fruit.htm (accessed on 21 October 2024).

26. Mukkun, L.; Simamora, A.V.; Lalel, H.J.; Pakan, P.D. Savanna biomass for cosmetics sources. In *Biomass-Based Cosmetics*; Springer: Singapore, 2024; pp. 85–112. [CrossRef]

27. Thakur, N.S.; Attar, K.; Gunaga, R.; Chauhan, R.S.; Hegde, H.; Balubhai, B.J. Anecdote about mythology and ethno-pharmacolgy of Asian palmyra palm (*Borassus flabellifer* L.). In Proceedings of the National Conference on Palmyra Palm, Gujarat, India, 7–8 January 2016. Available online: https://www.researchgate.net/publication/298788493_Anecdote_about_mythology_and_ethno-pharmacolgy_of_Asian_palmyra_palm_Borassus_flabellifer_L/citations (accessed on 21 October 2024).

28. Thevamirtha, C.; Balasubramaniyam, A.; Srithayalan, S.; Selvakumar, P.M. An insight into the antioxidant activity of the facial cream, solid soap and liquid soap made using the carotenoid extract of Palmyrah (*Borassus flabellifer*) fruit pulp. *Ind. Crop. Prod.* **2023**, *195*, 116413. [CrossRef]

29. Gnanaselvam, V.; Alakolanga, A.; Lochana, E.A.; Balasubramaniyam, A.; Thangavel, K. Formulation and standardization of face wash using Palmyrah pulp and aloe vera gel. In Proceedings of the International Research Conference of Uva Wellassa University (IRCUWU2020), Badulla, Sri Lanka, 29–30 July 2020. Available online: https://www.researchgate.net/publication/348621362_Formulation_and_Standardization_of_Face_Wash_using_Palmyrah_Pulp_and_Aloe_Vera_Gel (accessed on 21 October 2024).

30. Risla, K.R.; Balasubramaniyam, A.; Nithiyananthan, S.; Perera, G.A.; Lochana, E.A.; Srivijeindran, S.; Samsankapil, V. Formulation of a natural liquid soap enriched with carotenoids from Palmyrah fruit pulp and rock salt. In Proceedings of the Vavuniya University International Research Conference Human Empowerment through Research Excellence, Virtual, 15 October 2021. Available online: https://www.researchgate.net/publication/361903232_formulation_of_a_natural_liquid_soap_enriched_with_carotenoids_from_palmyrah_fruit_pulp_and_rock_salt (accessed on 21 October 2024).

31. Reju Ayur. Under Eye Gel with Ice Apple & Cucumber. (n.d.). Available online: https://www.rejuayur.com/products/under-eye-gel-with-ice-apple-cucumber?srsltid=AfmBOoqmWoi-PX8-kWrY27-YOdkZGJ9ds98Dkp4sBvEmigsIjXMfw-CQ (accessed on 21 October 2024).

32. Arunan, J.M.; Robikka; Gunasekara, N.; Sivaji, M. Development of a natural toothpaste from Palmyrah pulp. In Proceedings of the 1st International Conference on Technological Research and Innovation, Palachcholai, Sri Lanka, 27 September 2023. Available online: https://www.researchgate.net/publication/375524983_Development_of_A_Natural_Toothpaste_from_Palmyrah_Pulp/citations (accessed on 21 October 2024).

33. The Jakarta Post. Bali's Miracle: Turning Wine into Hand Sanitizer—National—The Jakarta Post. 2020. Available online: https://www.thejakartapost.com/news/2020/04/09/balis-miracle-turning-wine-into-hand-sanitizer.html (accessed on 21 October 2024).

34. Cocoon Original. An Giang Palmyra Palm Sugar Body Polish. (n.d.). Available online: https://cocoonoriginal.com/products/palmyra-palm-sugar-body-polish (accessed on 21 October 2024).

35. Monichan, S.; Thevamirtha, C.; Thanapaul, R.J.; Selvakumar, P.M. Palmyraculture: An insight into the nano medicines from Palmyra palm (*Borassus flabellifer* L.). *Acta Sci. Med. Sci.* **2021**, *5*, 143–153.

36. Ngoc, L.T.; Moon, J.Y.; Lee, Y.C. Plant extract-derived carbon dots as cosmetic ingredients. *Nanomaterials* **2023**, *13*, 2654. [CrossRef] [PubMed]

37. Goyal, N.; Jerold, F. Biocosmetics: Technological advances and future outlook. *Environ. Sci. Pollut. Res.* **2021**, *30*, 25148–25169. [CrossRef] [PubMed]

38. Singh, V.K. Biocosmetics—A Trend. *PharmaTutor*. 22 June 2023. Available online: https://www.pharmatutor.org/pharmapedia/biocosmetics-a-trend (accessed on 21 October 2024).

39. Dailin, D.J.; Rithwan, F.; Azelee, N.I.; Zainan, N.; Low, L.Z.; Zaidel, D.N.; El Enshasy, H. Trends in bio-based cosmetic ingredients. In *Biomass-Based Cosmetics*; Springer: Singapore, 2024; pp. 27–47. [CrossRef]

40. Jamkhande, P.G.; Suryawanshi, V.A.; Kaylankar, T.M.; Shailesh, L.P. Biological activities of leaves of ethnomedicinal plant, *Borassus flabellifer* Linn. (Palmyra palm): An antibacterial, antifungal andantioxidant evaluation. *Bull. Fac. Pharm. Cairo Univ.* **2016**, *54*, 59–66. [CrossRef]

41. Paschapur, M.S.; Patil, M.B.; Kumar, R.; Patil, S.R. Evaluation of anti-inflammatory activity of ethanolicextract of *Borassus flabellifer* L. male flowers(inflorescences) in experimental animals. *J. Med. Plants Res.* **2009**, *3*, 49–54.

42. Mariselvam, R.; Ighnachimuthu, S.J.; Selvakumar, P.M. Review on the nutraceutical values of Borassus flabelifer Linn. *J. Pharm. Drug Res.* **2020**, *3*, 268–275. Available online: https://www.scitcentral.com/article/14/1071/latestupdates.php (accessed on 21 October 2024).

43. Bharadvaja, N.; Gautam, S.; Singh, H. Natural polyphenols: A promising bioactive compounds for skin care and cosmetics. *Mol. Biol. Rep.* **2022**, *50*, 1817–1828. [CrossRef]

44. Gomez-Molina, M.; Albaladejo-Marico, L.; Yepes-Molina, L.; Nicolas-Espinosa, J.; Navarro-León, E.; Garcia-Ibañez, P.; Carvajal, M. Exploring phenolic compounds in crop by-products for cosmetic efficacy. *Int. J. Mol. Sci.* **2024**, *25*, 5884. [CrossRef]

45. Cosmetics Info. *Beta-Carotene*. 2023. Available online: https://www.cosmeticsinfo.org/ingredient/beta-carotene/ (accessed on 21 October 2024).

46. Aparecida Sales de Oliveira Pinto, C.; Elyan Azevedo Martins, T.; Miliani Martinez, R.; Batello Freire, T.; Valéria Robles Velasco, M.; Rolim Baby, A. *Vitamin E in Human Skin: Functionality and Topical Products*; IntechOpen: London, UK, 2021. [CrossRef]

47. Enna, C.D.; Dyer, R.F. The Histomorphology of the elastic tissue system in the skin of the human hand. *Hand* **1979**, *11*, 144–150. [CrossRef] [PubMed]

48. Wang, K.; Jiang, H.; Li, W.; Qiang, M.; Dong, T.; Li, H. Role of vitamin C in skin diseases. *Front. Physiol.* **2018**, *9*, 819. [CrossRef]

49. Gref, R.; Deloménie, C.; Maksimenko, A.; Gouadon, E.; Percoco, G.; Lati, E.; Desmaële, D.; Zouhiri, F.; Couvreur, P. Vitamin C-squalene bioconjugate promotes epidermal thickening and collagen production in human skin. *Sci. Rep.* **2020**, *10*, 16883. [CrossRef]

50. Darr, D.; Combs, S.; Dunston, S.; Manning, T.; Pinnel, S. Topical vitamin C protects porcine skin from ultraviolet radiation-induced damage. *Br. J. Dermatol.* **1992**, *127*, 247–253. [CrossRef] [PubMed]

51. Scherer Santos, J.; Diniz Tavares, G.; Nogueira Barradas, T. *Vitamin E and Derivatives in Skin Health Promotion*; IntechOpen: London, UK, 2021. [CrossRef]

52. Bocheva, G.; Slominski, R.M.; Slominski, A.T. The impact of vitamin D on skin aging. *Int. J. Mol. Sci.* **2021**, *22*, 9097. [CrossRef] [PubMed]

53. Shunatona, B. Dermatologists Explain Why Vitamin D is so Crucial for Healthy Skin. *Byrdie*. 1 May 2020. Available online: https://www.byrdie.com/vitamin-d-for-skin-4783626 (accessed on 21 October 2024).

54. Kahina. Which Skincare Products Contain Vitamin K? *Typology Paris*. 16 April 2024. Available online: https://us.typology.com/library/vitamin-k-where-to-find-it-in-cosmetics (accessed on 21 October 2024).

55. Rud, M. The Benefits of Vitamin K in Skincare, According to Derms. *Byrdie*. 17 June 2020. Available online: https://www.byrdie.com/vitamin-k-for-skin-4801332 (accessed on 21 October 2024).

56. Lautenschlaeger, H. (n.d.). Vitamins in Cosmetics. Hautschutz—Hautpflege—Dermatika—Willkommen. Available online: https://dermaviduals.de/english/publications/special-actives/vitamins-in-cosmetics.html (accessed on 21 October 2024).

57. Rodrigues, P.D.; Corrêa, A.G.; Baffi, M.A.; Pasquini, D. Potential applications of hemicellulose. In *Handbook of Biomass*; Springer: Singapore, 2023; pp. 1–31. [CrossRef]

58. Ariyanta, H.A.; Santoso, E.B.; Suryanegara, L.; Arung, E.T.; Kusuma, I.W.; Azman Mohammad Taib, M.N.; Hussin, M.H.; Yanuar, Y.; Batubara, I.; Fatriasari, W. Recent progress on the development of lignin as future ingredient biobased cosmetics. *Sustain. Chem. Pharm.* **2023**, *32*, 100966. [CrossRef]

59. Cosmetics Info. Pectin. 2023. Available online: https://www.cosmeticsinfo.org/ingredient/pectin/ (accessed on 21 October 2024).

60. Cosmetics Info. Glucose. 2023. Available online: https://www.cosmeticsinfo.org/ingredient/glucose/ (accessed on 21 October 2024).

61. Cosmetics Info. Sucrose. 2023. Available online: https://www.cosmeticsinfo.org/ingredient/sucrose/ (accessed on 21 October 2024).

62. Cosmetics Info. Fructose. 2023. Available online: https://www.cosmeticsinfo.org/ingredient/fructose/ (accessed on 21 October 2024).

63. Hebbar, J.V. Toddy palm (Asian Palmyra palm) Uses, Research, Medicines, Side Effects. *Easy Ayurveda*. 28 April 2022. Available online: https://www.easyayurveda.com/2017/10/04/toddy-palm-asian-palmyra-palm-tala/#interaction_with_medicines_supplements (accessed on 21 October 2024).
64. Działo, M.; Mierziak, J.; Korzun, U.; Preisner, M.; Szopa, J.; Kulma, A. The potential of plant phenolics in prevention and therapy of skin disorders. *Int. J. Mol. Sci.* **2016**, *17*, 160. [CrossRef] [PubMed]
65. Kuldeep, S.A.; Vijayakumar, V.; Segaruban, S.; Sinnathurai, Y.; Hossain, M.S.; Begum, K.; Paul, J.; Selvakumar, P.M. An insight into the molecules and materials of the haustorium of the Asian Palmyra palm (*Borassus flabellifer*) and their nutraceutical values. *Food Nutr. Chem.* **2023**, *1*, 115. [CrossRef]
66. Ravetti, S.; Clemente, C.; Brignone, S.; Hergert, L.; Allemandi, D.; Palma, S. Ascorbic Acid in Skin Health. *Cosmetics* **2019**, *6*, 58. [CrossRef]
67. Sarpooshi, H.R.; Mortazavi, F.; Vaheb, M.; Tabarayee, Y. The effects of topical vitamin C solution on burn wounds granulation: A randomized clinical trial. *J. Biomed.* **2016**, *1*, e8301. [CrossRef]
68. Miyazaki, A.; Taki, T.; Takeichi, T.; Kono, M.; Yagi, H.; Akiyama, M. Darier disease successfully treated with a topical agent containing vitamin A (retinyl palmitate), vitamin E, and urea. *J. Dermatol.* **2022**, *49*, 779–782. [CrossRef]
69. Giannella, R.A. Infectious enteritis and Proctocolitis and bacterial food poisoning. In *Sleisenger and Fordtran's Gastrointestinal and Liver Disease*; WB Saunders: Philadelphia, PA, USA, 2010; pp. 1843–1887.e7. [CrossRef]
70. Tunit, P.; Thammarat, P.; Okonogi, S.; Chittasupho, C. Hydrogel containing *Borassus flabellifer* L. male flower extract for antioxidant, antimicrobial, and anti-inflammatory activity. *Gels* **2022**, *8*, 126. [CrossRef]
71. Mitsuwan, W.; Sornsenee, P.; Romyasamit, C. *Lacticaseibacillus* spp., probiotic candidates from Palmyra palm sugar, possess antimicrobial, and anti-biofilm activities against methicillin-resistant *Staphylococcus aureus*. *Vet. World* **2022**, *15*, 299–308. [CrossRef]
72. Methicillin-Resistant *Staphylococcus aureus* (MRSA), Methicillin-Resistant *Staphylococcus aureus* (MRSA) Basics. 2024. Available online: https://www.cdc.gov/mrsa/about/index.html (accessed on 21 October 2024).
73. Rani, V.P.; Mirabel, L.; Priya, S.; Nancy, A.A.; Kumari, G.M. Phytochemical, antioxidant and antibacterial activity of aqueous extract of *Borassus flabellifer* (L.). *Int. J. Sci. Res. Sci. Technol.* **2021**, *4*, 405–410.
74. Artnarong, S.; Payap, M.; Jaruwan, M. Isolation of yeast and acetic acid bacteria from palmyra palm fruit pulp (*Borassus flabellifer* Linn.). *Int. Food Res. J.* **2016**, *23*, 308–1314. Available online: https://www.researchgate.net/publication/302584085_Isolation_of_yeast_and_acetic_acid_bacteria_from_palmyra_palm_fruit_pulp_Borassus_flabellifer_Linn (accessed on 21 October 2024).
75. Elhage, K.G.; St. Claire, K.; Daveluy, S. Acetic acid and the skin: A review of vinegar in dermatology. *Int. J. Dermatol.* **2021**, *61*, 804–811. [CrossRef] [PubMed]
76. Reshma, M.; Jacob, J.; Syamnath, V.; Habeeba, V.; Dileep Kumar, B.; Lankalapalli, R.S. First report on isolation of 2,3,4-trihydroxy-5-methylacetophenone from Palmyra palm (*Borassus flabellifer* Linn.) syrup, its antioxidant and antimicrobial properties. *Food Chem.* **2017**, *228*, 491–496. [CrossRef]
77. Wilson, G. Introducing escherichia coli (*E. coli*) Part 1. *The Secret Life of Skin*. 23 November 2020. Available online: https://thesecretlifeofskin.com/2020/11/10/introducing-e-coli/ (accessed on 21 October 2024).
78. Wang, C.J.; Song, Y.; Li, T.; Hu, J.; Chen, X.; Li, H. Mycobacterium smegmatis skin infection following cosmetic procedures: Report of two cases. *Clin. Cosmet. Investig. Dermatol.* **2022**, *15*, 535–540. [CrossRef]
79. Mastine, M.L.; Zajac, K. Staphylococcus Simulans. *Mechanisms of Pathogenicity*. 15 December 2021. Available online: https://mechpath.com/2021/11/18/staphylococcus-simulans/ (accessed on 21 October 2024).

cosmetics

Article

Adaptogen Technology for Skin Resilience Benefits

Andrea Cavagnino [1,*], Lionel Breton [2], Charline Ruaux [3], Celeste Grossgold [3], Suzy Levoy [4], Rawad Abdayem [4], Romain Roumiguiere [4], Stephanie Cheilian [4], Anne Bouchara [4], Martin A. Baraibar [1] and Audrey Gueniche [4]

[1] OxiProteomics SAS, 94000 Créteil, France; martin.baraibar@oxiproteomics.fr
[2] Cilia Consulting, 78000 Versailles, France; btwocg@gmail.com
[3] L'Oréal, PRADA Beauty, 92300 Levallois, France; charline.ruaux@loreal.com (C.R.); celeste.grossgold@loreal.com (C.G.)
[4] L'Oréal R&I, 94550 Chevilly La Rue, France; suzy.levoy@loreal.com (S.L.); rawad.abdayem@loreal.com (R.A.); romain2.roumiguiere@loreal.com (R.R.); stephanie.cheilian@loreal.com (S.C.); anne.bouchara@loreal.com (A.B.); audrey.gueniche@loreal.com (A.G.)
* Correspondence: andrea.cavagnino@oxiproteomics.fr

Abstract: (1) Background: Skin undergoes constant changes, providing capabilities to repair and renovate its constituents once damaged and a fundamental shield to contrast environmental stress. Nevertheless, environmental stressors may overcome the skin's protective potential inducing premature aging and accelerating the appearance of anaesthetic age-related skin aspects. Ultraviolet radiation (UVR) and pollutants (particulate matters, PAHs) contribute to skin aging and functional decline inducing harmful oxidative modifications of macromolecules and stress-related skin disorders. Innovative approaches to preserve skin are needed. (2) Methods: Skin keratinocytes were treated (or not) with a combination of ingredients (*Lactobacillus plantarum* extract, *Withania somnifera* root extract and *Terminalia ferdinandiana* fruit extract; "MIX") in the presence or absence of stress (oxidative stress or pollution). The effects of the MIX adaptogen technology on (a) cellular resilience, (b) the regulation of cellular functions and (c) regeneration of skin were disclosed through expression proteomics and bioinformatics analyses first, and then through focused evaluations of protein carbonylation as a hallmark of oxidative stress' deleterious impact and mitochondrial activity. (3) Results: The deleterious impact of stressors was evidenced, as well as the beneficial effects of the MIX through (a) mitochondrial activity preservation, (b) the "vigilance" of the NRF2 pathway activation, (c) NADPH production and protein homeostasis improvements, (d) preserving skin regeneration function and I the contrasting stress-induced oxidation (carbonylation) of mitochondrial and nuclear proteins. (4) Conclusions: The effects of the MIX on increasing cell adaptability and resilience under stress suggested a beneficial contribution in precision cosmetics and healthy human skin by acting as an adaptogen, an innovative approach that may be employed to improve resistance to harmful stress with a potential favourable impact on skin homeostasis.

Keywords: adaptogens; skin benefits; oxidative stress; pollution; skin disorders; mitochondrial function; protein carbonylation; proteomics

Citation: Cavagnino, A.; Breton, L.; Ruaux, C.; Grossgold, C.; Levoy, S.; Abdayem, R.; Roumiguiere, R.; Cheilian, S.; Bouchara, A.; Baraibar, M.A.; et al. Adaptogen Technology for Skin Resilience Benefits. *Cosmetics* **2023**, *10*, 155. https://doi.org/10.3390/cosmetics10060155

Academic Editor: Paulraj Mosae Selvakumar

Received: 30 September 2023
Revised: 1 November 2023
Accepted: 1 November 2023
Published: 10 November 2023

1. Introduction

Skin is the largest human organ providing a fundamental shield to contrast environmental stress. Cutaneous aging is a continuous, complex and cumulative biological process influenced by intrinsic (mainly genetic) and extrinsic factors. At midlife, these factors contribute to approximately 50% of alterations in skin function [1].

Physiologically, skin undergoes constant changes, with the capability to repair and renovate its constituents once damaged. Although the skin maintains a functional biological barrier against external aggressions, environmental stressors may overcome the skin's protective potential. Besides environmental factors, biological and chronological aging also hamper the skin's abilities to maintain homeostasis and repair, contributing to the aging process. Among environmental factors, ultraviolet radiation (UVR) is a primary element of skin impairment, inducing oxidative stress and harmful genetic modifications. Other factors, such as pollutants (particulate matters, volatile organic compounds (VOCs) and polycyclic aromatic hydrocarbons (PAHs)), also contribute to skin aging and decline the functionality of the skin [2–7].

Oxidative stress has been identified as a common and prevalent mechanism for cellular tissue damage and dysfunctions in an ample variety of age-related diseases and skin disorders, revealing, additionally, its interplay with inflammatory processes at both the tissue and organismal levels [8,9]. A major mechanism by which ambient PM exerts its detrimental effects is through the generation of oxidative stress. In particular, polycyclic aromatic hydrocarbons (PAHs) adsorbed on the surface of suspended particulate matter (PM) in air of urban areas can activate xenobiotic metabolism, ultimately inducing the production of reactive oxygen species (ROS) through the formation of quinone derivates and superoxide anion radicals. PAH metabolites contribute to the excessive generation of reactive oxygen species (ROS), causing oxidative stress. In addition, UV irradiation damages skin through the photosensitized production of ROS, mainly driven by UV-A (320–400 nm), which can penetrate to the deeper sections of the epidermis, reaching the dermis. Furthermore, several PAHs can induce strong oxidative stress and UV-A exposure and consequent photoactivation.

Among the oxidative modification of macromolecules, protein carbonylation has become a comprehensive warning and major hallmark of severe oxidative damage, protein dysfunction and stress-related disorders [10], being a harmful irreversible modification. This modification at the molecular level plays havoc with cellular/tissue functions in the aging organism, increasing alongside the organisms' life [11].

Carbonylation can negatively affect, or totally abrogate, protein catalytic functions and may trigger the formation of potentially cytotoxic protein aggregates.

The nuclear factor erythroid-2-related factor 2 (Nrf2)-mediated antioxidant response pathway plays a central role in the cellular reduction–oxidation homeostasis, promoting a youthful cellular phenotype. This evolutionarily conserved pathway acts firstly as the transcription of cytoprotective genes, resulting in cellular cryoprotection and in the physiological cellular resilience to stress exposures (antioxidant, prosurvival, anti-inflammatory and macromolecular damage repair) [12,13].

Another well-known cellular mechanism implicated in cellular resilience and having a prosurvival role is the production of NADPH. In this pathway, glucose-6-phosphate dehydrogenase (G6PD) has a key role assuring the production of ribose and reducing equivalent nicotinamide adenine dinucleotide phosphate (NADPH), being vital components for the synthesis of many biological building blocks, such as nucleic and fatty acids. NADPH is well known for its remarkable role in the maintenance of antioxidant defences promoting health and life-span [14,15].

Moreover, protein carbonylation has been spotted as a potential mechanism contributing to mitochondrial dysfunction [16]. Mitochondria can produce up to 95% of all adenosine triphosphate (ATP) in the cell, sustain critical metabolic processes [17–20] and play a crucial role in inducing cell differentiation, sustaining proliferation, maintaining cell survival and triggering apoptosis, key processes for the development and maintenance of skin homeostasis and health [21–23]. Mitochondria are highly susceptible to detrimental environmental factors [24–27], harming its function and its genetic patrimony, resulting in mitochondrial DNA (mtDNA) deletions, perturbing the electron flow and energy production [28–30]. Mitochondrial dysfunction increases as age progresses. Additionally, wound healing and hair growth are part of the physiological functions of the skin, where mitochondria support optimal regeneration and differentiation, respectively.

Playing an important role in cellular metabolism and cell life-span, mitochondria assure the generation of cellular "energy storage" molecules (adenosine triphosphate) through oxidative phosphorylation (OXPHOS). Under unfavourable conditions, reactive oxygen species (ROS) are generated during OXPHOS as a by-product of aerobic respiration.

Perturbations within the OXPHOS complexes generate more reactive oxygen species (ROS), less adenosine triphosphate and the mismanagement of mitochondrial dynamics with the consequent accumulation of oxidative damage. Moreover, the lower concentration of adenosine triphosphate restricts the activity of mitochondrial ATP-dependent proteases, which are less efficient or even not able to degrade all carbonylated proteins generated from the elevated ROS level. Reactive oxygen species can react and modify DNA, RNA, carbohydrates, proteins and lipids, triggering cellular and tissular dysfunction and promoting aging [31,32].

Biological adaptation refers to the ability of living organisms to adjust their biological functions to suit their external environment. This concept was first introduced by Laborit in 1976 [33], who proposed that the ability of a living being to survive and reproduce is based on its capacity to adapt to external conditions. In other words, biological adaptation involves the successful adjustment of an organism's physiological processes to changes in its environment, enabling it to survive and thrive [34]. The human body is continuously exposed to a range of environmental stressors, such as pollution, temperature fluctuations, UV radiation, tobacco smoke and screen light [35]. The skin, acting as a barrier between the internal body and the external environment, also plays a critical role in the adaptation of humans to these stressors.

While the cosmetic industry has largely focused on developing products that enhance the appearance of the skin, there is growing interest in identifying ingredients that can also support the skin's natural functions and ability to adapt to internal and external stressors. Adaptogens, in particular, have gained attention for their potential to help the skin maintain a healthy balance and resilience.

The definition of adaptogens has constantly evolved sinIe it was coined [36]. The dictionary definition is as follows: "natural substances used in herbal medicine to normalize and regulate the systems of the body".

The increase in research and scientific evidence to understand the pharmacological and molecular mechanisms of action of these active ingredients has demonstrated that the term adaptogen is related to a physiological process, the adaptation to environmental challenges, which is a multistep process that includes diverse mechanisms of extracellular and intracellular interactions [36–39].

Adaptogens are mediators that help plants to persist in stress conditions, defined also as substances causing the "state of unspecific resistance" of an organism. The term adaptogens comprises families of herbal and natural products able to promote the adaptability and survival of living organisms under stress conditions, resulting in a decreased sensitivity to stressors and/or prolonged phase of resistance (stimulatory effect). These compounds are mild stress mimetics at low doses, activating adaptive stress-response signalling pathways to cope with severe stress and have been used in traditional or alternative medicines for preventing premature aging and to maintain good health and vitality [36–42].

If the beneficial properties of adaptogens have been studied often with oral intake, some studies using topical applications in nonhuman models have also proven their efficacy, notably with ashwagandha [43–45].

It is important to note that certain antioxidant treatments, if used incorrectly, can have negative effects on the skin that are just as harmful as those caused by stress. For instance, treatments that are too concentrated or applied at the wrong time can have adverse effects on the skin. Additionally, some treatments may target the wrong biological pathways, leading to unintended consequences.

To avoid such deleterious effects, it is crucial to adopt an approach that is tailored to the skin's unique needs, ensuring that antioxidant treatments provide maximum benefits without causing any harm.

Preserving skin cells from oxidative damage and mitochondrial damage can help mitigate the cumulative injuries caused by environmental factors. Innovative approaches contrasting environmental aggressions that accelerate the aging process may employ compounds resulting in adaptogenic effects.

A mix of active ingredients was selected based on the understanding that the skin's ability to adapt to stressors is facilitated by its three primary functions: protection, regulation and regeneration. These functions work together to maintain the skin's health and resilience. The mix of ingredients includes:

- An extract of *Withania somnifera* roots (or Ashwagandha), an adaptogen plant with a long history of use in Ayurvedic medicine for promoting longevity and slowing aging;
- A ferment of *Lactobacillus plantarum*, known for its beneficial effects on the skin's barrier function;
- A superfruit extract of the Kakadu plum (*Terminalia ferdinandiana*), which is the richest natural source of vitamin C.

Together, these active ingredients were thoughtfully combined to create a technology ("Adapto.GN Smart Technology") that can help increase the skin's ability to adapt to modern life stressors of the current generation, GN.

By supporting the skin, the active ingredients can help optimize and enhance the skin's functions.

The objective of our investigation was firstly focused on the characterizations of the effects of a mix compounds (*Lactobacillus plantarum* extract, *Withania somnifera* root extract and *Terminalia ferdinandiana* fruit extract) on human keratinocytes exposed to stress conditions (urban pollution and oxidative stress) through an expression proteomics phenotypical analysis. The obtained observations then inspired further evaluations on the potentially beneficial effects of the mix of compounds (1) on contrasting stress-induced protein carbonylation as a biomarker of oxidative damage and (2) the beneficial effects against stress exposure mitochondrial injury resulting in preserved mitochondrial function.

2. Materials and Methods

2.1. Cell Culture and Treatments

Human primary keratinocytes were seeded in multiwell plates, cultured in an SFM medium (Gibco, Strasburg, France) and incubated in optimal conditions at 37 °C, a humified atmosphere and 5% CO_2. The stress exposures consisted of (1) 30 min of contact with a solution of 100 µM H_2O_2 or (2) contact with particulate matters (1 µg/cm^2, 1 h of contact; Ref. ERM-CZ100; The item (ERM-CZ100) is supplied and certified by the "European Reference Materials" from the Joint Research Centre of the European Commission in Geel, Belgium, and is distributed by Sigma-Aldrich, Merck. European Reference Material with certified content in PAHs and the presence of >25% of PM2.5-like particles on cumulative distribution), followed by UV-A irradiation (dose of 2 J/cm^2 at a wavelength peak of 365 nm). A fresh solution of active ingredients (the MIX (*Adapto.GN technology* ^{TM})), composed of *Lactobacillus plantarum* extract 0.5%, *Withania somnifera root* extract 1.0% and *Terminalia ferdinandiana* fruit extract 0.4%, was obtained through the direct solubilization of each ingredient in the culture medium. The MIX solution was incubated with the cells for 72 h of contact prior to stress exposure or in basal conditions (in the absence of stress). The active mix solution was previously evaluated on cellular viability with an MTS (tetrazolium inner salt) viability assay, excluding cytotoxic effects at the tested concentrations.

2.2. Expression Proteomics Analysis and Ingenuity Pathway Analysis (IPA)

Right after the treatments and/or stress exposure to H_2O_2, the cells were harvested, snap-frozen in liquid nitrogen and stored at −80 °C until analyses. The cellular pellets (n = 3 replicates per experimental group) were submitted to protein extraction by using a buffer solution of 6M Urea, 2M Thiourea, 1% SDS (sodium dodecyl sulphate) and 1% deoxycholate (Sigma-Aldrich, Merck, Darmstadt, Germany). The protein quantification was then realized by using the Bradford assay [46], and the profiles and qualities of protein extractions were evaluated using SDS-PAGE. Fifty micrograms of proteins from each sample was typically digested and desalted through anion-exchange liquid chromatography (LC). A nanoscale LC was effectuated on a C18 column (UltiMate^{TM} 3000 RSLCnano System; Thermo Fisher Scientific, Illkirch-Graffenstaden, France). Mass spectrometry (MS) analyses were performed on Tribrid Eclipse (Thermo Fisher Scientific) coupled to an Orbitrap-HCD-Ion Trap mass spectrometer (Thermo Fisher Scientific). Protein identifications and relative quantifications were conducted by using Protein Discoverer 2.4.0 software, MASCOT and SEQUEST on the SwissProt human protein database. Intergroup comparisons and statistics were performed by using PERSEUS software. The protein abundance ratio values were then converted to fold-change values, where the negative inverse $(-1/x)$ was taken for values between 0 and 1 (e.g., an abundance ratio of 0.5 was converted in fold change -2), while values greater than 1 were not affected. Graphs and a principal component analysis (PCA) were generated with Prism (GraphPad software version 10, San Diego, CA, USA). The PCA was calculated using all protein abundance values (Log2) on PC1 and PC2. The data mining analysis of identified and grouped proteins in relation with their biological functions was then realized using IPA (Ingenuity Pathway Analysis, Qiagen, Courtaboeuf, France). Proteins from the dataset that were associated with biological functions in the Ingenuity Knowledge Base were considered for the analysis (minimum threshold of selection: >1.3-fold change of upregulation or <-1.3 of downregulation and p-value < 0.1). Right-tailed Fisher's exact test was used to calculate a p-value determining the probability that each biological function, canonical pathway or upstream regulator assigned to that specific subdataset was due to chance alone. The functional analysis was restricted to human taxon and skin functions (focused on keratinocytes, fibroblasts, adipocytes, melanocytes, epidermis, dermis and skin over the Ingenuity Knowledge Base reference set) [47].

2.3. Mitochondrial Protein and Nuclear Protein (Histone-Enriched) Fraction Preparation

Right after the treatments and/or stress exposure to H_2O_2, the cells were harvested, snap-frozen in liquid nitrogen and stored at $-80\ °C$ until analyses. The cellular pellets (n = 5 replicates per group; per protein fraction) were independently processed and subjected to protein extraction using an adapted method based on a sequential lysis of cell membranes by increasing the detergent strength of the lysis buffers [48]. Reducing conditions were used in the overall process of extraction/enrichment in order to avoid unwanted oxidative reactions. The collected supernatants concerning organelle proteins (mitochondrial) or nuclear (histone enriched) proteins were precipitated with the trichloroacetic acid (TCA)/acetone method and then suspended in OxiProteomics buffer for solubilization. The quantification of the extracted proteins was performed using the Bradford method and the samples were equally distributed for a protein carbonylation level analysis.

2.4. Protein Carbonylation Analysed

The evaluation was performed on 5 replicates per experimental group, per each analysis: (1) organelles/mitochondrial proteins; (2) nuclear histone-enriched proteins. Right after the stress exposure, the cells were harvested, snap-frozen in liquid nitrogen and stored at $-80\ °C$. At the moment of the analysis, the carbonylated proteins were extracted from each sample using an optimized lysis buffer and then labelled with a specific fluorescent probe [49]. After labelling, the proteins were resolved onto 4–20% gradient SDS-PAGE, then fixed to the gel and evidenced with differential fluorescence scanning at an emission wavelength of 647 nm. The total proteins were poststained with SyproRuby^{TM} (Invitrogen, Strasbourg, France). Digital image acquisitions of carbonyl and the total proteins were performed using the iBright system (Thermo Fisher Scientific). Image processing and the densitometric analysis were performed using "ImageJ" software [50].

The signal of the carbonylated proteins (fluorescence units) was normalized with respect to the signal obtained with the total proteins for each sample to obtain the carbonyl score for each sample: carbonyl score (*sample X*) = carbonylated proteins (fluorescence signal; *sample X*)/total proteins (fluorescence signal; *sample X*). The statistics were obtained by using GraphPad Prism software.

2.5. Mitochondrial Function Analysis

The analysis was performed on n = 5 replicates per experimental group. Right after the stress exposures (H_2O_2 or pollution) and/or cell treatments with the active MIX, the mitochondrial function evaluation was conducted through an in situ treatment of cells with a Mitotracker® probe, following the provider's instructions (Red CMXRos; Invitrogen). The control group did not receive any treatment or exposure to stress. The image collection of specific fluorescence signals was performed with the epi-fluorescence imaging system (EVOS M5000; Thermo fisher). The image analysis was effectuated with ImageJ software [50]. The levels of fluorescent probe accumulation in active mitochondria were obtained by integrating the specific intensity of a signal normalized on the surface of the evaluation and N of the nuclei.

3. Results and Discussion

3.1. MS (Mass Spectrometry)-Based Proteomic Analysis

The effects of a combination of active ingredients (MIX) were evaluated on skin cells (human primary keratinocytes) exposed (or not) to oxidative stress (H_2O_2) through a proteomic analysis. This approach allowed us to describe and characterize the changes in the expression of thousands of proteins supporting a better understanding of the beneficial protection against stress. Through a MS (mass spectrometry)-based proteomic analysis, we investigated the association between biological pathways, phenotypes, skin cell treatments and exposure to stress supporting precision cosmetics and promoting a healthy human skin.

In our study, "upregulated" and "downregulated" proteins referred to the proteins that showed an increase or decrease in their expression levels, respectively, when comparing two or more conditions. To quantify these changes in the protein expression, we employed a metric called 'fold change'. Fold change is a ratio that represents how much a quantity has changed between an initial and a subsequent measurement. In our context, a fold change greater than one indicated an increase or upregulation in protein expression, while a fold change of less than one indicated a decrease or downregulation. For instance, a fold change of 2 would suggest that the protein expression doubled (upregulated) in the subsequent measurement compared to the initial one, whereas a fold change of 0.5 would indicate that the expression halved (downregulated).

More than 2400 proteins were identified and quantified per the experimental conditions (Figure 1). Among them, proteins with fold-change expression ratios versus the control group higher or lower than the control group or stress group (+1.3- or −1.3-fold change values) were considered, respectively, as upregulated (in red) or downregulated (in green, Figure 1).

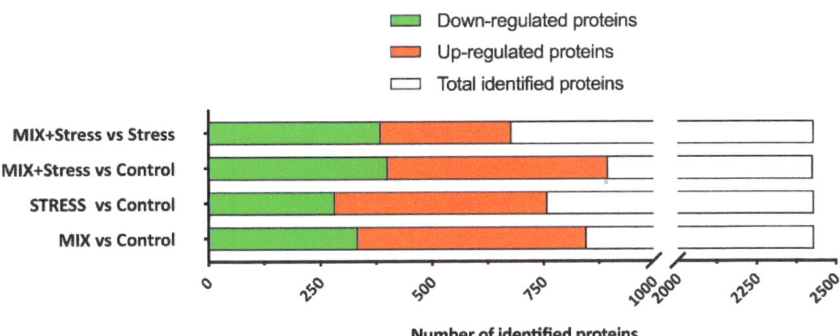

Figure 1. Stacked bar representation of significantly differentially expressed proteins in treated conditions versus "Ctrl" (basal conditions) or versus "Stress". The number of upregulated proteins with fold change expression ratios higher than 1.3 are represented in the red bar, while the number of proteins downregulated are shown in green (fold change ratio lower than −1.3). The total number of identified proteins in the experimental dataset is shown as a white bar (cumulated distribution = 2432 proteins).

The biplot representation (Figure 2 showed a homogenous positioning of the vectors (loadings, red dots) of each replicate belonging from the same group (generating clusters), and a relative difference in the positioning between the groups (correlation in this space) describing both intragroup (replicates n = 3) homogeneity and intergroup differential positioning.

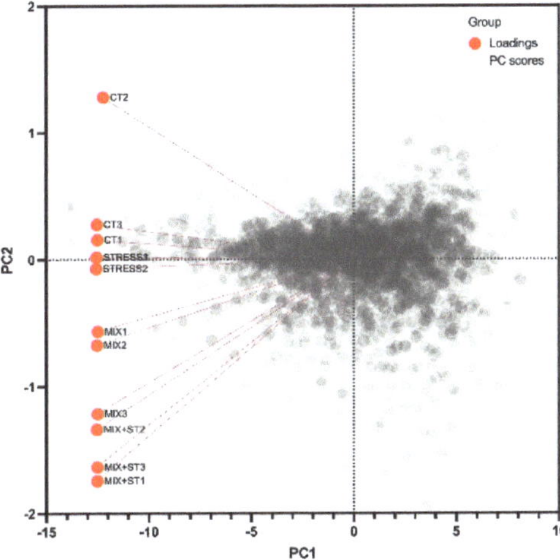

Figure 2. The results of the PCA were visualized though a biplot scaled representation of the loadings (each replicate for experimental group) and PC scores. Proportion of variance PC1 96.26%, PC2 0.81%.

The following results were evaluated according to skin functions: (1) regulation, (2) regeneration and (3) cellular protection/resilience. Among the identified proteins, the expression levels of "ADP/ATP translocase 1", an antiporter that mediates the import of ADP (adenosine diphosphate) into the mitochondrial matrix for ATP (adenosine triphosphate) synthesis and the export of ATP out to fuel the cell resulted affected by stress (H_2O_2) exposure (downregulated; fold change −2.09, Table 1), were assessed. The expression of ANT1 carrying mutations caused a wide range of mitochondrial abnormalities clinically characterized by exercise intolerance, ptosis and muscle weakness [51]. Adenine nucleotide translocases (ANTs) are highly abundant proteins in mitochondria, comprising up to 1%

of the total mitochondrial protein content [52]. In normal physiological conditions, the consumption of oxygen by mitochondria has been closely linked to the production of ATP. The proper functioning of ANT is crucial for maintaining healthy mitochondrial activity. One of the primary functions of ANT is to facilitate the exchange of ATP and ADP across the inner mitochondrial membrane, which is necessary for both ATP synthesis and the maintenance of a normal mitochondrial membrane potential ($\Delta\psi$m). Additionally, ANT also serves as an uncoupler, which is important for protecting mitochondria from excessive reactive oxygen species (ROS) production that can occur due to an elevated $\Delta\psi$m. These dual roles of ANT are critical for preserving mitochondrial health and function [53]. The pretreatment of cells with the MIX (MIX + Stress group) resulted in its significant upregulation (+2.37-fold change) when compared to the stressed group, contrasting the impact of stress exposure and preserving the levels of a key protein involved in mitochondrial activity.

Table 1. Subset of proteins significantly ($p < 0.05$) up- or downregulated in "Stress" (compared to control) and in "Mix + Stress" condition (compared to stress). Only the proteins showing a relative fold change (vs. ctrl) >1.3 in the group STRESS were included. The fold change ratios were presented as negative values for downregulation and positive values for upregulation.

Accession Number	Description	MIX (in Basal Conditions)		Stress (H_2O_2)		MIX + Stress	
		Fold Change (vs. Ctrl)	*p*-Value	Fold Change (vs. Ctrl)	*p*-Value	Fold Change (vs. Stress)	*p*-Value
P12235	ADP/ATP translocase 1	+1.14	0.525	−2.09	0.039	+2.37	0.007
Q9BY77	Polymerase delta-interacting protein 3	−1.08	0.780	−2.63	0.007	+2.43	0.012
Q9UQE7	Structural maintenance of chromosome protein 3	−1.87	0.416	−6.13	0.006	+5.53	0.020

Moreover, polymerase delta-interacting protein 3 contributed to an enhanced translational efficiency of spliced over nonspliced mRNAs. This protein recruited activated ribosomal protein S6 kinase beta-1 I/RPS6KB1 to the newly synthesized mRNA antiporter [54]. Oxidative stress (H_2O_2) negatively affected the expression of polymerase delta-interacting protein 3 (fold change variation versus control −2.63, Table 1), while the presence of the active MIX significantly contrasted the stress-induced variation (fold change +2.43 versus stress), suggesting a potentially improved and preserved protein homeostasis.

The structural maintenance of chromosome protein 3 is a central component of a complex required for chromosome cohesion during the cell cycle, which is coupled to DNA replication and involved in DNA repair [55], resulted in a significant downregulation (−6.13-fold change variation, Table 1) upon stress. The presence of the MIX of active ingredients (MIX + Stress) resulted in its significant upregulation (+5.53-fold change) when compared to the stressed group, preserving this important component of genetic maintenance and stability from the impact of oxidative stress exposure.

Interestingly, the levels of ADP/ATP translocase 1, polymerase delta-interacting protein 3 and structural maintenance of chromosome protein 3 were not significantly affected when the MIX was in contact with the keratinocytes in the basal conditions (absence of stress).

Then, the omics (proteome) dataset was analysed using IPA (Ingenuity Pathway Analysis; Qiagen) software to determine the biological significance of the observed experimental changes in the protein expression. Based on the integration of single protein expression results, the biological functions relevant to the skin expected to be activated or inhibited, given the observed protein expression profile, were identified as (a) major canonical pathways implicated in key aspects of cellular homeostasis (regulatory functions), (b) activated or inhibited upstream regulators (regulatory and protective functions, and (c) whether significant downstream biological processes increased or decreased (regeneration and resilience; Table 2).

Table 2. Relevant canonical pathway, upstream analysis and biofunction analysis originated from the comparison between the conditions "MIX (basal conditions)" or "Stress" or "MIX + Stress" versus the condition "Ctrl". A positive z-score indicated a predicted activation and a negative z-score indicated a predicted inactivation of the enriched pathway or biofunction in relation to the group of binary comparison. NaN, a z-core, could not be calculated. The overlap *p*-value identified pathways or biological functions that were statistically significantly overlapping between the dataset proteins and the proteins that were included in the specific pathway or biological function.

Analysis	MIX (in Basal Condition) (vs. Ctrl)		Stress (H_2O_2) (vs. Ctrl)		MIX + Stress (vs. Ctrl)	
	z-Score	*p*-Value	z-Score	*p*-Value	z-Score	*p*-Value
Canonical Pathway Oxidative Phosphorylation	2.309	<0.001	NaN	0.48	2.449	0.035
Upstream Analysis Nuclear factor, erythroid 2 like 2 (NRF2)	NaN	<0.001	NaN	/	2.236	<0.001
Biofunction Formation (Regeneration) of the skin	−1.091	<0.001	−1.715	<0.001	NaN	<0.001

The oxidative phosphorylation pathway, a key process in active mitochondria [56], was activated by the treatments with the combination of the MIX in the presence or absence of stress. This effect was not attended with the only exposure to stress (H_2O_2). This beneficial effect on mitochondria resulting from the activation of oxidative phosphorylation pathway was related to mitochondrial energy generation and production.

A predicted activation of the upstream regulator NRF2 was disclosed upon treatment with the MIX in the presence of stress underlying the effects on the regulation and protection of skin functions increasing cellular resilience to stress. In the nucleus, Nrf2 bound to the antioxidant response element (ARE) in the upstream promoter region of antioxidant genes and initiated their transcription, playing a vital role in response to oxidative stress, cytoprotection, being a recognized guardian of health span and a gatekeeper of species longevity [57]. The presence of the MIX in stress conditions showed beneficial effects on the "vigilance" of the Nrf2 signalling pathway resulting from activated antioxidant mechanisms (observed significant upregulation of five key proteins: the upregulation of phosphogluconate dehydrogenase [58], a key enzyme that produces NADPH, (PDG, fold change +1.48 vs. Ctrl), glucose-6-phosphate dehydrogenase [59] (G2PD; +1.39), malic enzyme 1 [60] (ME1; +1.39), transketolase [61] (TKT; +1.39), which connects the pentose–phosphate pathway to glycolysis and transaldolase 1 [62] (TALDO1; +1.39), an important enzyme for the balance of metabolites in the pentose–phosphate pathway. To note, transaldolase deficiency is associated with cutis laxa/wrinkled skin.

In basal conditions, the MIX treatment also showed a significant upregulation of PDG (+1.35), G2PD (+1.40) and malic enzyme 1 (ME1; +1.47). G2PD is responsible for the first step in the pentose–phosphate pathway, which, in a series of chemical reactions, converts glucose (a type of sugar found in most carbohydrates) to another sugar, ribose-5-phosphate. A dysfunction in the pentose–phosphate pathway can lead to a reduced production of NADPH (nicotinamide adenine dinucleotide phosphate) [59]. ME1 is a cytosolic protein that catalyses the conversion of malate to pyruvate, while concomitantly generating NADPH from NADP [60]. NADPH is one of the major reducing powers required for lipid production and protecting the cell against oxidative stress [63].

The biofunction formation (regeneration) of skin was predicted as slightly inhibited upon the basal treatment with the MIX (Z-score -1.091), being highly inhibited upon the exposure to the stress (Z-score -1.715), while not being affected in the presence of both the MIX and stress. These behaviours were aligned with the adaptogenic effects of the MIX, which exhibited slightly negative effects on the regeneration of the skin at basal conditions (mild stress-mimicking effects), but activated the adaptive stress response to protect and cope with the highly deleterious effects of stress, thus, maintaining a functional regeneration function of the skin comparable to the reference control conditions (not exposed to stress).

To note, among the differentially expressed proteins belonging to the so-called "formation of skin regeneration", Cornifin-A (SPRR1A) was upregulated upon the MIX treatment of the cells in the absence (basal conditions; fold change +6.37 vs. Ctrl) and in the presence of stress (fold change +10.12 vs. Ctrl), but not in the presence of stress only (-1.18 vs. Ctrl). Cornifin-A is a structural constituent of the skin epidermis that can participate widely in the construction of cell envelopes in cornifying epithelia characterized by either increased thickness or a requirement for extreme flexibility [64].

3.2. Carbonylation of Proteins

In order to confirm these promising results focused on key cellular functions, the effects of the MIX were evaluated on skin cells (primary keratinocytes) exposed (or not) to oxidative stress (H_2O_2) or pollution stress exposure (PM + UVA irradiation) through (a) a targeted evaluation of protein carbonylation levels of nuclear (histone-enriched) proteins and mitochondrial fraction and (b) a functional evaluation of mitochondrial activity. These investigations aimed to disclose the potential of the MIX to protect skin cells from stress-induced deleterious effects.

Carbonylated proteins (Oxi-Proteome [7,65]) were quantified upon detection, while the mitochondrial function was assessed by evaluating the membrane potential as a key indicator of mitochondrial activity through the accumulation of a selective fluorophore in active mitochondria. Increased levels of carbonylated (damaged) proteins were observed both in mitochondria and nuclear fraction upon H_2O_2 or pollution (PM + UVA) insults (Figure 3). In addition, the mitochondrial proteome fraction showed a higher susceptibility to oxidative damage than the nuclear proteome fraction for pollution stress exposure. The presence of the combination of active ingredients (MIX) showed a significant preservation of mitochondria-enriched proteome and nuclear (histones enriched) proteome from both H_2O_2-induced oxidative and pollution-induced damages, preventing the accumulation of carbonylated (oxidatively damaged) proteins. The alone presence of the MIX did not show any significant changes on the mitochondrial protein carbonylation basal levels. Interestingly, the presence of the MIX in basal conditions increased the carbonylation ($p < 0.01$) levels of nuclear proteins and showed protection upon stress conditions, again, in line with the action as stress mimetics in basal conditions and activating the adaptive stress response signalling pathways to contrast the deleterious effects of stress.

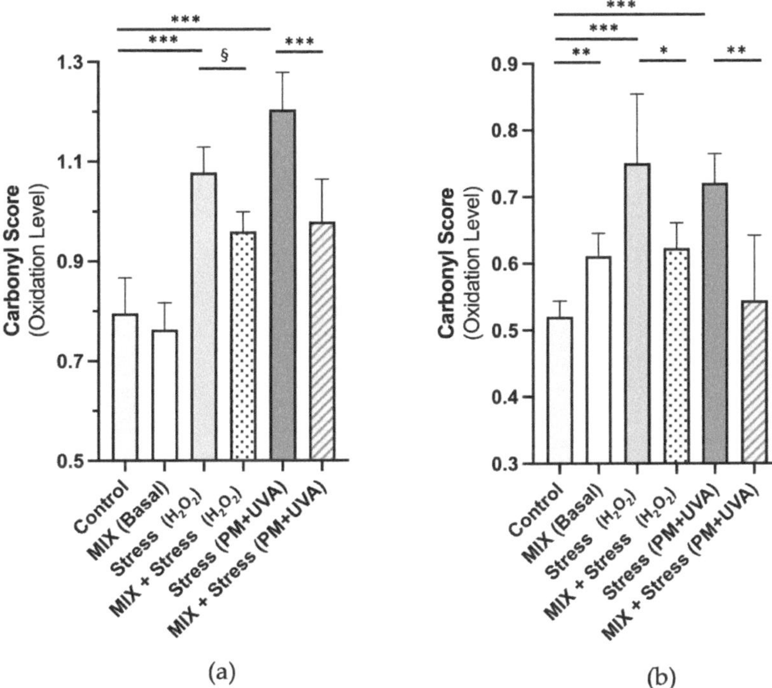

Figure 3. (**a**) Protein carbonylation of the mitochondrial purified fraction and (**b**) of the nuclear (histone-enriched) fraction. The levels of protein carbonylation for each experimental condition are reported as vertical bar graph representation of the mean and standard deviation. *** p-value < 0.001 ** p-value < 0.01 * p-value < 0.05 § p-value < 0.1—ANOVA and Tukey's post hoc multicomparison statistical analyses.

3.3. Mitochondrial Activity

Decreased levels of selective probe accumulation in the function of the mitochondrial membrane potential (essential for mitochondrial activity) were observed upon stress exposure (Figure 4). During the incubation with cells, the specific fluorescent probe passively diffused across the plasma membrane and selectively accumulated in active mitochondria in relation to their membrane potential, an essential component of mitochondrial function. The process in which the potential difference across the mitochondrial membrane was reduced is defined as mitochondrial depolarization, and it reflects a loss of mitochondrial function or activity. Upon both stress exposures (H_2O_2 or pollution), a significant decreased membrane potential was observed (reduced cumulation of the fluorescent probe in active mitochondria). The presence of the MIX showed a significant cellular resilience resulting in the protection of mitochondrial activity in the presence of pollution stress (PM + UVA; Figure 4).

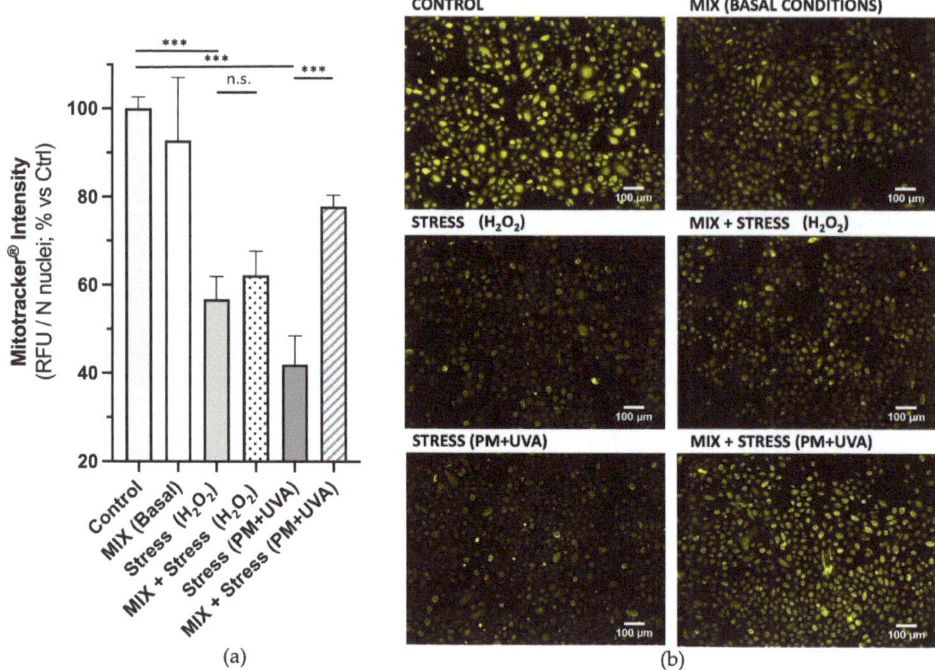

Figure 4. (**a**) Mitochondrial activity. The levels of fluorescent probe (Mitrotracker®) accumulation in active mitochondria are reported as vertical bar representation of mean values for each experimental group $+/-$ standard deviation from the mean. The intensity of fluorescence was normalized with the number of nuclei per each analysed image (detection and quantification from DAPI nuclear staining collected images). (**b**) In situ visualization of mitochondrial polarization (yellow), functional mitochondria accumulate more efficiently the probe (high intensity of the signal, yellow). Mitochondria injury and depolarization led to lower levels of probe accumulation (decreased fluorescence signal). The specific fluorescence emission signal for the fluorescent probe included in active mitochondria was obtained by using different and specific excitation and emission wavelengths (Ex: 542/20 Em: 593/40). *** *p*-Value < 0.001 n.s.—not significantly different—ANOVA and Tukey's post hoc multi-comparisons statistical analyses.

4. Conclusions

Taken together, these results suggested that the presence of the MIX of active ingredients (*Lactobacillus plantarum* extract, *Withania somnifera* root extract and *Terminalia ferdinandiana* fruit extract) improved the response to negative stress consequences through (1) preserving mitochondrial activity, (2) activating the "vigilance" of NRF2 pathway, (3) improving NADPH production and protein homeostasis and (4) preserving skin regeneration under pollution exposure by boosting cellular antioxidant mechanisms.

Moreover, the MIX showed a significant increase in cellular resilience resulting in the protection of both mitochondria-enriched proteome and nuclear (histones enriched) proteome from H_2O_2- and pollution-induced oxidative damage (carbonylation).

Both effects indicated that the MIX acted as an adaptogen, improving the cellular resistance to harmful stresses with a potential favourable impact on skin health homeostasis.

Based on the obtained results on carbonylation levels of nuclear proteins, a mode of action such as eustressors (i.e., good stressors) or as mild stress mimetics was evidenced in the basal conditions, associated then to a stress-protective response against both stressors (H_2O_2 and pollution), confirming their action as adaptogens.

This study showed the effects of the MIX on increasing cell adaptability and resilience under stress, thus, as adaptogen technology, suggesting a beneficial contribution in precision cosmetics and healthy human skin targeting the adverse consequences of environmental stress.

Author Contributions: A.C. coguided, conducted the analysis, interpreted the results and wrote the paper. L.B. coguided, interpreted the results and wrote the paper. C.R., C.G., S.C., A.B., S.L., R.R., R.A. and M.A.B. coguided, supervised the study and read the paper. A.G. coguided, interpreted the results, wrote and reviewed the paper. All authors have read and agreed to the published version of the manuscript.

Funding: This research received no external funding.

Institutional Review Board Statement: Not applicable.

Informed Consent Statement: The study has been conducted complying strictly with the ethical rules for the donation and use of human samples for research use only.

Data Availability Statement: Data are contained within the article.

Acknowledgments: We thank Philippe Bastien (L'Oréal R&I) for the insightful discussions on statistics.

Conflicts of Interest: The authors belongs to either L'OREAL groups (Charline Ruaux, Celeste Grossgold, Suzy Levoy, Rawad Abdayem, Romain Roumiguiere, Stephanie Cheilian, Anne Bouchara, Audrey Gueniche), Oxiproteomics company (Andrea Cavagnino, Martin Baraibar) or Cilia company (Lionel Breton). The authors declare that the research was conducted in the absence of any commercial or financial relationships that could be construed as a potential conflict of interest. The funders had no role in the design of the study; in the collection, analyses or interpretation of data; in the writing of the manuscript; or in the decision to publish the results. The authors declare no conflict of interest.

References

1. WHO. *Guidelines on Hand Hygiene in Health Care: First Global Patient Safety Challenge Clean Care is Safer Care*; World Health Organization—WHO: Geneva, Switzerland, 2009.
2. Fussell, J.C.; Kelly, F.J. Oxidative contribution of air pollution to extrinsic skin ageing. *Free Radic. Biol. Med.* **2020**, *151*, 111–122. [CrossRef] [PubMed]
3. Slominski, A.T.; Zmijewski, M.A.; Skobowiat, C.; Zbytek, B.; Slominski, R.M.; Steketee, J.D. Sensing the environment: Regulation of local and global homeostasis by the ski's neuroendocrine system. *Adv. Anat. Embryol. Cell Biol.* **2012**, *212*, 1–115.
4. Birch-Machin, M.A.; Russell, E.V.; Latimer, J.A. Mitochondrial DNA damage as a biomarker for ultraviolet radiation exposure and oxidative stress. *Br. J. Dermatol.* **2013**, *169* (Suppl. 2), 9–14. [CrossRef] [PubMed]
5. Hudson, L.; Bowman, A.; Rashdan, E.; Birch-Machin, M.A. Mitochondrial damage and ageing using skin as a model organ. *Maturitas* **2016**, *93*, 34–40. [CrossRef]
6. Dezest, M.; Le Bechec, M.; Chavatte, L.; Desauziers, V.; Chaput, B.; Grolleau, J.L.; Descargues, P.; Nizard, C.; Schnebert, S.; Lacombe, S.; et al. Oxidative damage and impairment of protein quality control systems in keratinocytes exposed to a volatile organic compounds cocktail. *Sci. Rep.* **2017**, *7*, 10707. [CrossRef]
7. Cavagnino, A.; Bobier, A.; Baraibar, M. The skin Oxi-Proteome as a molecular signature of exposome stress. *H&PC Today* **2019**, *14*, 4.
8. Zuo, L.; Prather, E.R.; Stetskiv, M.; Garrison, D.E.; Meade, J.R.; Peace, T.I.; Zhou, T. Inflammaging and Oxidative Stress in Human Diseases: From Molecular Mechanisms to Novel Treatments. *Int J Mol Sci.* **2019**, *20*, 4472. [CrossRef]
9. Fedorova, M.; Bollineni, R.C.; Hoffmann, R. Protein carbonylation as a major hallmark of oxidative damage: Update of analytical strategies. *Mass Spectrom. Rev.* **2014**, *33*, 79–97. [CrossRef]
10. Nyström, T. Role of oxidative carbonylation in protein quality control and senescence. *EMBO J.* **2005**, *24*, 1311–1317. [CrossRef]
11. Levine, R.L. Carbonyl modified proteins in cellular regulation, aging, and disease. *Free Radic. Biol. Med.* **2002**, *32*, 790–796. [CrossRef]
12. Zhang, H. Oxidative stress response and Nrf2 signaling in aging. *Free Radic. Biol. Med.* **2015**, *88 Pt B*, 314–336. [CrossRef]
13. Lavigne, E.G.; Cavagnino, A.; Steinschneider, R.; Breton, L.; Baraibar, M.A.; Jäger, S. Oxidative damage prevention in human skin and sensory neurons by a salicylic acid derivative. *Free Radic. Biol. Med.* **2022**, *181*, 98–104. [CrossRef] [PubMed]
14. Yang, H.C. What has passed is prolog: New cellular and physiological roles of G6PD. *Free Radic. Res.* **2016**, *50*, 1047–1064. [CrossRef]
15. Nobrega-Pereira, S. G6PD protects from oxidative damage and improves healthspan in mice. *Nat. Commun.* **2016**, *7*, 10894. [CrossRef] [PubMed]

16. Frohnert, B.I.; Bernlohr, D.A. Protein carbonylation, mitochondrial dysfunction, and insulin resistance. *Adv. Nutr.* **2013**, *4*, 157–163. [CrossRef] [PubMed]

17. McCormack, J.G.; Halestrap, A.P.; Denton, R.M. Role of calcium ions in regulation of mammalian intramitochondrial metabolism. *Physiol Rev.* **1990**, *70*, 391–425. [CrossRef]

18. He, W.; Newman, J.C.; Wang, M.Z.; Ho, L.; Verdin, E. Mitochondrial sirtuins: Regulators of protein acylation and metabolism. *Trends Endocrinol. Metab.* **2012**, *23*, 467–476. [CrossRef]

19. Tzameli, I. The evolving role of mitochondria in metabolism. *Trends Endocrinol. Metab.* **2012**, *23*, 417–419. [CrossRef]

20. Birsoy, K.; Wang, T.; Chen, W.W.; Freinkman, E.; Abu-Remaileh, M.; Sabatini, D.M. An Essential Role of the Mitochondrial Electron Transport Chain in Cell Proliferation Is to Enable Aspartate Synthesis. *Cell* **2015**, *162*, 540–551. [CrossRef]

21. Vakifahmetoglu-Norberg, H.; Ouchida, A.T.; Norberg, E. The role of mitochondria in metabolism and cell death. *Biochem. Biophys. Res. Commun.* **2017**, *482*, 426–431. [CrossRef]

22. Noguchi, M.; Kasahara, A. Mitochondrial dynamics coordinate cell differentiation. *Biochem. Biophys. Res. Commun.* **2018**, *500*, 59–64. [CrossRef]

23. Chandel, N.S. Mitochondria as signaling organelles. *BMC Biol.* **2014**, *12*, 34. [CrossRef]

24. Slominski, A.T.; Zmijewski, M.A.; Semak, I.; Kim, T.K.; Janjetovic, Z.; Slominski, R.M.; Zmijewski, J.W. Melatonin, mitochondria, and the skin. *Cell Mol. Life Sci.* **2017**, *74*, 3913–3925. [CrossRef] [PubMed]

25. Panich, U.; Sittithumcharee, G.; Rathviboon, N.; Jirawatnotai, S. Ultraviolet Radiation-Induced Skin Aging: The Role of DNA Damage and Oxidative Stress in Epidermal Stem Cell Damage Mediated Skin Aging. *Stem. Cells Int.* **2016**, *2016*, 7370642. [CrossRef] [PubMed]

26. Victorelli, S.; Passos, J.F. Reactive Oxygen Species Detection in Senescent Cells. *Methods Mol. Biol.* **2019**, *1896*, 21–29.

27. Bocheva, G.; Slominski, R.M.; Slominski, A.T. Neuroendocrine Aspects of Skin Aging. *Int. J. Mol. Sci.* **2019**, *20*, 2798. [CrossRef] [PubMed]

28. Krutmann, J.; Schroeder, P. Role of mitochondria in photoaging of human skin: The defective powerhouse model. *J. Investig. Dermatol. Symp. Proc.* **2009**, *14*, 44–49. [CrossRef]

29. Naidoo, K.; Hanna, R.; Birch-Machin, M.A. What is the role of mitochondrial dysfunction in skin photoaging? *Exp. Dermatol.* **2018**, *27*, 124–128. [CrossRef]

30. Cabrera, F.; Ortega, M.; Velarde, F.; Parra, E.; Gallardo, S.; Barba, D.; Soto, L.; Peña, G.; Pedroza, L.A.; Jorgensen, C.; et al. Primary allogeneic mitochondrial mix (PAMM) transfer/transplant by MitoCeption to address damage in PBMCs caused by ultraviolet radiation. *BMC Biotechnol.* **2019**, *19*, 42. [CrossRef]

31. Moller, I.M. Plant mitochondria and oxidative stress: Electron Transport, NADPH Turnover, and Metabolism of Reactive Oxygen Species. *Annu. Rev. Plant Physiol. Plant Mol. Biol.* **2001**, *52*, 561–591. [CrossRef]

32. Sreedhar, A.; Aguilera-Aguirre, L.; Singh, K.K. Mitochondria in skin health, aging, and disease. *Cell Death Dis.* **2020**, *11*, 444. [CrossRef] [PubMed]

33. Laborit, H. *Éloge de la Fuite*; Robert Laffont: Paris, France, 1976.

34. Simonet, G. Le concept d'adaptation: Polysémie interdisciplinaire et implication pour les changements climatiques. *Nat. Sci. Sociétés* **2009**, *17*, 392–401. [CrossRef]

35. Passeron, T.; Krutmann, J.; Andersen, M.L.; Katta, R.; Zouboulis, C. Clinical and biological impact of the exposome on the skin. *J. Eur. Acad. Dermatol. Venereol.* **2020**, *34* (Suppl. 4), 4–25. [CrossRef] [PubMed]

36. Panossian, A.G.; Efferth, T.; Shikov, A.N.; Pozharitskaya, O.N.; Kuchta, K.; Mukherjee, P.K.; Banerjee, S.; Heinrich, M.; Wu, W.; Guo, D.A.; et al. Evolution of the adaptogenic concept from traditional use to medical systems: Pharmacology of stress- and aging-related diseases. *Med. Res. Rev.* **2021**, *41*, 630–703. [CrossRef]

37. Panossian, A.; Wikman, G.; Wagner, H. Plant adaptogens III. Earlier and more recent aspects and concepts on their mode of action. *Phytomedicine* **1999**, *6*, 287–300. [CrossRef]

38. Panossian, A.; Gabrielian, E.; Wagner, H. On the mechanism of action of plant adaptogens with particular reference to cucurbitacin R diglucoside. *Phytomedicine* **1999**, *6*, 147–155. [CrossRef]

39. Panossian, A. Understanding adaptogenic activity: Specificity of the pharmacological action of adaptogens and other phytochemicals. *Ann. N. Y. Acad. Sci.* **2017**, *1401*, 49–64. [CrossRef]

40. Panossian, A.; Seo, E.J.; Efferth, T. Novel molecular mechanisms for the adaptogenic effects of herbal extracts on isolated brain cells using systems biology. *Phytomedicine* **2018**, *50*, 257–284. [CrossRef]

41. Lazarev, N.V. General and specific in action of pharmacological agents. *Farmacol. Toxicol.* **1958**, *21*, 81–86.

42. Lazarev, N.V.; Ljublina, E.I.; Ljublina, M.A. State of nonspecific resistance. *Patol. Fiziol. Exp. Terapia.* **1959**, *3*, 16–21.

43. Yadav, B.; Bajaj, A.; Saxena, M.; Saxena, A.K. In Vitro Anticancer Activity of the Root, Stem and Leaves of Withania Somnifera against Various Human Cancer Cell Lines. *Indian J. Pharm. Sci.* **2010**, *72*, 659–663. [CrossRef]

44. Ilayperuma, I.; Ratnasooriya, W.D.; Weerasooriya, T.R. Effect of Withania somnifera root extract on the sexual behaviour of male rats. *Asian J. Androl.* **2002**, *4*, 295–298. [PubMed]

45. Saleem, S.; Muhammad, G.; Hussain, M.A.; Altaf, M.; Bukhari, S.N.A. Withania somnifera L.: Insights into the phytochemical profile, therapeutic potential, clinical trials, and future prospective. *Iran. J. Basic Med. Sci.* **2020**, *23*, 1501–1526. [PubMed]

46. Bradford, M.M. A rapid and sensitive method for the quantitation of microgram quantities of protein utilizing the principle of protein-dye binding. *Anal. Biochem.* **1976**, *72*, 248–254. [CrossRef] [PubMed]

47. Krämer, A.; Green, J.; Pollard, J., Jr.; Tugendreich, S. Causal analysis approaches in Ingenuity Pathway Analysis. *Bioinformatics* **2014**, *30*, 523–530. [CrossRef]
48. Baghirova, S.; Hughes, B.G.; Hendzel, M.J.; Schulz, R. Sequential fractionation and isolation of subcellular proteins from tissue or cultured cells. *MethodsX* **2015**, *2*, 440–445. [CrossRef]
49. Baraibar, M.A.; Ladouce, R.; Friguet, B. Proteomic quantification and identification of carbonylated proteins upon oxidative stress and during cellular aging. *J. Proteom.* **2013**, *92*, 63–70. [CrossRef]
50. Schneider, C.A.; Rasband, W.S.; Eliceiri, K.W. NIH Image to ImageJ: 25 years of image analysis. *Nat. Methods* **2012**, *9*, 671–675. [CrossRef]
51. Kawamata, H.; Tiranti, V.; Magrané, J.; Chinopoulos, C.; Manfredi, G. adPEO mutations in ANT1 impair ADP-ATP translocation in muscle mitochondria. *Hum. Mol. Genet.* **2011**, *20*, 2964–2974. [CrossRef]
52. Brand, M.D.; Pakay, J.L.; Ocloo, A.; Kokoszka, J.; Wallace, D.C.; Brookes, P.S.; Cornwall, E.J. The basal proton conductance of mitochondria depends on adenine nucleotide translocase content. *Biochem. J.* **2005**, *392 Pt 2*, 353–362. [CrossRef]
53. Kim, E.H.; Koh, E.H.; Park, J.Y.; Lee, K.U. Adenine nucleotide translocator as a regulator of mitochondrial function: Implication in the pathogenesis of metabolic syndrome. *Korean Diabetes J.* **2010**, *34*, 146–153. [CrossRef] [PubMed]
54. Folco, E.G.; Lee, C.S.; Dufu, K.; Yamazaki, T.; Reed, R. The proteins PDIP3 and ZC11A associate with the human TREX complex in an ATP-dependent manner and function in mRNA export. *PLoS ONE* **2012**, *7*, e43804. [CrossRef] [PubMed]
55. Sumara, I.; Vorlaufer, E.; Gieffers, C.; Peters, B.H.; Peters, J.M. Characterization of vertebrate cohesin complexes and their regulation in prophase. *J. Cell Biol.* **2000**, *151*, 749–762. [CrossRef]
56. Vercellino, I.; Sazanov, L.A. The assembly, regulation and function of the mitochondrial respiratory chain. *Nat. Rev. Mol. Cell Biol.* **2022**, *23*, 141–161. [CrossRef] [PubMed]
57. Matsumaru, D.; Motohashi, H. The KEAP1-NRF2 System in Healthy Aging and Longevity. *Antioxidants* **2021**, *10*, 1929. [CrossRef]
58. Chayen, J.; Howat, D.W.; Bitensky, L. Cellular biochemistry of glucose 6-phosphate and 6-phosphogluconate dehydrogenase activities. *Cell Biochem. Funct.* **1986**, *4*, 249–253. [CrossRef] [PubMed]
59. Stanton, R.C. Glucose-6-phosphate dehydrogenase, NADPH, and cell survival. *IUBMB Life* **2012**, *64*, 362–369. [CrossRef]
60. White, K.; Someya, S. The roles of NADPH and isocitrate dehydrogenase in cochlear mitochondrial antioxidant defense and aging. *Hear. Res.* **2023**, *427*, 108659. [CrossRef]
61. Kochetov, G.A.; Solovjeva, O.N. Structure and functioning mechanism of transketolase. *Biochim. Biophys. Acta* **2014**, *1844*, 1608–1618. [CrossRef]
62. Moriyama, T.; Tanaka, S.; Nakayama, Y.; Fukumoto, M.; Tsujimura, K.; Yamada, K.; Bamba, T.; Yoneda, Y.; Fukusaki, E.; Oka, M. Two isoforms of TALDO1 generated by alternative translational initiation show differential nucleocytoplasmic distribution to regulate the global metabolic network. *Sci. Rep.* **2016**, *6*, 34648. [CrossRef]
63. Xu, M.; Ding, L.; Liang, J.; Yang, X.; Liu, Y.; Wang, Y.; Ding, M.; Huang, X. NAD kinase sustains lipogenesis and mitochondrial metabolismthrough fatty acid synthesis. *Cell Rep.* **2021**, *37*, 110157. [CrossRef] [PubMed]
64. Rajagopalan, P.; Jain, A.P.; Nanjappa, V.; Patel, K.; Mangalaparthi, K.K.; Babu, N.; Cavusoglu, N.; Roy, N.; Soeur, J.; Breton, L.; et al. Proteome-wide changes in primary skin keratinocytes exposed to diesel particulate extract-A role for antioxidants in skin health. *J. Dermatol. Sci.* **2019**, *96*, 114–124. [CrossRef] [PubMed]
65. Lourenço Dos Santos, S.; Baraibar, M.A.; Lundberg, S.; Eeg-Olofsson, O.; Larsson, L.; Friguet, B. Oxidative proteome alterations during skeletal muscle ageing. *Redox Biol.* **2015**, *5*, 267–274. [CrossRef] [PubMed]

Communication

The Natural *Centella asiatica* Extract Acts as a Stretch Mark Eraser: A Biological Evaluation

Cloé Boira [1,*], Marie Meunier [1], Marine Bracq [1], Amandine Scandolera [1] and Romain Reynaud [2]

[1] Givaudan Active Beauty, Science and Technology, 51110 Pomacle, France
[2] Givaudan Active Beauty, Science and Technology, 31000 Toulouse, France
[*] Correspondence: cloe.boira@givaudan.com; Tel.: +33-51008011

Abstract: Stretch marks are far from exclusively appearing on pregnant women and appear whenever the body experiences rapid growth. Collagen fibres are altered in the dermis, which is associated with a loss of orientation, and the elastic network is disrupted, leading to a fibrotic organisation. This results in epidermal tearing that produces skin lesions. *Centella asiatica* (CAST) is a well-known medicinal plant rich in active triterpenic molecules and traditionally used to treat wounds and help skin repair. The aim of this study was to evaluate CAST extract as a natural way to solve stretch mark concerns and understand its mechanism of action. Fibroblast proliferation based on scratch assay model and their gene expression by RT-qPCR was first evaluated. At the ex vivo level, elastin fibres were quantified by immunofluorescence. The orientation of the collagen fibres and their occupation of the dermis were analysed after Sirius red staining and specific software analysis. We showed that CAST stimulated fibroblast proliferation and reduced extracellular matrix degradation and fibrosis. On a stretch-marked skin explant, CAST increased the occupation of collagen fibres and elastin production. Based on the mechanisms behind the formation of stretch marks, CAST restored the dermis network by optimising fibre organisation for a visible skin remodelling effect.

Keywords: *Centella asiatica*; stretch marks; skin repair; striae

Citation: Boira, C.; Meunier, M.; Bracq, M.; Scandolera, A.; Reynaud, R. The Natural *Centella asiatica* Extract Acts as a Stretch Mark Eraser: A Biological Evaluation. *Cosmetics* **2024**, *11*, 15. https://doi.org/10.3390/cosmetics11010015

Academic Editor: Paulraj Mosae Selvakumar

Received: 4 December 2023
Revised: 19 January 2024
Accepted: 21 January 2024
Published: 23 January 2024

1. Introduction

Stretch marks, also called striae, are a form of tissue damage on the skin due to an excessive stretching of the dermis [1,2]. Stretch marks are commonly associated with pregnancy and affect between 60% and 90% of women during their pregnancy [3]. In reality, they occur as the result of tearing of the dermis during periods of rapid growth of the body or some body parts. Therefore, around 27% of adolescents are concerned by stretch marks [3]. As they are caused by the sudden expansion of the skin, especially in areas where fat is most likely to be stored in our body, obesity may lead to the apparition of stretch marks. Stretch marks are thus usually found on the breasts, thighs, and belly, especially near the navel, upper arms, underarms, and lower back, both in women and men [4].

Stretch marks affect the dermis by preventing the fibroblasts from organising collagen fibres to keep up with the skin's stretching. The collagen bundles are altered, lose their orientation, and are organised into a fibrotic structure, while the elastic network is also disrupted [5]. As a result, the skin appears less firm, less deformable, and less plastic. This normally leads to epidermal tearing, which can produce a lesion in the form of a visible stretch mark. Stretch marks could be classified into six different types depending on their appearance and epidemiology; *striae atrophicans* (thinned skin), *striae gravidarum* (following pregnancy), *striae distensae* (stretched skin), *striae rubrae* (red), *striae albae* (white), *striae nigra* (black), and *striae caerulea* (dark blue) [4]. Over time, stretch marks tend to atrophy and lose pigmentation, depending on how recent they are on the skin. Thus, hyper-pigmented stretch marks related to an acute stage are characterised by the initial erythematous, and a chronic stage is characterised by a hypo-pigmented and atrophic lesion [6].

Current treatment involves invasive methods such as laser therapy, light therapy, collagen injection, laser lipolysis, radiofrequency techniques, and microdermabrasion [1,6,7]. A topical solution may be used, and among these, a formulation with CAST has been described to improve stretch marks, but its mechanism of action is poorly described [8–11]. CAST is one of the most known traditional medicinal plants. It has long been used to promote skin repair and wound healing, and it is a medicinal remedy present in the Ayurvedic system and traditional Chinese medicine [12]. It grows in cultivated fields in Madagascar and other Asian and African countries. It expresses, in its tiny and round leaves, as secondary metabolites, some biologically active triterpenic molecules (asiaticoside, asiatic acid and madecassic acid) [13]. CAST is a promising candidate to solve stretch mark concerns as it enhances cell production and matrix components and improves tensile strength [12]. Clinical studies have shown that the application of a formulation with CAST significantly improved stretch mark appearance [9,14]. We have also previously conducted a clinical study on 54 women divided into two groups (the control group applying a placebo formula without CAST and a group using a formulation with CAST [15]). The volunteers used the product three times per day for one month. After four weeks of using the CAST formulation, skin thickness at the edge and in the centre of the stretch marks was significantly increased in comparison to the placebo, such as skin vascularisation and elasticity for a visual reduction in stretch marks appearance [15]. Following this preliminary work, we wanted to explore the mechanism of action of CAST further, and the aim of this study was to unveil potential unknown pathways that can be impacted by CAST at the in vitro and ex vivo levels. Transcriptomic analysis was performed in order to identify potential new actors in stretch mark-related fibrosis, such as CGTF. Modulation of its expression by CAST could be the result of improvement of the dermal matrix, improving the quality of the dermis and thus reducing stretch mark damages.

2. Materials and Methods

The aim of this study was to evaluate the active ingredient obtained from *Centella asiatica* (supplied by Givaudan Active Beauty, Argenteuil, France) as a new way to solve stretch mark concerns.

Centella CAST (INCI: asiaticoside, asiatic acid, madecassic acid) was obtained by Indena Spa, Viale Ortles 12, Milan, Italy. The product is obtained through the extraction, fractionation, and recombination of terpenes typical of *Centella asiatica*, which are the major bioactive constituents. These include asiaticoside, in which a trisaccharide moiety is linked to the aglycone asiatic acid, madecassoside, and madecassic acid. *Centella asiatica* is sourced in Madagascar, and leaves are hand collected from spontaneous growth during their growing season, harvested, and naturally dried. After a preliminary extraction of Centella soft extract, the product is concentrated and purified to the first fraction of asiaticoside. Discarded solutions are concentrated, the pH is adjusted, and the fraction is purified to obtain the madecassic and asiatic acid fractions. The asiaticoside fraction and the acids fraction are then mixed to match the specifications. This natural extract has a very fixed composition and is composed of asiaticoside (about 10–16%), asiatic acid (about 6–9%) and madecassossic acid (about 10–14%).

2.1. Wound Healing on Fibroblast

Fibroblasts (NIH-3T3, ATCC CRL-1658) were maintained in supplemented DMEM medium (Gibco®, Life Technologies, Carlsbad, CA, USA) containing 10% fetal bovine serum (FBS), (Hyclone, Logan, UT, USA) and 1% antibiotics (Sigma-Aldrich, St. Louis, MO, USA) (DMEM complete medium) at 37 °C 5% CO_2. NIH-3T3 (10^5 cells/well) were seeded in 96-well Essen ImageLock plates (Essen BioScience, Ann Arbor, MI, USA) and grown to 80% confluence in a CO_2-humidified incubator. After 24 h, the scratch was made using the 96-pin WoundMaker (Essen BioScience, Ann Arbor, MI, USA). Fibroblasts were treated with CAST at 1, 5, 10, 25, or 50 µg/mL. FBS 10% was used as a positive control. Wound images were taken every hour for 36 h, and the data were analysed using the integrated metric "relative wound density" from the live-content cell imaging system IncuCyte HD (Essen BioScience, Ann Arbor, MI, USA). The experiments were performed in triplicate wells.

2.2. Transcriptomic Analysis on Fibroblast

Normal human dermal fibroblasts (NHDFs) freshly isolated from non-stretch-marked or from stretch-marked areas from the same donor were seeded at 300,000 cells per well in 6-well plates and cultivated in DMEM medium (Gibco®, Life Technologies, Carlsbad, CA, USA) supplemented by FBS with 10% and 1% antibiotics (Sigma-Aldrich, St. Louis, MO, USA). After 48 h of culture, NHDFs were rinsed two times with PBS (Gibco®, Life Technologies, Carlsbad, CA, USA) and allowed to rest in FCS-free medium overnight before stimulation. Cells were stimulated with CAST extract at 10 µg/mL versus untreated condition. After 24 h of stimulation, total RNA was extracted using the extract-all method (Fisher Scientific, Hampton, NH, USA). RNA quality was controlled, and reverse transcription was performed to obtain cDNA using the Verso cDNA kit (Thermo Fisher Scientific, Waltham, MA, USA). RT-qPCR was performed on specific pre-coated plates (Applied Biosystems, Foster City, CA, USA) designed to study the transcriptomic expression of different genes involved in fibrosis and matrix remodelling with 10 ng of cDNA per well using CFX96 Touch (Biorad, Hercules, CA, USA) and Universal Taqman mix (Quantabio, Beverly, MA, USA). CTGF: Unique Assay ID qHsaCED0002044, FAK: Unique Assay ID qHsaCED0001879, MMP1: Unique Assay ID qHsaCID0017039, MMP7: Unique Assay ID qHsaCID0011537. The relative quantification (RQ) of gene expression was calculated according to POP4 (Ribonuclease/MRP subunit) (POP4: qHsaCID0015127) and B2M (Beta-2-Microglobulin) (B2M: qHsaCID0015347) housekeeping genes. The data are expressed in fold change relative to normal fibroblasts or to untreated stretch-marked fibroblasts.

2.3. Collagen Network Analysis on Skin Explant

Skin explants were obtained according to ethical and regulatory rules and under the agreement of the participants. Skin explants with stretch marks from a 60-year-old female volunteer donor (Biopredic International, Saint Grégoire, France) were treated topically with CAST at 0.5% for 5 days. Each day, treatment and medium (Genoskin, Toulouse, France) were renewed, and explants were incubated at 37 °C 5% CO_2.

INCI placebo: AQUA/WATER, CETYL ALCOHOL, GLYCERYL STEARATE, PEG-75 STEARATE, CETETH-20, STEARETH-20, ISODECYL NEOPENTANOATE, PHENOXYETHANOL.

After 5 days, skin biopsies were fixed in formalin (Sigma-Aldrich, St. Louis, MO, USA) for 48 h and then dehydrated overnight by automated dehydration using Histocore Pearl (Leica, Wetzlar, Germany). After dehydration, skin explants were included in paraffin (Leica, Wetzlar, Germany), and skin sections of 4 µm were obtained with a microtome (Leica, Wetzlar, Germany). Paraffin was removed thanks to xylene baths (Sigma-Aldrich, St. Louis, MO, USA) followed by ethanol (VWR International, Radnor, PA, USA) dehydration. Skin slices were stained with a ready-to-use Sirius red colouration (Labo Modern, Gennevilliers, France) before mounting on a thin glass slide. Dermis structure was observed using a bright field, and collagen fibre orientation was analysed in polarised light. Under polarised light, collagen I fibres were detected using the red channel of a slide scanner (Olympus, Tokyo, Japan). The quantity of collagen I was quantified using ImageJ 1.53k software on the whole dermis. Furthermore, the images were oriented with the epidermal part up at 90° of the x-axis. The images were then processed to remove a 200 µm thickness from the outer-most stratum corneum to eliminate the papillary dermis for the orientation analysis of collagen fibres. The conditioned images were processed by a segmentation algorithm, allowing the analysis to focus only on the collagen fibres. Based on a plane parallel to the epidermis, collagen I fibres were quantified at 10° intervals in order to evaluate collagen I occupation in the dermis.

2.4. Elastin Immunofluorescence on Skin Explant

The skin explants were cultivated and treated as previously described. Paraffin was removed thanks to xylene baths followed by ethanol dehydration. Antigenic sites were revealed with a citrate buffer (Sigma-Aldrich, St. Louis, MO, USA). Then, skin sections were BSA-saturated (Sigma-Aldrich, St. Louis, MO, USA) for 30 min, followed by immunostaining to specifically detect elastin (1/75e, Santa-Cruz, Dallas, TX, USA). After overnight incubation of the primary antibody at 4 °C, skin sections were washed and incubated for 1 h with secondary antibody (Anti-mouse, 1/500e, Abcam, Cambridge, UK). The staining was observed using fluorescent microscopy (Zeiss, Axio observer, Iena, Germany). Relative fluorescence was quantified for all images using ImageJ 1.53k software.

2.5. Statistical Analysis

For all studies, a Shapiro-Wilk test was used to verify whether the raw data followed the Gaussian Law. In the case of normally distributed data, the mean values were compared with untreated conditions or normal skin conditions (no stretch marks) using an unpaired t-student test. In the case of non-normally distributed data, a Mann–Whitney U test was used. For all statistical tests, we considered significant results as follows: $\# p < 0.1$, $* p < 0.05$, $** p < 0.01$ and $*** p < 0.001$.

3. Results

3.1. Centella asiatica Effect on Fibroblast Migration

NIH-3T3 monolayer cell cultures were wounded and then treated with non-cytotoxic concentrations (1, 5, 10, 25, and 50 µg/mL) of CAST. NIH-3T3 fibroblast migration was observed following CAST treatment, even at the lower concentration tested. After 36 h, a wound closure of up to 85% was induced by CAST at a tested concentration between 1 and 25 µg/mL. However, significant inhibition of basal wound healing in NIH-3T3 cells was demonstrated at the higher concentration tested (50 µg/mL) of CAST (Figure 1). Interestingly, after 24 h of incubation, better wound closure was observed on cells treated with CAST at 10 µg/mL comparatively with FBS at 10% (positive reference), as observed in illustrative pictures and graphs. Our results demonstrated that CAST initiated a more rapid cell migration process for better wound-healing activity.

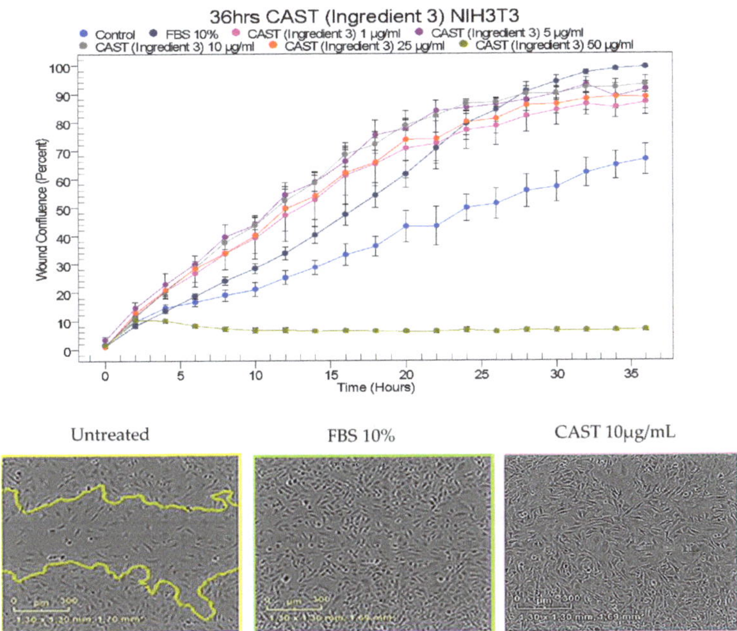

Figure 1. Effect of CAST on NIH-3T3 in wound healing. Wound closure kinetics after 36 h of treatment with FBS 10% (positive control) and CAST at 1, 5, 10 and 25 µg/mL starting just after the injury (**top**). Representative pictures of healing after 24 h following the treatment with FCS 10% (positive control) and CAST at 10 µg/mL (**bottom**).

3.2. Anti-Fibrotic Effect of Centella asiatica

Stretch-marked fibroblasts were treated with CAST at 10 µg/mL for 24 h. An RT-qPCR analysis was performed with a focus on matrix preservation and fibrosis markers. In comparison to normal fibroblasts, stretch-marked fibroblasts significantly increased fibrosis markers by 63% and 66% regarding the expression of CTGF and PTK2, respectively. Regarding matrix remodelling, MMP1 and MMP7 were significantly increased by 873% and 1230%, respectively, in comparison to normal fibroblasts. After CAST treatment, fibrosis and matrix degradation markers were significantly decreased by 128%, 58%, 149%, and 50% for CTGF, PTK2, MMP1, and MMP7, respectively, in comparison to the stretch-marked fibroblasts (Table 1).

Table 1. Gene expression (fold change) on NHDFs from a non-stretch-marked donor or stretch-marked donor. Fibroblasts with stretch marks were treated with CAST at 10 µg/mL. Unpaired *t*-test was performed to analyse the comparison with non-stretch marks or untreated conditions with # $p < 0.1$, * $p < 0.05$, ** $p < 0.01$ and *** $p < 0.001$.

	Genes	**Fold Change Stretch Marks versus Non-Stretch-Marked**	**Fold Change CAST versus Stretch Marks**
Fibrosis	CTGF (CCN2)	1.6 *	0.44 **
	PTK2 (FAK)	1.7 #	0.63 **
Extra Cellular Matrix	MMP1	9.7 *	0.40 **
	MMP7	13.3 ***	0.67 *

On stretch-marked skin explants, the fibrosis is visible in the bright field after Sirius red staining, and the accumulation of collagen I deposits was revealed by an intense red staining and an anarchic orientation of collagen bundles. The treatment with CAST at 0.5% reduced the fibrosis, and collagen bundles were well organised, as observed by the less impacted and better-defined papillary and reticular dermis organisations in Figure 2C. Moreover, the special Sirius red staining highlights the natural birefringence of collagen fibres when exposed to polarised light. Collagen type I would show a yellow–red colour, and the quantification showed a significant increase in collagen for the stretch-marked area in comparison to the non-stretch-marked region by 10%. After treatment with CAST at 0.5%, the quantity of collagen I was significantly reduced by 10% in comparison to untreated stretch marks. The placebo showed a strong impact on collagen and reduced it by 18% in comparison to stretch marks. CAST was able to increase the amount of collagen I in comparison to placebo, reaching a normal collagen quantity (Figure 2).

This special Sirius red staining highlights the configuration of the collagen, as well as the heterogeneity of the direction of the fibres in the tissues. Image analysis after polarised light acquisition allows the quantification of the number of collagen bundles in each direction, starting from a plane parallel to the epidermis. Stretch marks significantly decreased the occupation of the collagen fibre in the dermis by 33% in comparison to the normal skin. CAST at 0.5% significantly increased their occupation by 49% in comparison to the stretch mark explants (Figure 3). In addition, stretch marks impacted the preferential direction of the fibres, especially between 20° and 60° relative to the epidermis. This preferential direction of the collagen bundles was restored by CAST at 0.5% (Figure 4).

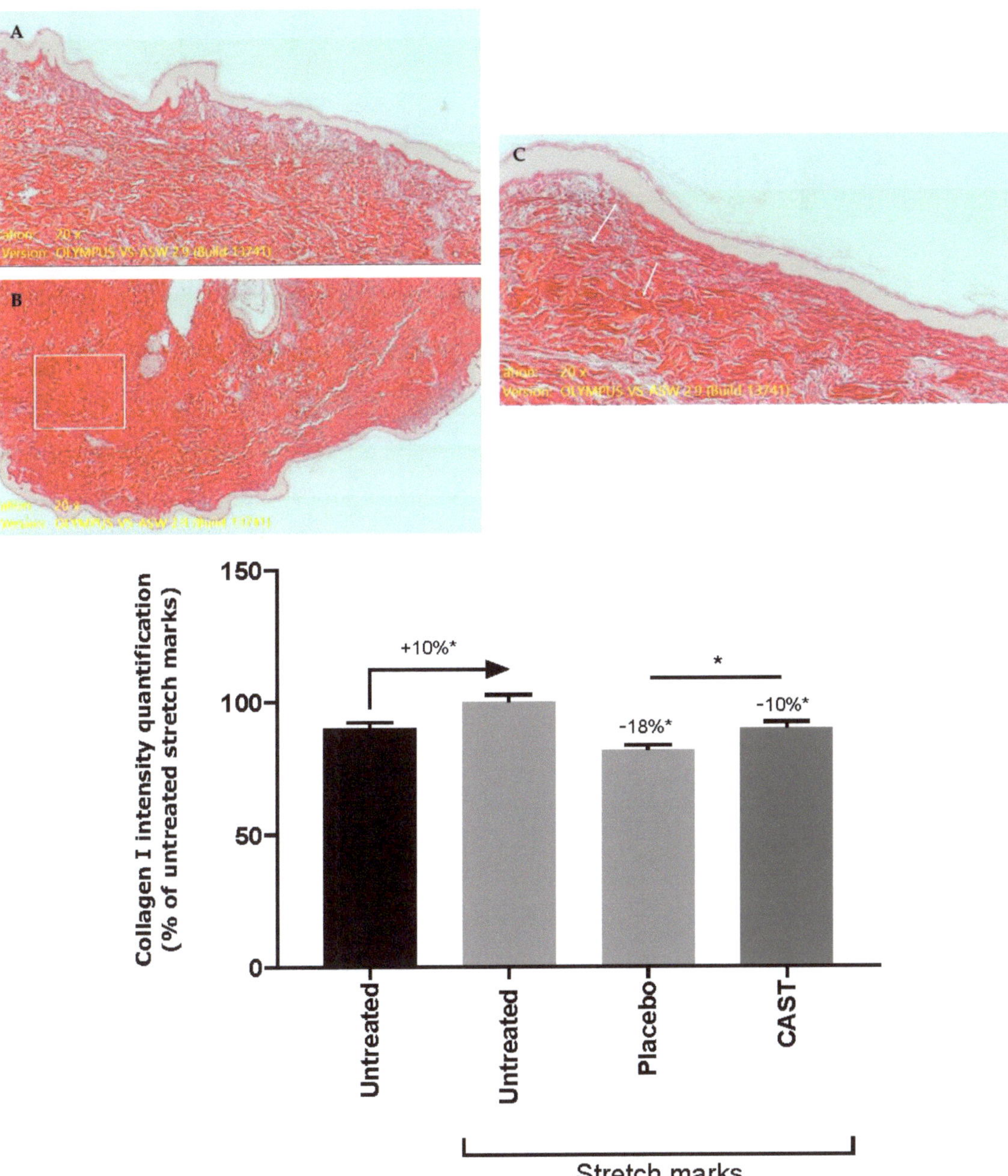

Figure 2. Effect of CAST on skin fibrosis. Dermis network visualisation in bright field (×20) after Sirius red staining (**top**). (**A**) Control; (**B**) untreated stretch mark skin explant (the square highlights a fibrotic organisation); (**C**) CAST 0.5% treatment (the arrows illustrate normal collagen bundles). Quantification of collagen I quantity in the whole dermis after Sirius red staining observed under polarised light (**bottom**). Mann-Whitney test was performed to analyse the comparison with untreated stretch mark conditions with * $p < 0.05$.

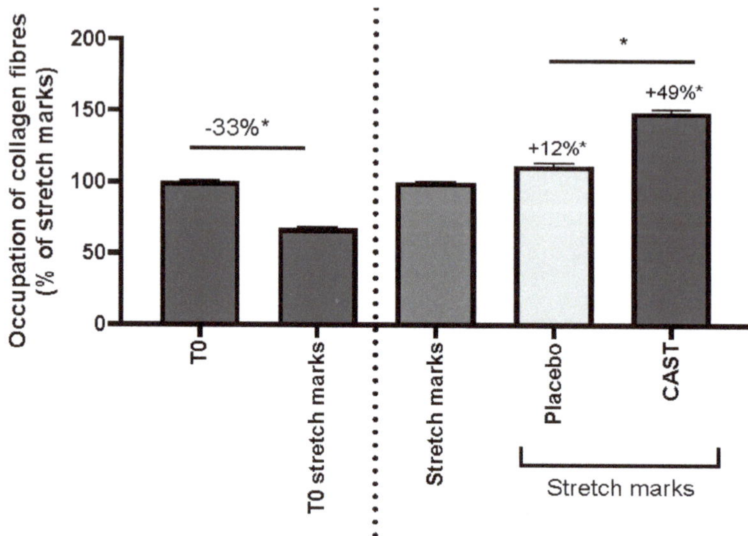

Figure 3. Collagen I fibres occupation in the dermis after Sirius red staining and analysis in polarized light. Mann–Whitney test with * $p < 0.05$.

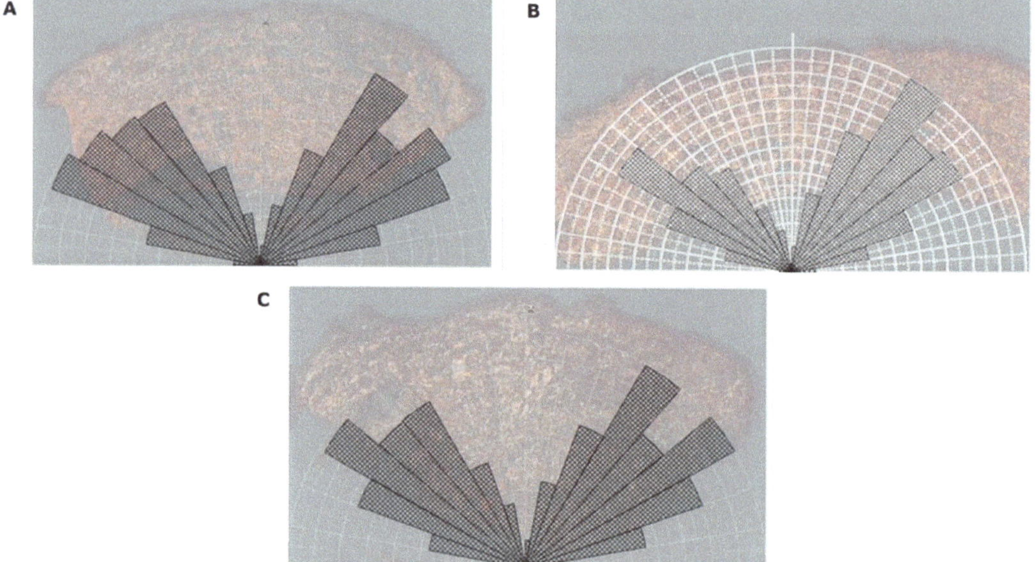

Figure 4. Collagen I orientation in the dermis. The fibres were quantified every $10°$ based on a parallel plane to the epidermis. (**A**) Non-stretch-marked skin, (**B**) untreated stretch-marked skin, and (**C**) stretch-marked skin explant treated with CAST at 0.5%.

3.3. *Centella asiatica Improves Skin Elasticity*

The elastin network was stained with immunofluorescence on human skin sections. The elastic fibres significantly increased by 11% in normal skin in comparison to stretch marks. After treatment with CAST 0.5%, elastin significantly increased by 37% in comparison to the untreated stretch-marked skin (Figure 5).

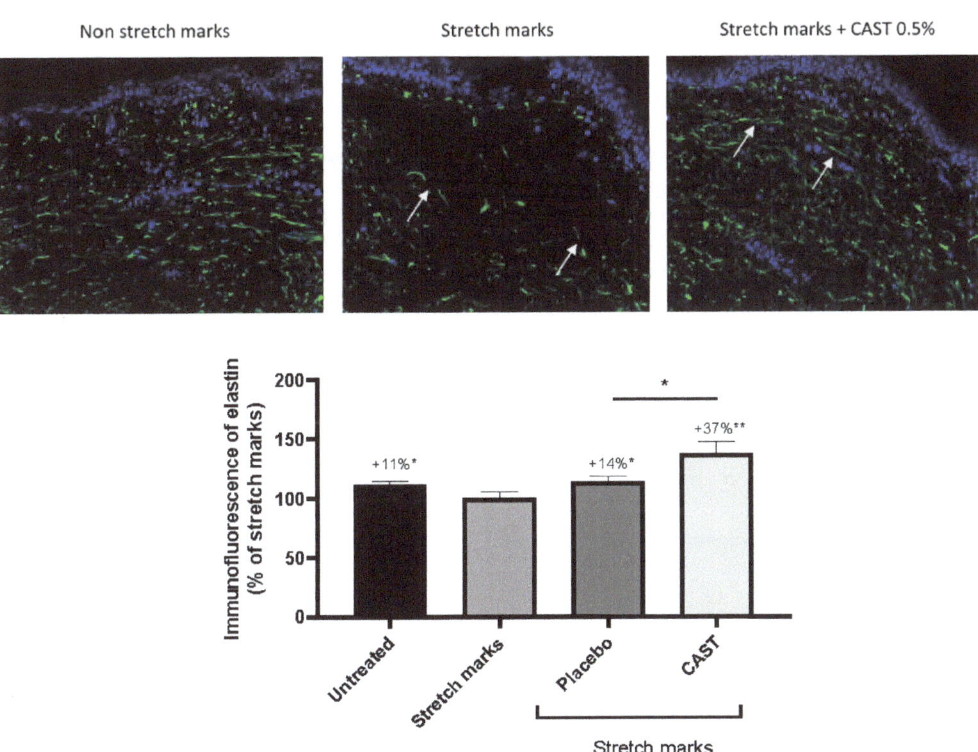

Figure 5. Elastin immunofluorescence visualisation and quantification. The arrows indicate, in the case of stretch marks, a damaged and short elastic fibre, whereas the treatment with CAST increased elastin content and fibril maturation. Mann–Whitney test with * $p < 0.05$, ** $p < 0.01$.

4. Discussion

Stretch mark concerns are mainly characterized by fibrotic skin. We showed that fibroblasts isolated from stretch-marked tissue expressed fibrosis markers (CTGF and PTK2), leading to an intense Sirius red staining on stretch-marked skin sections. CAST has been described to promote in vitro fibroblast migration, matrix components, and collagen synthesis [13,16]. In 2023, Leite Diniz et al. described tissue regeneration, cell migration, and wound repair processes that were mediated by CAST extract, where, among the active compounds, asiatic acid is especially involved in this activity [17]. Interestingly, during our study, we confirmed that CAST enhanced fibroblast migration, but on stretch-marked skin explants, the quantity of collagen I was not impacted by the treatment. We hypothesized that the effect of CAST at the in vitro level may be different than the effect on ex vivo skin explants depending on the physio-pathological environment and expression of complex markers. For example, a previous study using animal models showed that CAST fastens skin healing by the formation of a thick epidermal layer but with a moderate formation of granulation tissues and collagen, which corroborates our observation in humans [11,18]. In the context of pulmonary fibrosis, CAST was able to decrease fibrosis by decreasing collagen I content as well [19]. However, in addition to the quantity of collagen, the orientation and maturation of the collagen beams are essential. Using three-dimensional video, a study has demonstrated that collagen bundles are disrupted in stretch marks with marked separation of bundles and disorganisation of collagen fibrils, leading to this fibrotic organisation [20]. By using specific software that quantifies collagen fibres in all directions, we showed that the occupation of the dermis by collagen bundles was significantly improved following treatment with CAST. Finally, regarding collagen, CAST was able to promote maturation and restore the orientation of the bundles without enhancing the quantity of collagen. The elastic fibre network is also disrupted in stretch marks associated with micro fragmentation, and newly formed immature elastic fibres are found in those gaps [20–23]. By increasing elastin synthesis, CAST is able to promote elastic fibre networks, and staining analysis has shown the restoration of long, thick, and mature elastic fibres. Indeed, CAST was able to reduce fibrotic gene expression and also matrix degradation via MMPs to prevent this anarchic dermis organisation. Among them, connective-tissue growth factor (CTGF) is upregulated in various types of physio-pathological fibrosis [24]. Interestingly, an upregulation of CTGF in a stretch mark context has never been described. During our work, we compared a gene expression profile from normal skin and stretch-mark skin, which demonstrated that stretch-mark-related fibrosis involved a strong overexpression of gene coding for CTGF. This CTGF upregulation, which is reversed by topical application of CAST, showed evidence of an improvement in fibrosis through the control of the CTGF pathway. Overexpression of metalloproteinase within a fibrosis context has already been described in the literature by Giannandrea et al. in 2014. They are key proteases for matrix remodelling, but upregulation is the result of fibrosis [25]. Likewise, our study has revealed that by promoting dermis organisation among the components, CAST works on the physio-pathological pathways of stretch marks.

5. Conclusions

Centella asiatica acts as a natural remodelling partner that is useful for solving stretch mark concerns. Our study has unveiled the mechanism of action behind the efficacy of CAST. This traditional key plant significantly reduced fibrotic markers and extracellular matrix degradation. By decreasing collagen I degradation, the occupation of the beams and their preferential direction were restored for a plumping effect in the centre of the stretch mark. CAST has also stimulated micro-circulation and tropoelastin synthesis to fully replenish the dermis network, which led to a previously published clinical improvement in skin elasticity and a visible erasure of stretch marks.

Author Contributions: Conceptualisation, C.B., M.M., A.S. and R.R.; methodology, C.B., M.M., A.S. and R.R.; investigation, C.B., M.M. and M.B.; writing—original draft preparation, C.B. All authors have read and agreed to the published version of the manuscript.

Funding: This research received no external funding.

Institutional Review Board Statement: Not applicable.

Informed Consent Statement: Not applicable.

Data Availability Statement: Previous clinical study available at https://doi.org/10.23736/S1128-9155.18.00460-0, accessed on 20 June 2018.

Acknowledgments: The authors would like to acknowledge Biomeca for their support of this study.

Conflicts of Interest: Authors Cloe Boira, Marie Meunier, Marine Bracq, Amandine Scandolera and Romain Reynaud were employed by the company Givaudan Active Beauty. The remaining authors declare that the research was conducted in the absence of any commercial or financial relationships that could be construed as a potential conflict of interest.

References

1. Ud-Din, S.; McGeorge, D.; Bayat, A. Topical management of striae distensae (stretch marks): Prevention and therapy of striae rubrae and albae. *J. Eur. Acad. Dermatol. Venereol.* **2016**, *30*, 211–222. [CrossRef]
2. Borrelli, M.R.; Griffin, M.; Ngaage, L.M.; Longaker, M.T.; Lorenz, H.P. Striae Distensae: Scars without Wounds. *Plast. Reconstr. Surg.* **2021**, *148*, 77–87. [CrossRef]
3. García Hernández, J.Á.; Madera González, D.; Padilla Castillo, M.; Figueras Falcón, T. Use of a specific anti-stretch mark cream for preventing or reducing the severity of striae gravidarum. Randomized, double-blind, controlled trial. *Int. J. Cosmet. Sci.* **2013**, *35*, 233–237. [CrossRef]
4. Oakley, A.M.; Patel, B.C. *Stretch Marks*; StatPearls Publishing: Treasure Island, FL, USA, 2023.

5. Veronese, S.; Picelli, A.; Zoccatelli, A.; Zadra, A.; Faccioli, N.; Smania, N.; Sbarbati, A. The pathology under stretch marks? An elastosonography study. *J. Cosmet. Dermatol.* **2022**, *21*, 859–864. [CrossRef]
6. Lokhande, A.; Mysore, V. Striae distensae treatment review and update. *Indian Dermatol. Online J.* **2019**, *10*, 380–395.
7. Liu, L.; Ma, H.; Li, Y. Interventions for the treatment of stretch marks: A systematic review. *Cutis* **2014**, *94*, 66–72.
8. Draelos, Z.D.; Gold, M.H.; Kaur, M.; Olayinka, B.; Grundy, S.L.; Pappert, E.J.; Hardas, B. Evaluation of an onion extract, *Centella asiatica*, and hyaluronic acid cream in the appearance of striae rubra. *Skinmed* **2010**, *8*, 80–86.
9. Rawlings, A.V.; Bielfeldt, S.; Lombard, K.J. A review of the effects of moisturizers on the appearance of scars and striae. *Int. J. Cosmet. Sci.* **2012**, *34*, 519–524. [CrossRef]
10. Farahnik, B.; Park, K.; Kroumpouzos, G.; Murase, J. Striae gravidarum: Risk factors, prevention, and management. *Int. J. Women's Dermatol.* **2017**, *3*, 77–85. [CrossRef]
11. Young, G.; Jewell, D. Creams for preventing stretch marks in pregnancy. *Cochrane Database Syst. Rev.* **1996**, *2*, CD000066. [CrossRef]
12. Sun, B.; Wu, L.; Wu, Y.; Zhang, C.; Qin, L.; Hayashi, M.; Kudo, M.; Gao, M.; Liu, T. Therapeutic Potential of *Centella asiatica* and Its Triterpenes: A Review. *Front. Pharmacol.* **2020**, *11*, 568032. [CrossRef]
13. Bylka, W.; Znajdek-Awiżeń, P.; Studzińska-Sroka, E.; Brzezińska, M. *Centella asiatica* in cosmetology. *Adv. Dermatol. Allergol.* **2013**, *1*, 46–49. [CrossRef]
14. Korgavkar, K.; Wang, F. Stretch marks during pregnancy: A review of topical preventi-on. *Br. J. Dermatol.* **2015**, *172*, 606–615. [CrossRef]
15. Togni, S.; Maramaldi, G.; Pagin, I.; Riva, A.; Eggenhoffner, R.; Giacomelli, L.; Cesarone, M.R.; Belcaro, G. Postpartum stretch marks treated with *Centella asiatica* cream: Report of efficacy from a pilot registry study. *Esperienze Dermatol.* **2018**, *20*, 23–26. [CrossRef]
16. Maquart, F.X.; Chastang, F.; Simeon, A.; Birembaut, P.; Gillery, P.; Wegrowski, Y. Triterpenes from *Centella asiatica* stimulate extracellular matrix accumulation in rat experimental wounds. *Eur. J. Dermatol.* **1999**, *9*, 289–296.
17. Leite Diniz, L.R.; Calado, L.L.; Duarte, A.B.S.; de Sousa, D.P. *Centella asiatica* and its metabolite Asiatic Acid: Wound healing effects and therapeutic potential. *Metabolites* **2023**, *13*, 276. [CrossRef]
18. Sh. Ahmed, A.; Taher, M.; Mandal, U.K.; Jaffri, J.M.; Susanti, D.; Mahmood, S.; Zakaria, Z.A. Pharmacological properties of *Centella asiatica* hydrogel in accele-rating wound healing in rabbits. *BMC Complement. Altern. Med.* **2019**, *19*, 213.
19. Dong, S.-H.; Liu, Y.-W.; Wei, F.; Tan, H.-Z.; Han, Z.-D. Asiatic acid ameliorates pulmo-nary fibrosis induced by bleomycin (BLM) via suppressing pro-fibrotic and inflammatory signaling pathways. *Biomed. Pharmacother.* **2017**, *89*, 1297–1309. [CrossRef]
20. Wang, F.; Calderone, K.; Do, T.; Smith, N.; Helfrich, Y.; Johnson, T.; Kang, S.; Voorhees, J.; Fisher, G. Severe disruption and disorganization of dermal collagen fibrils in early striae gravidarum. *Br. J. Dermatol.* **2018**, *178*, 749–760. [CrossRef]
21. Johnson, T.M.; Lowe, L.; Brown, M.D.; Sullivan, M.J.; Nelson, B.R. Histology and Physiology of Tissue Expansion. *J. Dermatol. Surg. Oncol.* **1993**, *19*, 1074–1078. [CrossRef]
22. Wang, F.; Calderone, K.; Smith, N.; Do, T.; Helfrich, Y.; Johnson, T.; Kang, S.; Voorhees, J.; Fisher, G. Marked disruption and aberrant regulation of elastic fibres in early striae gravidarum. *Br. J. Dermatol.* **2015**, *173*, 1420–1430. [CrossRef] [PubMed]
23. Basile, F.V.; Basile, A.V.; Basile, A.R. Striae Distensae After Breast Augmentation. *Aesthetic Plast. Surg.* **2012**, *36*, 894–900. [CrossRef] [PubMed]
24. Yanagihara, T.; Tsubouchi, K.; Gholiof, M.; Chong, S.G.; Lipson, K.E.; Zhou, Q.; Scallan, C.; Upagupta, C.; Tikkanen, J.; Keshavjee, S.; et al. Connective-Tissue Growth Factor contributes to TGF-β1-induced lung fibrosis. *Am. J. Respir. Cell Mol. Biol.* **2022**, *66*, 260–270. [CrossRef]
25. Giannandrea, M.; Parks, W.C. Diverse functions of matrix metalloproteinases during fibrosis. *Dis. Model. Mech.* **2014**, *7*, 193–203. [CrossRef]

Article

Unlocking the Potential: A Comprehensive Analysis of the Technological Properties and Consumer Perception of Shampoo Enriched with Patchouli Extract and Allantoin

Ugnė Žlabienė [1], Erlita Bartkutė [2] and Jurga Bernatonienė [1,2,*]

[1]   Institute of Pharmaceutical Technology, Lithuanian University of Health Sciences, Sukilėlių pr. 13, LT-50162 Kaunas, Lithuania; ugne.cizauskaite@lsmuni.lt
[2]   Department of Drug Technology and Social Pharmacy, Lithuanian University of Health Sciences, Sukilėlių pr. 13, LT-50162 Kaunas, Lithuania; erlita.bartkute@stud.lsmu.lt
*   Correspondence: jurga.bernatoniene@lsmuni.lt

Abstract: Amidst a growing shift towards eco-friendly choices in personal care products, the challenge of formulating herbal shampoos with efficacy comparable to synthetic counterparts persists. This study investigates the potential of incorporating patchouli extract and allantoin as additives in anti-dandruff shampoo formulations, assessing their impact on the technological properties of the product. With limited research on their efficacy, our investigation contributes valuable insights to the development of effective and consumer-friendly shampoos targeting dandruff concerns. Physicochemical characteristics (pH, surface tension, texture) were evaluated, alongside specific quality assessments such as wetting time, dirt dispersion, foaming, and cleaning action, in in vivo consumer research. Shampoo formulations incorporating 0.5% Patchoul'Up™ and 1% allantoin exhibited acceptable properties. However, the addition of plant-derived ingredients resulted in a beneficial decrease in surface tension (5.87%). Nevertheless, a decrease in cohesiveness (18%) over a 5-month period resulted in rheological changes, indicating potential instability ($p < 0.05$). While the consumer evaluation aligns with laboratory findings, continuous research is essential to ensure stability and validate the anti-dandruff potential of the formulation, both in vitro and in vivo. This involves expanding the number of volunteers, with a specific focus on individuals experiencing dandruff concerns, to assess the shampoo's efficacy and impact on diverse user experiences.

Keywords: patchouli extract; allantoin; texture analysis; shampoo; formulation; consumer evaluation

Citation: Žlabienė, U.; Bartkutė, E.; Bernatonienė, J. Unlocking the Potential: A Comprehensive Analysis of the Technological Properties and Consumer Perception of Shampoo Enriched with Patchouli Extract and Allantoin. *Cosmetics* **2024**, *11*, 53. https://doi.org/10.3390/cosmetics11020053

Academic Editor: Paulraj Mosae Selvakumar

Received: 22 February 2024
Revised: 21 March 2024
Accepted: 26 March 2024
Published: 2 April 2024

1. Introduction

Shampoos, initially devised for scalp cleansing, have evolved into versatile products that cater to diverse hair types, care preferences, and scalp conditions. Typically formulated as aqueous solutions or emulsions, shampoos comprise a mixture of surfactants (synthetic or natural), cleansing and foaming agents, along with other excipients (viscosity controllers, emollients, preservatives, etc.) and active ingredients [1,2]. The surfactants play a crucial role in both the cleaning and lathering abilities of shampoos and contribute to their skin tolerance [3].

Modern consumers, driven by concerns about synthetic chemicals impacting long-term hair health and the environment, are increasingly inclined toward herbal shampoos [4]. The perception of safety and minimal side effects associated with natural-origin products contributes to the rising popularity of herbal shampoos [1,2]. Numerous medicinal plants, known for their beneficial effects on hair, are commonly used in shampoo formulations [5]. However, creating an herbal shampoo that competes favorably with synthetic ones in terms of foaming and detergency remains a challenging task [6]. Considering the diverse phytochemical composition of plant extracts, it becomes crucial to assess not only their therapeutic effects but also the changes in technological properties resulting from their incorporation into formulations. This is particularly significant when dealing with active ingredients, where compatibility with the base formulation can be challenging. To address this, various shampoo compositions were prepared and evaluated, encompassing formulations with and without active ingredients.

Patchouli extracts and essential oil, widely used in various industries (pharmaceutical, food, perfume, cosmetic), contain bioactive ingredients, including sesquiterpenes, known for potential health benefits [7]. Despite its traditional uses, patchouli is recognized for its antimicrobial and anti-fungal properties, making it valuable for managing various skin and scalp conditions, including fungal infections, acne, dermatitis, and even dandruff [8,9]. However, these qualities have not been extensively proven in shampoo formulations, with limited available research [7]. To our knowledge, there are no available data on the innovative ingredient Patchoul'Up™—a 100% upcycled active ingredient crafted through green fractionation from distilled patchouli leaves, which was used throughout the research.

Allantoin, a plant-derived substance, stimulates cell proliferation, promoting internal and external wound healing. Allantoin has been incorporated into various cosmetic products, including shampoos, to improve skin health by reducing inflammation and irritation [10]. It dissolves the intercellular cement that holds the corneocytes together, promoting natural desquamation of the stratum corneum, increasing skin smoothness and water binding to the intercellular matrix and keratin [10,11]. These properties make it well suited for an anti-dandruff shampoo.

Lemon essential oil contains primarily monoterpenoids, with D-limonene being a key compound believed to confer anti-inflammatory properties. These beneficial chemical constituents contribute to the significant role of *Citrus limon* in both the food and cosmetics industries [12]. In shampoos, it acts as a natural cleanser suitable for normal-to-oily hair types. Its acidic nature tightens hair follicles, promoting healthier growth and reducing hair fall while enhancing hair shine [13,14]. Research has shown that lemon peel extract exhibits potent antimicrobial activity against *Candida albicans*, *Aspergillus*, and *Aspergillus flavus*, suggesting potential benefits against dandruff. Furthermore, lemon juice and lemon peel powder effectively inhibit the growth of *Malassezia furfur*, a fungus responsible for causing dandruff [15,16].

Dandruff, a common dermatological condition affecting 5% of the population, is characterized by excessive scaling of scalp tissue [1,17]. To address the individuals dealing with dandruff who prefer using natural ingredients rather than opting for medicinal shampoos, this research aims to formulate a stable herbal shampoo containing allantoin, patchoul'Up™, with the potential to benefit individuals dealing with dandruff concerns. Also, this comprehensive evaluation aims not only to shed light on the potential impact of plant extracts on therapeutic efficacy but also the technological characteristics of the shampoo base.

2. Materials and Methods

2.1. Materials

Lemon fruit essential oil was purchased from "ACappella naturals" (distributor UAB "Aromika", Kaunas, Lithuania). Allantoin, potassium sorbate, sodium benzoate, and citric acid were procured from "Sigma-Aldrich" (Taufkirchen, Germany). Cocamidopropyl betaine and glyceryl cocoate were purchased from "Dragonspice Naturwaren" (Reutlingen, Germany). Decyl glucoside and sodium cocoyl isethionate were obtained from "Berg + Schmidt" (Hamburg, Germany). Patchoul'Up™ was purchased from "Givaudan Active Beauty" (Jawa Barat, Indonesia). D-panthenol was bought from "ThermoFisher" (Karlsruhe, Germany). Indian ink was purchased from "Daravija" (Ariogala, Lithuania) manufactured by "Koh-In-Noor" (České Budějovice, Czech Republic). Parfum was obtained from the local distributor "ELL" (Vilnius, Lithuania), and purified water was made in the Lithuanian University of Health Sciences laboratory according to Ph. Eur. 01/2008:0008.

2.2. Methods

2.2.1. Shampoo Formulation

The formulation process started by dissolving D-panthenol in purified water, followed by the addition of allantoin and Patchoul'Up™ if necessary (Table 1). Subsequently, preservatives (potassium sorbate and sodium benzoate) were added and stirred until fully dissolved. The anionic detergent sodium cocoyl isethionate was incorporated and mixed until complete dissolution. Next, the non-ionic detergent decyl glucoside was added, followed by the amphoteric detergent cocamidopropyl betaine. Glyceryl cocoate was introduced, followed by the addition of lemon essential oil and parfum. Finally, a few drops of a 50% citric acid solution were added to achieve the desired pH.

Table 1. Compositions of formulated shampoos.

Ingredients	Compositions (%)			
	1	**2**	**3**	**4**
Sodium cocoyl isethionate	10.0	10.0	10.0	10.0
Cocamidopropyl betaine	10.0	10.0	10.0	10.0
Decyl glucoside	5.0	5.0	5.0	5.0
Glyceryl cocoate	5.0	5.0	5.0	5.0
D-panthenol	1.0	1.0	1.0	1.0
Potassium sorbate	0.1	0.1	0.1	0.1
Sodium benzoate	0.1	0.1	0.1	0.1
Parfum	1.0	1.0	1.0	1.0
Patchoul'Up™	-	0.5	0.5	0.5
Lemon essential oil	0.1	0.1	0.1	0.1
Allantoin	-	0.5	1	1.5
50% citric acid solution	*	*	*	*
Purified water	67.7	66.7	66.2	66.2

* as much as needed to reach the appropriate pH.

2.2.2. Determination of pH Value

A 10% aqueous solution of shampoo was made carefully to avoid foam formation and filtered through a paper filter with 20–25 μm pore size (DP 411, Albet-Hahnemuehle S.L., Barcelona, Spain) [18]. The pH was measured at room temperature ($22 \pm 2\,°C$) using a pH meter (inoLab®pH/ION 7320, Berlin, Germany).

2.2.3. Determination of Dirt Dispersion

To initiate the experiment, two drops of herbal shampoo were introduced into a wide-mouthed Falcon tube filled with 10 mL of distilled water. Following this, a single drop of India ink was incorporated. The Falcon tube was securely covered and subjected to ten vigorous shakes. The discernible presence of ink within the resulting foam was assessed using qualitative descriptors: None, Light, Moderate, or Heavy [19].

2.2.4. Determination of Wetting Time

The wetting time was determined by observing the duration it took for a canvas paper to fully submerge into 1% aqueous solution of shampoo. The size of the canvas paper was 0.44 g, with a diameter of approx. 2.54 cm. The canvas paper disc was placed on the surface of the sample solution and the stopwatch was used to measure the time it took for the paper to submerge entirely [19].

2.2.5. Determination of Surface Tension

Throughout the experiment, 10% aqueous shampoo solutions were consistently utilized. The experiment was conducted at room temperature, employing a Traube stalagmometer with a diameter of 0.62 mm. Equal volumes of test samples were filled into the stalagmometer, secured in a stand, and allowed to descend gradually until reaching the mark on the tube [18]. The number of drops was counted and the surface tension was calculated using Equation (1).

$$R2 = \frac{(W3 - W1)N1}{(W2 - W1)N2} \times R1,$$ (1)

where W_1—weight of empty beaker (g); W_2—weight of beaker with distilled water (g); W_3—weight of beaker with test solution (g); N_1—no. of drops of distilled water; N_2—no. of drops of shampoo solution; R1—surface tension of distilled water at room temperature (72.75 mN/m); R2—surface tension of shampoo solution (mN/m).

2.2.6. Determination of Foaming Ability and Foam Stability

The foaming ability was assessed using a modified cylinder shake method. In total, 50 mL of the 1% aqueous solution of shampoo was poured into a 250 mL graduated measuring cylinder, covered with parafilm, and inverted ten times. The total volume of the foam content was then recorded. The recordings were taken again after 1, 2, 3, and 4 min as well to account foam stability [19].

2.2.7. Detergency Test

The cleansing efficacy of the developed shampoo was assessed through the slightly modified washing-of-wool-yarn technique [18]. Initially, 5 g of wool yarn, pre-moistened with sunflower oil, was added into a cylinder containing 200 mL 35 °C purified water along with 1 g of the test sample (shampoo or control solution). The cylinder was subjected to shaking for approximately 4 min, with around 50 shakes per minute. Following the shaking process, the water was drained from the cylinder, and the yarn threads were dried and weighed; 35 °C purified water was used as the control solution. The cleansing action was determined using Equation (2).

$$DP = 100 \left(1 - \left(\frac{T}{C} \right) \right)$$ (2)

where DP—detergency power (%); T—weight of yarn after shaking with shampoo (g); C—weight of yarn after shaking with water (g).

2.2.8. Texture Analysis

It was performed using the back extrusion test by the Ta.Xtplus Texture Analyzer (Stable Micro Systems, Vienna, Austria). During this test, four parameters were determined: firmness (g), consistency (g·s), cohesiveness (g), and viscosity index (g·s). The back extrusion rig, denoted as A/BE, consisted of a sample container filled with the developed shampoo, positioned on the analyzer platform beneath a disc plunger. The compression test was initiated by lowering the test probe into the sample at a speed of 3 mm/s, causing the product to extrude up and around the disc's edge. Once a trigger force of 5 g was reached, the disc plunger initiated the deformation of the sample to a predetermined distance (5 mm), following which the probe returned to its initial position [20].

2.2.9. Consumer Evaluation

The research was granted approval (No. BEC-FF-52) by the Lithuanian University of Health Sciences Center for Bioethics after evaluating the protocol. Ten students from the Lithuanian University of Health Sciences, serving as volunteers, participated in the study. The inclusion criteria comprised being healthy, proficient in Lithuanian and English, aged above 18, without a medical history of scalp conditions or diseases (except for dandruff), and no known allergic reactions to the shampoo ingredients. Informed consent was obtained from all subjects involved in the study, acknowledging the objectives and methods of the research. Participants refrained from using any hair or scalp care products 12 h before and during the study. Each participant received 30 g of shampoo for hair washing. To assess the quality of the formulated shampoo, all volunteers responded to specific questions outlined in the questionnaire (Appendix A).

2.2.10. Microscopic Evaluation

The microscopic analysis was carried out using an optical microscope (BMS 739960) equipped with a digital USB camera (Breukhoven, Netherlands) for processing the image (400× magnification fold). The diameter of droplets in each sample was measured using previously calibrated BMS 2.0 software.

2.2.11. Statistical Analysis

Data are presented as mean $\pm$ SD. Statistical analysis was performed by one-way and two-way analysis of variance (ANOVA), followed by Dunnett's post-test using the software package Prism 6.0 (GraphPad Software Inc., La Jolla, CA, USA). A value of $p < 0.05$ was taken as the level of significance.

3. Results and Discussion

Given the diverse phytochemical composition of plant extracts and the complexities associated with integrating various active ingredients, it is imperative to assess not only their therapeutic effects but also the resulting changes in technological properties. This becomes particularly crucial when dealing with active ingredients or unknown composition plant extracts, where compatibility with the base formulation can pose challenges due to the lack of scientific research data. As part of our investigation, we meticulously prepared several shampoo compositions, including the shampoo base and various formulations containing investigated active ingredients (Table 1).

3.1. Physical Appearance and Microscopic Evaluation

In the assessment of the physical appearance, the developed formulation underwent evaluation for clarity, color, and odor. Shampoo formulas were crafted with varying concentrations of allantoin powder, specifically 0.5%, 1%, and 1.5%. Becker et.al determined that it is safe to use up to 2% allantoin in cosmeceuticals [10]. Visual examination revealed that the addition of 1.5% allantoin resulted in a less homogeneous shampoo, while 0.5% and 1% concentrations produced a more uniform and consistent product.

In Figure 1, microscopic evaluation confirmed that the droplet size of shampoo with 1.5% allantoin concentration was over 10–12% higher compared to shampoos with lower allantoin concentrations ($p < 0.05$). Consequently, a shampoo formula containing 1% allantoin was selected for further evaluation, as it is anticipated to showcase superior moisturizing and keratolytic benefits.

3.2. Evaluation of pH

Continuing with the evaluation, the pH levels of both the shampoo base and the formulated shampoo containing allantoin and patchouli extract were assessed. Although there is not a universally defined recommendation for the pH of shampoo, it is crucial to ensure it is suitable for use on the scalp. The scalp's pH is generally around 5.5, and the hair shaft's pH is approximately 3.67. Research by M. Dias et al. suggests that an alkaline pH may increase the negative electrical net charge on the hair fiber surface, leading to increased friction between fibers, potential cuticle damage, and fiber breakage. It is a well-established fact that lower pH in shampoos can contribute to reduced frizz by minimizing negative static electricity on the fiber surface. However, it is imperative that the shampoo ingredients collectively result in a final pH of around 5.5 to prevent any damage to the scalp [21]. On the other hand, AlQuadeib B. et al. suggest that the acceptable pH range for hair shampoo should be 5.0–7.0 [22]. In our research, we attempted to achieve a pH of approximately 5.8 by adding a 50% citric acid solution. The pH levels of the developed shampoos remained mildly acidic and exhibited insignificant variation throughout a span of 5 months (Figure 2).

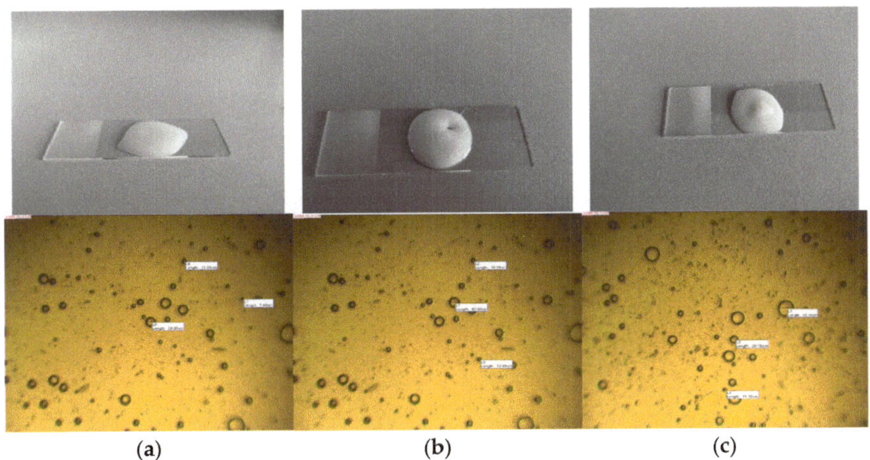

Figure 1. Physical appearance and microscopic evaluation (400× fold) of shampoos containing various concentrations of allantoin: (**a**) 0.5%; (**b**) 1.0%; (**c**) 1.5%.

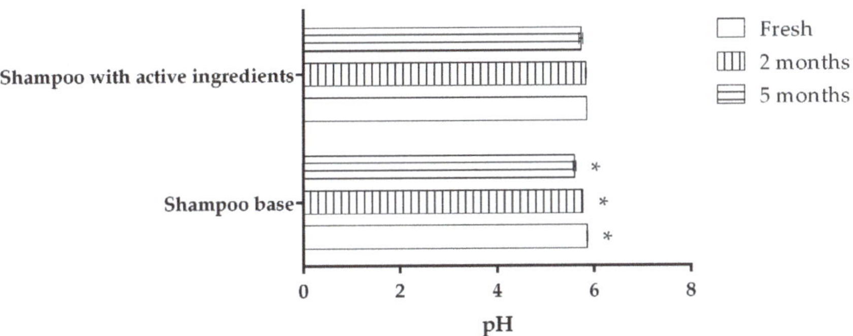

Figure 2. pH variation in shampoo formulations over time. * $p > 0.05$ vs. shampoo with active ingredients.

Similar investigations on herbal shampoos, which adjusted the pH to be below 7, also reported stable pH levels over time [18,19].

3.3. Evaluation of Wetting Time

In the context of our results, we evaluated the wetting time, a critical parameter hinging on surfactant concentration, commonly employed to assess its efficacy. The wetting process involves intricate interactions influenced by factors like surface tension, diffusion, concentration, and the nature of the surface being wetted [6,19,23]. Specifically, the wetting time for the shampoo with active ingredients was 74.5 ± 1.025 s, while the comparison shampoo exhibited a wetting time of 75.32 ± 1.459 s. Despite a slight increase in wetting time, statistical analysis deemed this change to be insignificant ($p > 0.05$). Consequently, the detergency difference between the two samples was considered inconsequential.

Comparatively, existing research indicates that some commercial shampoo brands, assessed using Drave's test, exhibit wetting times ranging from 12.7 ± 5.0 s to 18.9 ± 8.2 s [22]. On the other hand, compositions with natural ingredients, evaluated using a paper canvas test, showed wetting times extending up to 120–187 ± 4 s [22,24]. These variations underscore the impact not only of the shampoo's chemical composition but also the subtle nuances in the wetting test technique, which can lead to significant changes in wetting time.

3.4. Evaluation of Dirt Dispersion

In evaluating shampoo effectiveness, an essential parameter is its ability to disperse dirt. A premium-quality shampoo should efficiently eliminate dirt without allowing its redeposition on the hair. Shampoos that lead to ink accumulation in the foam are generally deemed of poor quality, as it becomes challenging to rinse away and may result in dirt accumulation on the hair [19]. To determine the dirt dispersion capability, both the shampoo with active ingredients and the comparison shampoo underwent testing. Remarkably, no ink was detected in the foam for either shampoo (Figure 3). This observation remained consistent when the test was repeated after 2 and 5 months. The enduring absence of ink in the foam suggests that the formulated shampoos effectively remove dirt without leaving any residue on the hair, aligning with the anticipated standards of high-quality shampoo. In comparing the dirt dispersion of the formulated shampoo with the findings from other researchers, it is evident that all marketed shampoos exhibit no dirt dispersion as well [22,25]. However, according to the data from studies conducted by A. Pradhan and A. Bhattacharyya or T. Malpani, experimental shampoos, particularly those incorporating herbal extracts, often leave light or moderate dirt residue in the foam. This may signal the need for formula refinement before marketing [4,19].

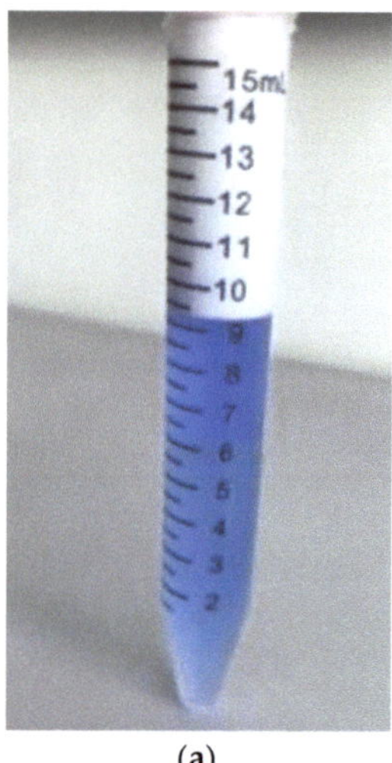

(a)

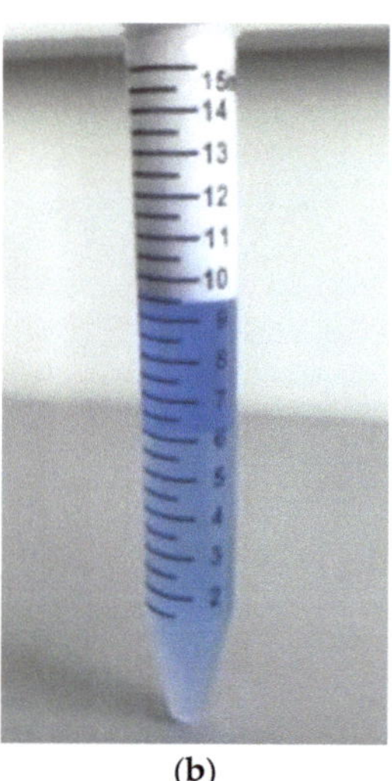

(b)

Figure 3. View of dispersion test tubes: (**a**) shampoo base; (**b**) shampoo with active ingredients.

3.5. Evaluation of Detergency Test Results

Moving on to the detergency test, the evaluation focused on one of the most crucial properties: the cleansing power of the shampoo. There is a consensus among cosmetic chemists that a shampoo should not be so powerful that it strips all natural secretions from the hair and scalp. The detergency ability of a normal hair shampoo is expected to be higher than that of dry-hair shampoos but lower than greasy-hair shampoos [3].

Upon analyzing the results of the investigated shampoo samples, it can be concluded that both shampoos exhibit a similar cleansing power, showing no statistically significant difference (Figure 4). The addition of allantoin and Patchoul'Up did not impact the cleansing power of the shampoo. Furthermore, the cleansing power remained statistically unchanged even after periods of 2 and 5 months ($p > 0.05$). In comparison, Dhayanithi et al. found that the cleansing power of some marketed shampoos ranged from 18 to 33% [26], indicating that our samples have a suitable cleaning action. However, in contrast to some other detergency studies, the investigated shampoos showed up to four times lower detergency ability when compared with the formulations assessed in those studies [3,25] ($p < 0.05$). This discrepancy might arise due to the lack of standardization in experimental detergency evaluation. Standardizing this process has proven difficult as there is no unanimous agreement on a standard soil, a reproducible soiling process, or the ideal amount of soil a shampoo should remove [3,26].

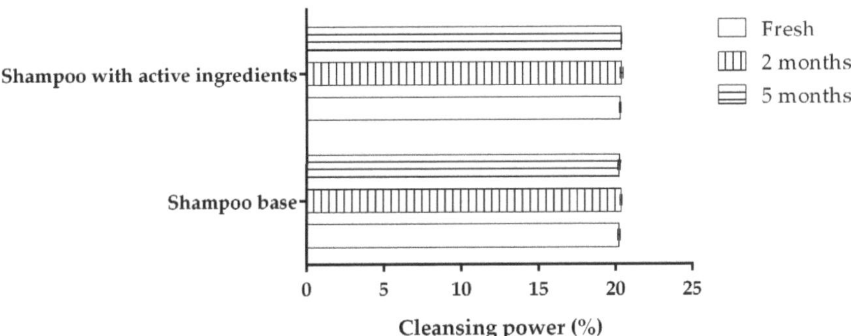

Figure 4. Cleansing power of investigated shampoo samples over time.

3.6. Evaluation of Surface Tension

Continued investigation focused on the surface tension of the shampoo samples, and the findings are illustrated in Figure 5. Detergents, which are integral components of shampoos, play a key role in reducing surface tension, thereby enhancing the cleaning efficacy of the shampoo [17]. Numerous studies suggest that an effective shampoo should be capable of decreasing the surface tension of pure water to approximately 32–40 mN/m [4,17]. The surface tension of the shampoo containing active ingredients was found to be 5.87% lower than that of the shampoo base ($p < 0.05$). This difference may be attributed to the inclusion of plant extracts and allantoin in the formulation. It is worth noting that the impact of additives on surface tension can vary; for instance, Farah et al. reported a significant increase in surface tension with the addition of patchouli oil (5%) to a fuel mixture [27]. Conversely, the incorporation of allantoin (0.2%) in bioadhesive gels resulted in a decrease in surface tension of 4.05 to 7.03% [16]. Regarding stability, there was no statistically significant difference in surface tension observed between the shampoos produced immediately and those tested after 2 and 5 months ($p > 0.05$). This suggests that the formulated shampoos maintained consistent surface tension values over the specified time frame, indicating stability in this particular aspect.

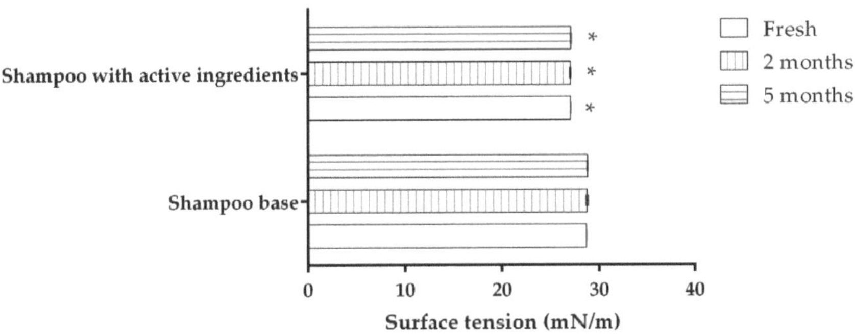

Figure 5. Surface tension of investigated shampoo samples over time. * $p < 0.05$ vs. shampoo base.

3.7. Evaluation of Foaming Ability and Foam Stability

In this section, this study delves into the investigation of foam volume and stability, recognizing the consumer's perception of foam as a crucial factor in the success of a shampoo, even though it does not have a direct correlation with cleansing efficacy [28]. It is expected for the shampoo to lather well and keep it stable for a certain period of time. Also, foaming ability strongly correlates with the content and type of surface active ingredients added to clean the sebum. The results of the formulated shampoo's foam volume and stability are presented in Figure 6.

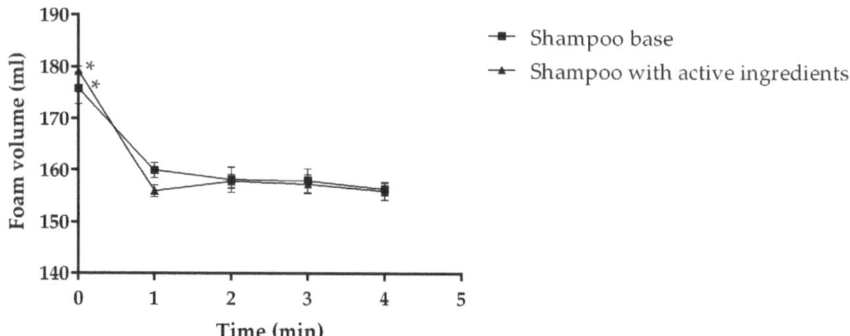

Figure 6. Foam volume of investigated shampoo samples over time. * $p < 0.05$ vs. 1, 2, 3, and 4 min.

The foam volume of the shampoo exhibited a slight 1.9% increase with the addition of active ingredients, which was deemed statistically insignificant ($p > 0.05$). Comparatively, a study involving shikakai and soapnut in shampoo formulation demonstrated enhanced foam characteristics, producing uniform, small-sized, compact, denser, and stable foam in comparison to the control, emphasizing the role of natural saponins found in these plant extracts [24]. A decline in foam volume was observed over a 4 min interval for both the shampoo without active ingredients (up to 11.9%, with a Pearson correlation coefficient of -0.8078) and the shampoo with allantoin and patchouli extract (up to 13%, with Pearson correlation coefficients determined to be -0.7094). Previous evaluations of marketed shampoos suggested no significant difference in foam volume after 5 min [22]. A. Pradhan and A. Bhattacharyya emphasized that foaming abilities are influenced by factors such as soaking time, temperature, and surfactant properties, contributing to variations in foam volume and stability among different compositions [4].

3.8. Evaluation of Texture Analysis Results

In this section, we have investigated the rheological properties of the shampoo samples. Texture analysis, a crucial aspect, not only aids in formulating a product that aligns with consumer expectations but also characterizes rheology, a key determinant when assessing the stability of semi-solid preparations. The back extrusion test was chosen, yielding four parameters: firmness (g), consistency (g·s), cohesiveness (g), and viscosity index (g·s). The software of the equipment facilitated graphical data presentation (Figure 7).

For the most part, parameters describing the texture of the shampoo with active ingredients did not exhibit statistically significant differences ($p > 0.05$) when compared to the shampoo without active ingredients. Notably, a more substantial difference was observed in cohesiveness: the addition of active ingredients led to an 18% reduction in cohesiveness, suggesting an enhanced likelihood of breakage ($p < 0.05$). These findings align with our previous studies, where plant materials were shown to not only enhance product quality but also impact technological parameters, including rheology [29].

A comparative analysis of the shampoo's texture with active ingredients over a 5-month period revealed statistically significant differences ($p < 0.05$) in consistency. The formulation immediately tested showed lower consistency after 2 months and 5 months, with reductions of 6.97% and 13.19%, respectively, indicating decreased resistance to flow or deformation and suggesting a less solid and cohesive structure. Concurrent changes were observed in cohesiveness, which increased from -56.13 ± 2.45 g to -36.80 ± 0.40 g over the 5-month period ($p < 0.05$). In contrast, firmness exhibited minimal variation over time. A noteworthy 24.25% increase in viscosity index was observed, indicating that the sample's viscosity changed less with temperature variations, a favorable outcome ($p < 0.05$). Comparing our results with texture analyses conducted by other researchers, such as L. Silva et al., the incorporation of botanical extracts similarly resulted in shampoos with decreased hardness, firmness, and shear. According to the researchers, this outcome enhances user experience and comfort during shampoo use, attributed to easier spreadability [30]. During the texture analysis, the non-Newtonian pseudoplastic behavior of the investigated samples was observed. This is a desired property of shampoo formulations because it makes it easier to spread and massage the product into the hair and scalp during application. This property enhances the spreadability and sensory experience of the shampoo, providing a smoother and more luxurious feel during use [31].

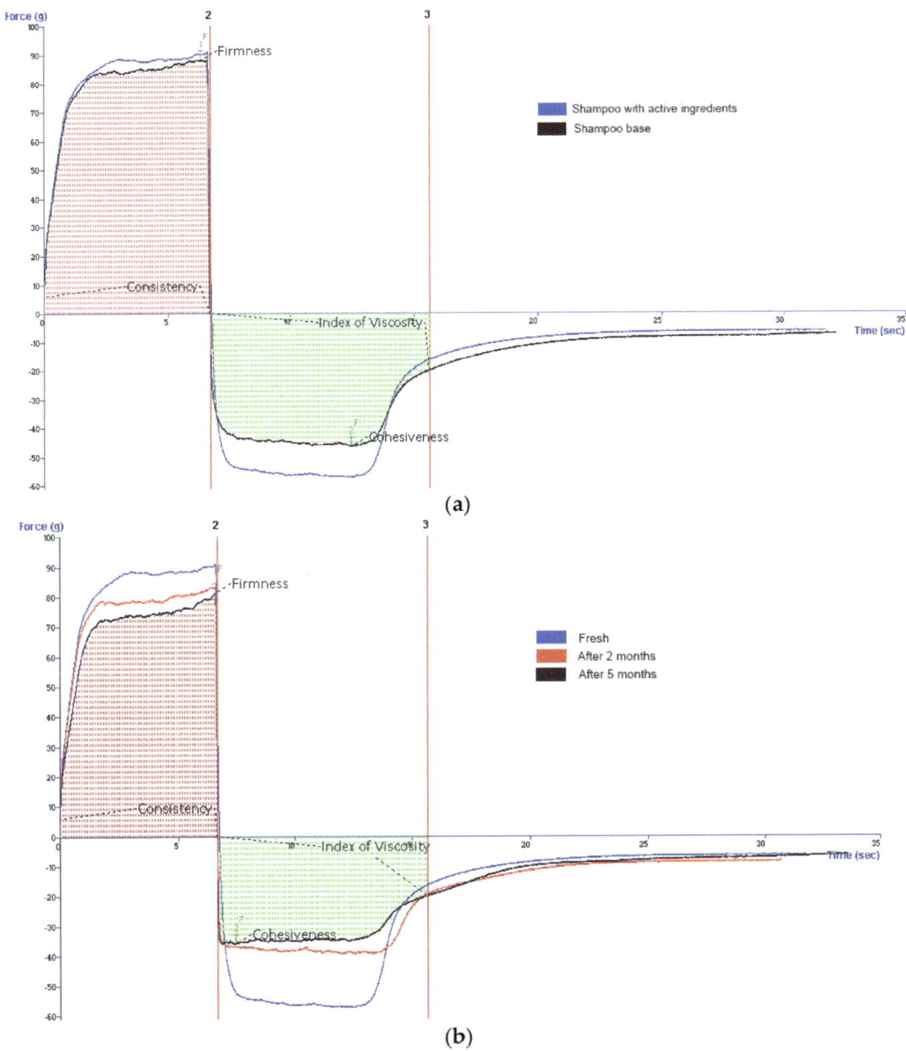

Figure 7. Graphical view of shampoo sample texture analysis: (**a**) comparison of texture properties between shampoo base and shampoo with active ingredients; (**b**) changes in the texture properties of shampoo with active ingredients over time.

3.9. Evaluation of Consumer Research Results

After formulating and rigorously assessing the shampoo's quality, consumer research was conducted with ten volunteers to evaluate whether it possessed the desired anti-dandruff properties and was acceptable for the consumers. The research protocol, subjected to the Center of Bioethics, received approval. Volunteers, aged 18 and above, without known allergic reactions to any shampoo ingredients, proficient in English and Lithuanian, were selected. Each participant refrained from using any hair care products 12 h before the experiment, receiving 30 g of a shampoo sample and a questionnaire (Appendix A). Respondents evaluated the formulated shampoo's quality by answering questionnaire questions, and the survey results are described below and presented in Figures 8 and 9.

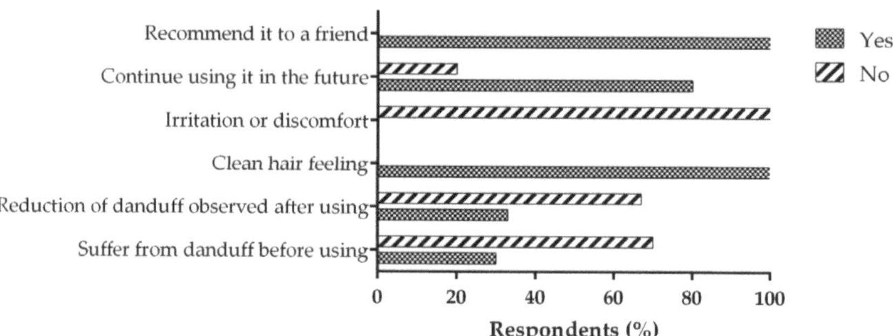

Figure 8. Shampoo evaluation by consumer: graphical questionnaire representation.

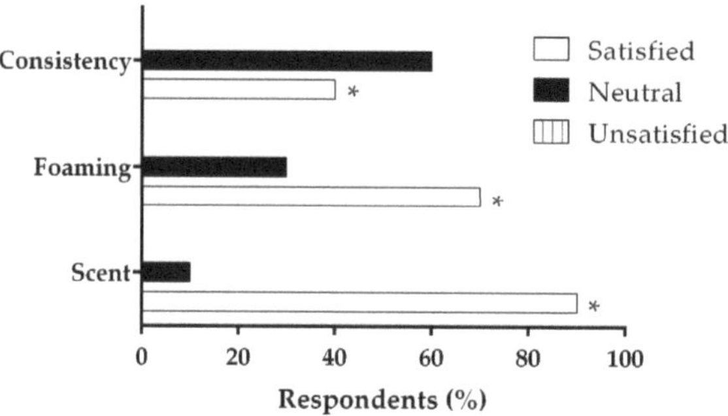

Figure 9. Shampoo sensory properties evaluation by consumer. * $p < 0.05$ vs. neutral.

The survey involved 10 participants, comprising 3 males and 7 females. Among the respondents, 30% reported experiencing dandruff, while 70% did not face this issue. Of the three participants with dandruff, only one observed a reduction in dandruff after using the shampoo, while the remaining two did not provide specific feedback on any reduction. As illustrated in Figure 8, an overwhelming 100% of the respondents reported that the shampoo left their hair feeling clean, and they would recommend it to a friend. In a consumer behavior study, it was found that 40% of individuals do not recommend the use of shampoos to others unless they find it genuinely useful [32]. This positive feedback suggests that the shampoo was well received among the surveyed individuals. None of the respondents experienced any irritation from using the shampoo. According to J.S Pahmar, every second respondent believes herbal shampoos are safer than non-herbal ones [32]. Overall, 80% of participants expressed their intention to continue using the shampoo, emphasizing their satisfaction. However, two individuals indicated that they would not continue using it.

In the survey, participants were provided with response options, including 'satisfied', 'neutral', and 'unsatisfied', to assess the scent, foaming, and texture of the shampoo. Figure 9 reveals that a significantly higher number of respondents expressed satisfaction with the scent of the shampoo compared to those who were neutral ($p < 0.05$), with none expressing dissatisfaction with the aroma.

Concerning the lather or foaming of the shampoo, 40% more respondents were satisfied ($p < 0.05$ vs. neutral), and none reported dissatisfaction. Additionally, 20% more participants adopted a neutral stance on the thickness of the shampoo ($p < 0.05$ vs. satisfied). Consumer preferences for hair shampoo are influenced by various factors, including usage experience, ethical and eco-friendly attributes, brand preference, product quality, and problem-solving ability [33].

4. Conclusions

Based on the comprehensive analysis of the obtained results, it can be deduced that the shampoo formulation incorporating 0.5% Patchoul'Up™ and 1% allantoin exhibits satisfactory physicochemical properties, including pH, texture, and surface tension. The product also demonstrates favorable characteristics in terms of wetting time, dirt dispersion, foam volume, and cleansing action. The consumer evaluation aligns with the laboratory findings, indicating that the shampoo was well received without causing irritations. Acknowledging the wide phytochemical composition of plant extracts, it is advisable to not only concentrate on their bioactivity but also evaluate the alterations in technological properties that arise upon their addition into the shampoo base. Continuous research is essential to enhance rheological and foam stability while adjusting consistency based on consumer evaluations. Further investigation is warranted to substantiate the anti-dandruff potential by expanding the study to include volunteers specifically experiencing scalp problems.

Author Contributions: Conceptualization, J.B. and E.B.; methodology, J.B.; validation, E.B. and J.B.; formal analysis, E.B. and U.Ž.; investigation, E.B.; resources, J.B.; data curation, E.B., U.Ž. and J.B.; writing—original draft preparation, E.B. and U.Ž.; writing—review and editing, U.Ž.; visualization, E.B. and U.Ž.; supervision, J.B.; project administration, J.B. All authors have read and agreed to the published version of the manuscript.

Funding: This research received no external funding.

Institutional Review Board Statement: This study was conducted in accordance with the Declaration of Helsinki, and approved by the Institutional Review Board (or Ethics Committee) of Lithuanian University of Health Sciences Center for Bioethics (No. BEC-FF-52, 2023/04/27) for studies involving humans.

Informed Consent Statement: Informed consent was obtained from all subjects involved in the study.

Data Availability Statement: All data are available upon request.

Conflicts of Interest: The authors declare no conflicts of interest.

Appendix A

QESTIONNAIRE
The purpose of this survey is to gather feedback on the effectiveness of herbal shampoo targeted for antidandruff. Your participation is voluntary, and all information collected will remain confidential. Your response will be analyzed in aggregate and used only for research purposes. If you agree to participate, please answer the questions honestly to the best of your ability.
By signing below, I acknowledge that I have read and understand the above information.
Signature   Date

1.   Gender
    i.   Male
    ii.   Female

2.   Do you suffer from dandruff? (If no, go to question 4)
    i.   Yes
    ii.   No

3. Did you observe any reduction in dandruff after using herbal shampoo?
   i. Yes
   ii. No

4. How would you rate the scent of the herbal shampoo?
   i. Satisfied
   ii. Neutral
   iii. Unsatisfied

5. How would you rate the lather/foaming of the herbal shampoo?
   i. Satisfied
   ii. Neutral
   iii. Unsatisfied

6. How would you rate the thickness or consistency of the herbal shampoo?
   i. Satisfied
   ii. Neutral
   iii. Unsatisfied

7. Did the herbal shampoo leave your hair feeling clean?
   i. Yes
   ii. No

8. Did you experience any irritation or discomfort while using the herbal shampoo?
   i. Yes
   ii. No

9. Would you continue to use this shampoo in the future?
   i. Yes
   ii. No

10. Would you recommend the herbal shampoo to a friend?
    i. Yes
    ii. No

References

1. Pierard-Franchimont, C.; Xhauflaire-Uhoda, E.; Pierard, G. Revisiting dandruff. *Int. J. Cosmet. Sci.* **2006**, *28*, 311–318. [CrossRef] [PubMed]
2. Mainkar, A.; Jolly, C. Formulation of natural shampoos. *Int. J. Cosmet. Sci.* **2001**, *23*, 59–62. [CrossRef]
3. Fazlolahzadeh, O.; Masoudi, A. Cosmetic evaluation of some Iranian commercial normal hair shampoos and comparision with new developed formulation. *Int. J. Pharmacogn.* **2015**, *2*, 259–265.
4. Pradhan, A.; Bhattacharyya, A. Shampoos then and now: Synthetic versus natural. *J. Surf. Sci. Technol.* **2014**, *30*, 59–76.
5. Halith, S.M.; Abirami, A.; Jayaprakash, S.; Karthikeyini, C.; Pillai, K.K.; Firthouse, P.M. Effect of Ocimum sanctum and Azadiracta indica on the formulation of antidandruff herbal shampoo powder. *Pharm. Lett.* **2009**, *1*, 68–76.
6. Al Badi, K.; Khan, S.A. Formulation, evaluation and comparison of the herbal shampoo with the commercial shampoos. *Beni-Suef Univ. J. Basic Appl. Sci.* **2014**, *3*, 301–305. [CrossRef]
7. Muhammad, S.; Utari, R.S.D.; Rahmatullah, M.; Fadhlurrahman, H.; Arie, F.M.; Amanda, T.; Erwan, F.; Lufika, R.D.; Ernawati, E. Innovation of Shampoo Bar From Natural Herbal Essential Oil of Aceh. *J. Patchouli Essent. Oil Prod.* **2022**, *1*, 18–21. [CrossRef]
8. Pandey, S.K.; Bhandari, S.; Sarma, N.; Begum, T.; Munda, S.; Baruah, J.; Gogoi, R.; Haldar, S.; Lal, M. Essential oil compositions, pharmacological importance and agro technological practices of Patchouli (*Pogostemon cablin* Benth.): A review. *J. Essent. Oil Bear. Plants* **2021**, *24*, 1212–1226. [CrossRef]
9. Ramya, H.; Palanimuthu, V.; Rachna, S. An introduction to patchouli (*Pogostemon cablin* Benth.)—A medicinal and aromatic plant: It's importance to mankind. *Agric. Eng. Int. CIGR J.* **2013**, *15*, 243–250.
10. Becker, L.C.; Bergfeld, W.F.; Belsito, D.V.; Klaassen, C.D.; Marks, J.G.; Shank, R.C.; Slaga, T.J.; Snyder, P.W.; Andersen, F.A. Final report of the safety assessment of allantoin and its related complexes. *Int. J. Toxicol.* **2010**, *29*, 84S–97S. [CrossRef]
11. Savić, V.L.; Nikolić, V.D.; Arsić, I.A.; Stanojević, L.P.; Najman, S.J.; Stojanović, S.; Mladenović-Ranisavljević, I.I. Comparative study of the biological activity of allantoin and aqueous extract of the comfrey root. *Phytother. Res.* **2015**, *29*, 1117–1122. [CrossRef] [PubMed]

12. Klimek-Szczykutowicz, M.; Szopa, A.; Ekiert, H. *Citrus limon* (Lemon) phenomenon—A review of the chemistry, pharmacological properties, applications in the modern pharmaceutical, food, and cosmetics industries, and biotechnological studies. *Plants* **2020**, *9*, 119. [CrossRef] [PubMed]
13. Madhusudhan, M.; Rao, M.K.; Radha, G.; Ganapathy, S. Use of traditional herbs for the formulation of herbal powdered shampoos and their evaluation. *Plant Arch.* **2021**, *21*, 845–856. [CrossRef]
14. Arora, P.; Nanda, A.; Karan, M. Shampoos based on synthetic ingredients vis-a-vis shampoos based on herbal ingredients: A review. *Int. J. Pharm. Sci. Rev. Res.* **2011**, *7*, 41–46.
15. Olakunle, O.; Joy, B.; Irene, O. Antifungal activity and phytochemical analysis of selected fruit peels. *J. Biol. Med.* **2019**, *3*, 040–043. [CrossRef]
16. Rathi, S.; Murarkar, K.; Chandak, A. Antimicrobial activity of natural herbal products against dandruff causing fungus and bacteria. *World J. Pharm. Res.* **2019**, *8*, 1460–1467.
17. Singh, A.; Saxena, A. Formulation and Evaluation of Herbal Anti-Dandruff Shampoo from Bhringraj Leaves. *Pharm. Pract. Res.* **2018**, *1*, 5–11.
18. Revansiddappa, M.; Sharadha, R.; Abbulu, K. Formulation and evaluation of herbal Anti-dandruff shampoo. *J. Pharmacogn. Phytochem.* **2018**, *7*, 764–767.
19. Malpani, T.; Jeithliya, M.; Pal, N.; Puri, P. Formulation and evaluation of Pomegranate based herbal shampoo. *J. Pharmacogn. Phytochem.* **2020**, *9*, 1439–1444.
20. Kasparaviciene, G.; Kalveniene, Z.; Pavilonis, A.; Marksiene, R.; Dauksiene, J.; Bernatoniene, J. Formulation and characterization of potential antifungal oleogel with essential oil of thyme. *Evid. Based Complement. Altern. Med.* **2018**, *2018*, 9431819. [CrossRef]
21. Dias, M.F.R.G.; de Almeida, A.M.; Cecato, P.M.R.; Adriano, A.R.; Pichler, J. The shampoo pH can affect the hair: Myth or reality? *Int. J. Trichol.* **2014**, *6*, 95. [CrossRef]
22. AlQuadeib, B.T.; Eltahir, E.K.; Banafa, R.A.; Al-Hadhairi, L.A. Pharmaceutical evaluation of different shampoo brands in local Saudi market. *Saudi Pharm. J.* **2018**, *26*, 98–106. [CrossRef] [PubMed]
23. Jolly, M. Evaluation of commercial herbal shampoos. *Int. J. Cosmet. Sci.* **2000**, *22*, 385–391. [CrossRef] [PubMed]
24. Vijayalakshmi, A.; Sangeetha, S.; Ranjith, N. Formulation and evaluation of herbal shampoo. *Asian J. Pharm. Clin. Res.* **2018**, *11*, 121–124.
25. Krunali, T.; Dhara, P.; Meshram, D.; Mitesh, P. Evaluation of standards of some selected shampoo preparation. *World J. Pharm. Pharm. Sci.* **2013**, *2*, 3622–3630.
26. Dhayanithi, S.; Hoque, E.; Pharm, B.; Pallavi, N.; Pn, K. Formulation and evaluation of herbal shampoo. *Natl. J. Pharm. Sci.* **2021**, *1*, 88–93.
27. Farhan, M.; Gamayel, A. An Addition of Ethanol and Patchouli Oil on Pertamax: The Case Study of Their Effect on Exhaust Gas Emission for 4-stroke Engine. *J. Tek. Mesin Mech. Xplore* **2023**, *3*, 87–95. [CrossRef]
28. D'Souza, P.; Rathi, S.K. Shampoo and Conditioners: What a Dermatologist Should Know? *Indian J. Dermatol.* **2015**, *60*, 248–254. [CrossRef] [PubMed]
29. Cizauskaite, U.; Marksa, M.; Bernatoniene, J. The optimization of technological processes, stability and microbiological evaluation of innovative natural ingredients-based multiple emulsion. *Pharm. Dev. Technol.* **2018**, *23*, 636–645. [CrossRef]
30. Silva, L.N.; Leite, M.G.A.; Costa, G.M.D.A.; Campos, P.M.B.G.M. Influence of botanical extracts in the texture profile of shampoo formulations. *Int. J. Phytocosmetics Nat. Ingred.* **2020**, *7*, 6. [CrossRef]
31. Kumar, A.; Mali, R.R. Evaluation of prepared shampoo formulations and to compare formulated shampoo with marketed shampoos. *Evaluation* **2010**, *3*, 025.
32. Parmar, J.S. Customer Behavior on the Purchase Decision and Brand Preference for Shampoos. *Apeejay J. Manag. Technol.* **2013**, *8*, 10.
33. Rawal, N.; Singh, G.S. A Study on Consumer Preference and Satisfaction Towards Various Brands of Hair Shampoo in Ahmedabad City. In *Data-Driven Decision Making for Long-Term Business Success*; IGI Global: Hershey, PA, USA, 2023; pp. 356–368.

Article

Formulation and Characterization of Non-Toxic, Antimicrobial, and Alcohol-Free Hand Sanitizer Nanoemulgel Based on Lemon Peel Extract

Faten Mohamed Ibrahim [1,*], Eman Samy Shalaby [2], Mohamed Azab El-Liethy [3], Sherif Abd-Elmaksoud [4], Reda Sayed Mohammed [5], Said I. Shalaby [6], Cristina V. Rodrigues [7], Manuela Pintado [7,*] and El Sayed El Habbasha [8]

1   Medicinal and Aromatic Plants Research Department, Pharmaceutical and Drug Industries Research Institute, National Research Centre, Cairo P.O. Box 12622, Egypt
2   Pharmaceutical Technology Department, Pharmaceutical and Drug Industries Research Institute, National Research Centre, Cairo P.O. Box 12622, Egypt; dremanshalaby_nrc@hotmail.com
3   Environmental Microbiology Lab., Water Pollution Research Department, National Research Centre, Dokki, Giza 12622, Egypt; mohamedazabr@yahoo.com
4   Environmental Virology Lab., Water Pollution Research Department, National Research Centre, Dokki, Giza 12622, Egypt; sherifnrc@yahoo.com
5   Pharmacognosy Department, Pharmaceutical and Drug Industries Research Institute, National Research Centre, Cairo P.O. Box 12622, Egypt; redamohammed2015@gmail.com
6   Complementary Medicine Department, National Research Centre, Cairo P.O. Box 12622, Egypt; saidshalaby7@gmail.com
7   CBQF—Centro de Biotecnologia e Química Fina—Laboratório Associado, Escola Superior de Biotecnologia, Universidade Católica Portuguesa, Rua Diogo Botelho 1327, 4169-005 Porto, Portugal; civrodrigues@ucp.pt
8   Field Crops Research Department, National Research Centre, Cairo P.O. Box 12622, Egypt; sayedhabasha@yahoo.com
*   Correspondence: fatenmibrahim@gmail.com (F.M.I.); mpintado@ucp.pt (M.P.)

Citation: Ibrahim, F.M.; Shalaby, E.S.; El-Liethy, M.A.; Abd-Elmaksoud, S.; Mohammed, R.S.; Shalaby, S.I.; Rodrigues, C.V.; Pintado, M.; Habbasha, E.S.E. Formulation and Characterization of Non-Toxic, Antimicrobial, and Alcohol-Free Hand Sanitizer Nanoemulgel Based on Lemon Peel Extract. *Cosmetics* **2024**, *11*, 59. https://doi.org/10.3390/cosmetics11020059

Academic Editor: Vasil Georgiev

Received: 27 February 2024
Revised: 29 March 2024
Accepted: 9 April 2024
Published: 12 April 2024

Abstract: Recently, hand sanitization has gained attention for preventing disease transmission. Many on-the-market convenient dermal sanitizers contain alcohol, which can be detrimental to the skin. Therefore, three nanoemulgel formulations (LN-F1, LN-F2, LN-F3) incorporating lemon peel extract (LE), and with various increasing concentrations of xanthan gum as a gelling agent and stabilizer, were developed and characterized as a novel alternative. All formulations showed non-Newtonian shear-thinning flow behavior, particle size values below 200 nm, and increasing zeta potential with higher xanthan gum concentrations. All nanoemulgel formulations exhibited greater in vitro phenolic compound release than free LE. LN-F2 (1.0% LE, 20.0% mineral oil, 20.0% Span 80, 4.0% Cremophor RH 40, 4.0% PEG 400, 0.5% xanthan gum, 50.5% dH$_2$O) was selected as the optimal formulation due to improved characteristics. LE and LN-F2 potential cytotoxicity was assessed on MA-104, showing no significant cellular morphological alterations up to 10 mg/mL for both samples. LN-F2 showed in vitro antimicrobial activity against *E. coli*, *S.* Typhimurium, *P. aeruginosa*, *S. aureus*, *L. monocytogenes*, and *C. albicans*, as well as antiviral activity against phiX 174, but no effect against rotavirus (SA-11). *In vivo*, LN-F2 presented a removal capacity of 83% to 100% for bacteria and 89% to 100% for fungi. These findings suggest that the formulated nanoemulgel holds potential as a safe and effective antiseptic, providing a viable alternative to commercial alcohol-based formulations.

Keywords: lemon peels; plant-based antimicrobial agents; bioactive compounds; sustainable nanoemulgel formulations; dermal sanitizer

1. Introduction

Nowadays, hand sanitizing has become part of most people's daily routine, primarily aimed at neutralizing various pathogens that can be transmitted through contaminated surfaces [1]. This prophylactic measure not only helps safeguard the user from potential infection by these pathogenic microorganisms but also contributes to preventing their transmission, thereby reducing the risk of infectious disease propagation [2]. Moreover, due to the global pandemic caused by SARS-CoV-2, this practice has been increasingly adopted worldwide, due to the continuous and on-the-go necessity for hand hygiene [3]. Therefore, dermal sanitizers that allow for direct application to the skin without the need for rinsing have been developed and optimized, aiming to enhance the convenience of hand disinfection. A hand sanitizer is typically characterized as a liquid gel or foam developed to eliminate viruses, germs, and various other microorganisms from the hands [3]. Moreover, it can be categorized into two main types: alcohol-based (ABHS) and alcohol-free hand sanitizer. Dermal alcohol-based products have been commonly used for their safety and effective antisepsis properties, attributed to their antimicrobial capacities [2]. However, the use of ABHS may have some drawbacks, such as causing skin dryness [2], which can compromise the skin's barrier with prolonged use. On the other hand, alcohol-free hand sanitizers are generally considered safer, as they are less abrasive to the skin's barrier [3].

Citrus limon belongs to the *Rutaceae* family and is one of the most important fruit crops, growing widely in tropical and subtropical regions. According to the Food and Agriculture Organization (FAO), the global production of citrus exceeded 140 million tons in 2019 [4]. Besides being consumed as fresh produce, citrus fruits undergo processing into juice, canned or dehydrated products, marmalades, jams, and flavoring agents. Moreover, approximately 50–60% of the fruit weight, including peels, seeds, and pulps, is discarded after processing. These by-products contain several compounds such as vitamins, minerals, phenolic compounds, terpenoids, and dietary fiber which have different bioactivities [5,6]. For instance, *C. limon* has been extensively cultivated for its alkaloids present in crude extracts obtained from different parts of the fruit, which have demonstrated antibacterial potential against clinically relevant bacterial strains (e.g., *Escherichia coli*, *Staphylococcus aureus*, and *Salmonella* spp.) [7]. Additionally, these fruits are also important sources of other compounds of interest, such as polyphenols, which are bioactive molecules widely found in plant species, affecting their morphology, growth, reproduction, and resistance to pathogens and environmental stresses. Flavonoids, the most common group of polyphenols in plants, play crucial roles in plant responses and exhibit a broad spectrum of biological activity such as antibacterial, antifungal, antidiabetic, anticancer, and antiviral properties [7,8]. The most abundant flavonoids identified in citrus are naringin, hesperidin, narirutin, and neo-hesperidin [9].

Traditional topical vehicles, including lotions, ointments, patches, and creams, are associated with disadvantages such as poor penetrability. Furthermore, they may exhibit undesirable physicochemical characteristics, with a lower spreading coefficient, requiring rubbing and resulting in a more difficult application. In addition, ensuring the stability of these formulations can also be an issue. However, it has been reported that topical nanosized formulations can offer a viable solution to these concerns by promoting drug delivery through the induction of temporary disruption in the highly organized lipid bilayer structure of the skin. These formulations are also advantageous due to their low irritation to the skin, high stability, and improved local activity [10,11]. For instance, nanoemulsions are characterized as transparent and thermodynamically stable systems with droplet sizes of 20–200 nm, composed of oils and surfactants usually in association with co-surfactants. Despite their many benefits, low viscosity and spreadability are the main disadvantages of topical nanoemulsion formulations. Based on this knowledge, attempts have been made to formulate nanoemulgels by combining nanoemulsions with gelling agents, aiming to achieve a formulation with non-staining, thixotropy, emollient, non-greasy properties and improved spreadability [12]. Moreover, for individuals sensitive to alcohol-based formulations, incorporating a less abrasive substance, such as lemon peel extract with antimicrobial activities, into a nanoemulgel formulation could present a promising opportunity for hand hygiene. Therefore, the objective of the present study is to develop and characterize a non-toxic and alcohol-free nanoemulgel based on an upcycled lemon peel extract for hand-sanitizing purposes. Moreover, this research aims to address literature gaps regarding sustainable and effective hand sanitization ingredients, which is essential for advances in the field of resistant-pathogen-targeting formulations, representing a viable alternative to commercial alcohol-based solutions.

2. Materials and Methods

2.1. Chemicals

The sorbitan monooleate (Span 80) was acquired from LobaChemie (Mumbai, Maharashtra, India). Xanthan gum (Rheocare® XGN) and polyethylene glycol hydrogenated castor oil (CREMOPHOR® RH 40), were obtained from BASF (Ludwigshafen am Rhein, Germany). Polyethylene glycol 400 (PEG 400) was obtained from Sigma-Aldrich (St. Louis, MO, USA). Liquid paraffin, potassium monohydrogen phosphate, and dipotassium hydrogen phosphate were purchased from El Nasr Pharmaceutical Chemicals (Cairo, Egypt). Trypticase soy broth and Mueller–Hinton agar were obtained from BBL (Darmstadt, Germany). Malt extract agar was obtained from Merck (Merck, St. Louis, MO, USA, Sigma-Aldrich, St. Louis, MO, USA). The standard plate count agar was obtained from Hach (Ames, IA, USA) All reagents were used as received.

2.2. Raw Materials and Sample Preparation

The lemons (*Citrus limon*) used to obtain the bioactive extract were freshly collected at the National Research Center Farm (Agricultural Production and Research Station, National Research Center, El Nubaria Province, El Behira Governorate). The lemon peels were then removed from the fruit and used as needed.

2.3. Microbial Strains' Growth Conditions

Escherichia coli (ATCC 25922), *Salmonella enterica* serovar Typhimurium (ATCC 14028), *Pseudomonas aeruginosa* (ATCC 10145), *Staphylococcus aureus* (ATCC 43300), *Listeria monocytogenes* (ATCC 35152), and *Candida albicans* (ATCC 10231) were cultured in Mueller–Hinton medium (BBL, Darmstadt, Germany). These microbial cultures were prepared by inoculating the medium with each test microbe and incubating at 35 °C for 24 h. Microbial cells were obtained by centrifuging each culture under sterile conditions at 4000 rpm for 15 min. The cells were washed with 20 mL of sterile phosphate-buffered saline (PBS) and centrifuged (4000 rpm for 5 min) until the supernatant became clear. The optical density for each suspension was recorded at 500 nm, and serial dilutions were carried out using appropriate aseptic techniques until the optical density was between 0.5 and 1.0. The number of colony-forming units was assessed to guarantee a concentration of 5.0×10^6 CFU/mL.

2.4. Cell Culture Growth Conditions

The MA-104 cell line (Holding Company for Biological Products & Vaccines VAC-SERA, Agouza, Giza, Egypt) was cultured in Dulbecco's modified Eagle's medium—high glucose (DMEM, Gibco™). The cell culture was incubated at 37 °C in a humidified incubator containing 5% CO_2, and the medium was replaced every 2 to 3 days until an 85% confluence of cells was achieved. After cell trypsinization using TrypLE Express (Gibco™), the cells were counted in a hemocytometer and seeded into 96-well plates (Greiner-Bio one, Frickenhausen, Germany) at a density of 5.0×10^4 cells/mL. The cells were then left to adhere for 24 h.

2.5. Lemon Peel Extraction Methodology

To develop the nanoemulgel formulation, a hydroethanolic extraction process was conducted to acquire a polyphenolic-rich extract from the peels of *Citrus limon* (LE). Briefly, the fresh peels were weighed and macerated for 24 h using 80% (*v/v*) ethanol in a 1:3 (*w/v*) ratio. The resulting mixture was then filtered and concentrated through vacuum-drying using a rotary evaporator at 45 °C. The obtained extract was stored at 4 °C in an amber bottle until further use.

2.6. Antimicrobial Activity of the Lemon Peel Extract (LE)

The antimicrobial activity of LE was assessed through minimum inhibitory concentration (MIC) and minimum microbicidal concentration (MMC) determination against *Staphylococcus aureus* (ATCC 6538, Gram-positive), *Escherichia coli* (ATCC 25922, Gram-negative), and *Candida albicans* (ATCC 10231, yeast). LE was diluted in dimethyl sulfoxide (DMSO) to obtain a stock solution of 5 mg/mL, followed by two-fold dilutions to achieve several testing concentrations. To a 96-well plate, 100 µL of each microbial suspension (5.0×10^6 CFU/mL) and 100 µL of each LE dilution were added per well. As a positive control of microbial growth, medium with only DMSO (without LE) was used. The inoculated microplates were loosely wrapped with Parafilm and incubated at 37 °C for 18–24 h.

The resazurin solution was prepared by dissolving 67.5 mg of resazurin dye in 10 mL of dH_2O, followed by vortex homogenization. The solution was sterilized by filtration using a membrane filter with a pore size of 0.22–0.45 µm. Afterward, 10 µL of this dye solution was added to each testing and control well, and an additional 4 h of incubation was carried out. The color changes within each well upon resazurin incubation were analyzed. The MIC value was defined as the lowest concentration of LE at which color change was observed.

Using an inoculation loop, samples were taken from each well corresponding to the two lowest concentrations right before the MIC value for each strain and spread on sterile nutrient agar plates. The plates were incubated at 35 °C for 24 h. MMC was taken as the lowest sample concentration that did not exhibit any microbial growth. Each test condition was performed in duplicate.

2.7. Preparation of Lemon-Extract-Based Nanoemulgel (LN) Formulations

The nanoemulgels loaded with lemon peel extract (LN) were prepared in two phases: phase A and phase B. Phase A was prepared by mixing LE, surfactants (Span 80 and Cremophor RH 40), and oils (PEG 400 and mineral oil) to obtain the nanoemulsion. This mixture was stirred at 1500 rpm using a mechanical stirrer and vortexed (2000 rpm, 5 min). For phase B, dH_2O was slowly added to xanthan gum and mechanically stirred until a homogenous polymeric soluble gel was formed. The extract-loaded emulsion (phase A) was then slowly added to the polymeric gel base (phase B) and then mixed under magnetic stirring for 5 min until a homogenous gel was obtained [13]. The final lemon peel extract content was 10 mg/g of nanoemulgel for all tested formulations. This extract concentration was chosen for its improved antimicrobial capacity. The composition of each LN formulation (LN-F1, LN-F2, and LN-F3) is detailed in Table 1.

Table 1. Composition of the developed nanoemulgel formulations (LN-F1, LN-F2, and LN-F3), based on *Citrus limon* peel extract (LE).

	Ingredients (g)	Formulations		
		LN-F1	**LN-F2**	**LN-F3**
	LE	0.100	0.100	0.100
	Mineral Oil	2.000	2.000	2.000
Phase A	Span 80	2.000	2.000	2.000
	Cremophor RH 40	0.400	0.400	0.400
	PEG 400	0.400	0.400	0.400
Phase B	Xanthan Gum	0.025	0.050	0.100
	Water (dH_2O)	5.075	5.050	5.000

2.8. Fourier Transform Infrared (FTIR) Spectroscopy Characterization of LE and LN Formulations

The chemical integrity and potential chemical interactions between LE and the components of LN formulations were assessed using FTIR spectrophotometry (JASCO 6100, Tokyo, Japan). Xanthan gum was mixed separately with potassium bromide and compressed (200 kg/cm^2, for 2 min) in a hydraulic press to form a compact disk. In contrast, the liquid samples (LE, LN formulations, and the remaining individual ingredients) were analyzed directly. All samples were scanned against a blank KBr pellet background in the spectral region between 4000 cm^{-1} and 400 cm^{-1} [14,15]. The resolution used was 4 cm^{-1}, and each spectrum was acquired by an average of 32 scans, in transmission mode.

2.9. Organoleptic Properties and Phase Separation Evaluation of LN Formulations

Firstly, the LN formulations' organoleptic properties, namely texture and color, were macroscopically evaluated. Also, to evaluate the occurrence of phase separation under extreme conditions, all LN formulations were subjected to centrifugation. Accordingly, 5 g of each nanoemulgel formulation was centrifuged at 5000 rpm for 30 min at room temperature [16–18]. Subsequently, the samples were examined regarding phase separation. Triplicates were made for this assessment.

2.10. Rheology Analysis of LN Formulations

Rheological measurements of the developed LN formulations were acquired using a parallel-plate rheometer (Physica MCR 301, Anton Paar GmbH, Graz, Austria. Measurements were conducted to determine shear viscosity as a function of shear rate. A total of 25 measurement points were acquired with shear rates ranging from 1 s^{-1} to 250 s^{-1} [19], at 25.0 $\pm$ 0.1 °C. The plot was obtained from the average of duplicates.

2.11. Particle Size, Polydispersity Index, Zeta Potential, and pH Analysis of LN Formulations

The physical characteristics of the LN nanoemulsions, including particle size (PS), polydispersity index (PDI), and zeta potential (ZP), were assessed at 25.0 $\pm$ 2.0 °C by diluting each formulation at a ratio of 1:100 (v/v) in dH$_2$O [19]. These analyses were conducted using photon correlation spectroscopy with a Malvern Zetasizer (ZS, Malvern Instruments, Ltd., Worcestershire, UK). Additionally, the pH of the LN formulations was assessed under the same temperature conditions using a pH meter Orion Versa Star^{TM}, Thermo Fisher Scientific, Waltham, MA, USA).

2.12. In Vitro Polyphenolic Content Release of LE and LN Formulations

The in vitro polyphenolic content release profile from LE and LN formulations was analyzed by dialysis [20]. An amount equivalent to 15 mg of extract for both LE and LN formulation testing was placed into the dialysis membrane (cellulose membrane, molecular weight cut-off of 12,000–14,000, Sigma-Aldrich Co., St. Louis, MO, USA). This procedure was conducted against 100 mL of acetate buffer (pH 5.5) [20] and a 10% ethanol (v/v) solution [21], in a shaking water bath (MemmertSV-1422, Memmert GmbH, Schwabach, Germany) at 32.0 $\pm$ 0.5 °C, 100 rpm. Samples were withdrawn every 2 h, up until the 24 h time point, to measure the polyphenol concentration spectrophotometrically, using the UV-2401 PC (Shimadzu Co., Kyoto, Japan). The release profile of the phenolic compounds from the selected formulations was compared to the raw lemon extract polyphenol solution containing the same extract concentrations. The cumulative percentage of phenolic compounds released was determined as the ratio between the quantity of total polyphenols released (C_f) and the initial quantity of total phenolic compounds inserted in the dialysis membrane (C_i) (Equation (1)). All measurements were carried out in triplicate. Different release mathematical models were employed to elucidate the release mechanism. This analysis included the zero order, first order, Higuchi model, and Korsmeyer–Peppas models [22].

$$\text{Polyphenols Release Quantity (\%)} = C_f/C_i \times 100 \qquad (1)$$

2.13. Identification and Quantification of LN-F2 Phenolic Content by LC-ESI-MS/MS

The phenolic content analysis of LN-F2 was performed by liquid chromatography–electrospray ionization–tandem mass spectrometry (LC-ESI-MS/MS) with an Exion LC AC system for separation and SCIEX Triple Quad 5500+ MS/MS system equipped with electrospray ionization (ESI) for detection. The separation was performed using a ZORBAX

Eclipse Plus C18 Column (4.6 × 100 mm, 1.8 µm). The mobile phases consisted of two eluents: (A) 0.1% (*v/v*) aqueous formic acid and (B) LC-grade acetonitrile. The mobile phase was programmed as follows: 0–1 min, 2% B; 1–21 min, 2–60% B; 21–25 min, 60% B; 25.01–28 min, 2% B at a flow rate of 0.8 mL/min and an injection volume of 3 µL. For MRM analysis of the selected phenolic compounds, negative ionization modes were applied: curtain gas, 25 psi; ion spray voltage, 4500 V; temperature source, 400 °C; ion gas source 1 and 2, 55 psi; declustering potential, 50 V; collision energy, 25 V; collision energy spread, 10%. This analysis was conducted in triplicate.

2.14. LN-F2 Transmission Electron Microscopy (TEM) Characterization

The selected nanoemulgel system (LN-F2) was characterized through transmission electron microscopy (TEM) (JEM-2100, JEOL Co., Tokyo, Japan) to evaluate its components' morphological attributes, including lamellarity, shape, and size.

2.15. LE and LN-F2 Cytotoxicity Evaluation by Inverted Light Microscopy Analysis

This test was conducted to evaluate the potential cytotoxic concentrations of LE and LN-F2 [23]. Concentrations ranging from 10 µg/100 µL to 1 mg/100 µL for each sample were prepared using Dulbecco's Modified Eagle's Media (DMEM). Additionally, MA-104 cells (2×10^5 cells/mL) were prepared in 96-well culture plates (Greiner-Bio one, Germany). After confluency, the medium was removed from the wells and replaced with 100 µL of each prepared testing concentration. As a negative control of cytotoxicity, 100 µL of DMEM without any sample was used. The cells were incubated at 37 °C in a humidified atmosphere containing 5% CO_2 for 72 h. Cell morphology was microscopically analyzed at 24, 42, and 72 h after sample incubation to evaluate any potential morphological alterations, such as loss of confluence, cell rounding and shrinking, cytoplasm granulation, and/or vacuolization [23].

2.16. LE and LN-F2 Bioactivity Assessment

2.16.1. In Vitro Antiviral Activity

The antiviral activity of LE and LN-F2 was analyzed against rotavirus SA-11 and the bacteriophage phiX 174. For this purpose, LE was prepared at 10 mg/mL, while LN-F2 was prepared at 1 and 10 mg/mL, using sterile dH_2O with 10% (*v/v*) dimethyl sulfoxide (DMSO).

Viruses' Propagation Conditions

Rotavirus was activated with 10 µg/mL trypsin before propagation on MA-104 cells. Centrifugation of the rotavirus was performed at $1000 \times g$ for 5 min in order to purify the sample by removing cell debris. After filtration through a 0.2 µm membrane, the supernatant was kept and used as a virus stock suspension, containing 10^6–10^7 TCID50/mL. The suspension was then stored at −80 °C until further use.

Bacteriophage phiX 174 was propagated in an *E. coli* DSM 13127 host, in 3.0% trypticase soy broth (TSA) (BD, Franklin Lakes, NJ, USA) containing 0.1% glucose, 2 mM $CaCl_2$, and 10 mg/mL thiamine, at 37 °C for 18 h. The bacteriophage was then harvested by centrifugation and filtration using the abovementioned procedure for rotavirus.

Viruses' Quantification

Virus quantification was performed on cells exposed to each virus, either untreated or after treatment with LE or LN-F2. Moreover, the 50% tissue culture infectious dose (TCID50), a parameter used to quantify and assess the infectivity of a virus in cells, defined as the virus solution concentration at which 50% of the cells show cytopathic effects (CPE), was also assessed during this assay.

The infectivity of the rotavirus SA-11 stock was activated by adding 10 µg/mL trypsin for 30 min at 37 °C. A 100 µL aliquot of the activated rotavirus SA-11 at a final concentration of 1.0×10^6 TCID50/mL was mixed with an equal volume of LE (10 mg/mL) or LN-F2 (containing 1 mg/mL or 10 mg/mL of LE) at the prepared concentrations. The different solutions containing rotavirus with each sample underwent an initial incubation at 37 °C, for 30 min, followed by mixing and subsequent incubation for an additional 60 min.

After MA-104 cells adhered onto 96-well plates, the cells were incubated with the activated rotavirus suspension, as a positive control of cytotoxicity, or with the various sample (10 mg/mL of LE, 1 mg/mL of LN-F2, or 10 mg/mL of LN-F2) solutions containing rotavirus, for 7 days, at 37 °C, under a humid atmosphere containing 5% CO_2. During the seven days of incubation, cells were microscopically monitored daily to check for CPE, and the TCID50 was analyzed. This analysis was performed in duplicate for each sample.

To assess the effect of the testing solutions on the phiX 174 bacteriophage, equal volumes of virus suspension and each sample were mixed and incubated for 30 min, at 37 °C, as described previously. The viral titer after each treatment was evaluated by standard plaque assay. Briefly, 0.9 mL of phage sample solution and 0.1 mL of *E. coli* culture solution were added to 0.6% molten top agar (BD, Franklin Lakes, NJ, USA), mixed, and poured on tryptic soy agar (TSA) (BD, Franklin Lakes, NJ, USA) plates. The plates were incubated at 37 °C overnight, and then the colony-forming units were counted. The virus inhibition after treatment with each sample was calculated as the difference between initial (V_i) and final (V_f) virus counts, according to Equation (2). This assessment was carried out in duplicate.

$$\text{Virus inhibition (\%)} = (V_i - V_f)/V_i \times 100 \qquad (2)$$

2.16.2. In Vitro and In Vivo LN-F2 Antimicrobial Activity
Zone of Inhibition (ZI) Analysis by Well and Disk Diffusion Assays

LN-F2 antimicrobial properties were assessed in vitro through the disk and well diffusion techniques, against Gram-positive bacterial strains: *Staphylococcus aureus* (ATCC 43300), *Listeria monocytogenes* (ATCC 35152); Gram-negative bacterial strains: *Escherichia coli* (ATCC 25922), *Salmonella enterica* serovar Typhimurium (ATCC 14028), *Pseudomonas aeruginosa* (ATCC 10145); and a yeast: *Candida albicans* (ATCC 10231). Each microbial strain was added to a tube containing trypticase soy broth (BBL, Germany). The inoculation tubes were kept at 37 °C for 24–48 h.

To carry out both tests, the inoculated microorganisms were cultivated in Mueller–Hinton agar plates, using sterile cotton swabs. For the disk diffusion method, sterile disks (6 mm in diameter) were submerged in the developed LN and then were aseptically transferred to the center of each cultured agar plate surface. For the well diffusion test, wells were excavated in the cultured Mueller–Hinton agar media plates, using sterile tips. Following that, 100 μL of the testing LN was added to the wells. Regarding both tests, all plates were then incubated for 24–48 h at 37 °C. After incubation, the zone of inhibition (ZI) was measured in mm, using a measuring ruler (HiMedia Co., Maharashtra, India).

In Vivo Assessment

The in vivo testing of the LN-F2 antimicrobial properties was carried out according to the protocols outlined by Public Health England [24] and Lambrechts et al. [25]. Briefly, samples were collected from 20 healthy volunteers, with ages ranging from 25 to 67 years and of different genders, both before and after the application of LN-F2 (2.0 g). A 5.0×5.0 cm^2 area of volunteers' hands was scrubbed using sterile swabs, which were then placed in a tube filled with 5 mL of sterile dH_2O. The same procedure was repeated after applying 2.0 mg of LN-F2 on the volunteers' hands, and the nanoemulgel was fully absorbed by the skin. The tubes were then vortexed for 30 s.

In accordance with the American Public Health Association's (APHA) recommended procedure [26], total bacterial and total fungal counts were determined for all collected samples. The total fungal counts were conducted using the pour plate method on malt extract agar (MEA), while total bacterial counts were determined through standard plate count (SPC) agar. For bacteria, the inoculation plates were incubated at 37 °C for 24–48 h, while fungi were incubated for 48–72 h. Microbial counts were estimated using the colony counter model SC6PLUS (Stuart, UK). The viable microbial cell density was calculated and expressed by CFU/cm^2. The removal efficiency (R%) of the total bacterial and fungal counts was calculated according to Equation (3).

$$R\ (\%) = (ib - ia)/ib \times 100 \qquad (3)$$

where ib represents microbial counts before the addition of the LN-F2, and ia represents microbial counts after the addition of the LN-F2.

2.17. Statistical Analysis

Experiments were conducted in duplicate or triplicate, and values were expressed as mean ± standard deviation (SD). IBM® SPSS® Statistics (version 25) was used to perform the statistical analysis, using the one-way ANOVA and post hoc tests for multiple comparisons. Statistical significance was established at p-value < 0.05. Data processing and plot development were carried out using Microsoft Excel 2016.

3. Results and Discussion

3.1. Lemon Peel Extract Antimicrobial Activity

After obtaining the lemon peel extract (LE), the potential in vitro antimicrobial activity was assessed through MIC and MMC determination.

The MIC value, representing the minimum concentration at which a compound inhibits the growth of a specific microorganism, is a standard measure of antimicrobial efficacy [27,28]. The MMC value, indicating the lowest concentration required to eliminate 99.9% of a microorganism, is also crucial in assessing antimicrobial effectiveness [29]. Therefore, to gain further insights into the potential antimicrobial capacity of the obtained LE, its MIC and MMC values were assessed against *E. coli* (Gram-negative), *S. aureus* (Gram-positive), and *C. albicans* (yeast). As shown by the results presented in Figure 1, the extract demonstrated low MIC values against all tested microorganisms, namely 156.25 ± 3.50 µg/mL for *S. aureus*, 78.13 ± 2.85 µg/mL for *E. coli*, and 19.53 ± 1.08 µg/mL for *C. albicans*. Regarding the MMC results, LE also presented relatively low values against *E. coli* (78.13 ± 1.91 µg/mL) and *C. albicans* (78.13 ± 1.57 µg/mL), while for *S. aureus*, those values were notably higher (625.00 ± 5.60 µg/mL). Nonetheless, for each microorganism, the MIC and MMC values were correlated and followed the same tendency.

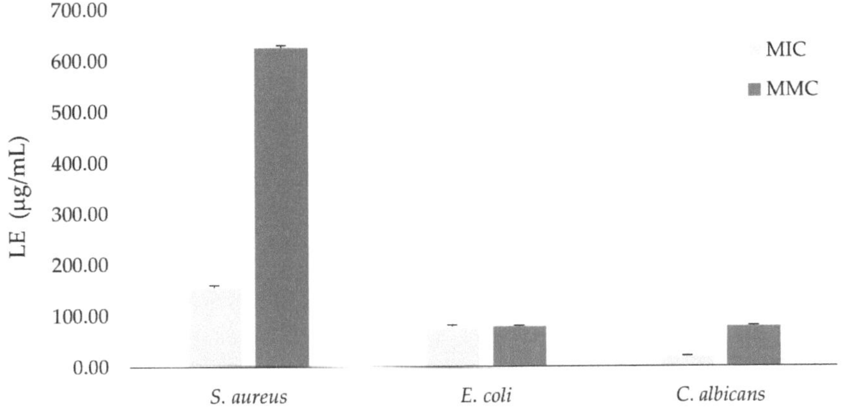

Figure 1. Minimum inhibition concentration (MIC) and minimum microbicidal concentration (MMC) of the lemon peel extract (LE) against *S. aureus*, *E. coli*, and *C. albicans*.

These findings align with existing literature, which also supports the antimicrobial efficacy of lemon peel extracts against several Gram-positive and Gram-negative bacteria, as well as *Candida* strains [30]. Henderson et al. [31] highlighted the antibacterial properties of a lemon peel extract against *Escherichia coli.*, *Staphylococcus aureus*, *Salmonella enteritidis*, and *Listeria monocytogenes*. Jonh et al. [32] also reported that employing different solvents during the extraction process might result in extracts with varying antimicrobial capacity, showing the impact of the extraction method on the antimicrobial effectiveness of the extract. For instance, the authors demonstrated that the methanolic extract showed higher activity against *S. flexneri*, *S. aureus*, and *E. coli*, while the acetone extract was only effective against *S. aureus* and *E. coli*, and the ethanolic extract presented higher antimicrobial capacity against *P. aeruginosa* and *S. enterica* serovar Typhimurium.

According to Saleem et al. [33], the LE antimicrobial effect might be mainly related to the presence of bioactive compounds such as phenolic compounds in the lemon peel. Overall, the extract exhibited a higher antimicrobial effect on *E. coli* and *C. albicans*, attributed to the ability of the bioactive compounds present in LE to attach to the outer membrane of Gram-negative bacteria, compromising its integrity and leading to the release of cytoplasmatic content and cell death [34]. Furthermore, Henderson et al. [31] explained that LE can have antimicrobial properties through the degradation of lipids and proteins constituting the microorganism's membranes, resulting in their disruption. Nonetheless, it has also been stated that the thicker peptidoglycan layer in the cell walls of Gram-positive bacteria can be infiltrated by some bioactive compounds found in lemon peels [35,36], which can explain LE's effectiveness against *S. aureus*.

3.2. Development of LE-Based Nanoemulgel Formulations

After assessing the antimicrobial activity of the LE, three different formulations (LN-F1, LN-F2, and LN-F3) were developed and subsequently characterized, aiming to achieve an optimal LE-based nanoemulgel (LN) (Figure 2) for potential hand-sanitizing applications. An LE extract concentration of 10 mg/g was chosen to develop all LN formulations to guarantee an effective antimicrobial activity. Regarding the chosen gelling agent, a variety of compounds, such as Carbopol, poloxamer, tragacanth, and HPMC have been applied in the production of nanoemulgels. Nonetheless, xanthan gum was selected for the development of these LN formulations, due to its many physical–chemical and mechanical advantages. Xanthan gum offers higher viscosity even at low concentrations, unlike other polysaccharides, and its elastic properties make it an ideal stabilizer. Also, xanthan gum is highly resistant to pH variations, meaning that it is stable in both alkaline and acidic conditions. Moreover, xanthan gum has an improved thermal stability compared to other water-soluble hydrocolloids, providing a better stability to the formulation at a wide temperature range [37]. Despite xanthan gum being reported in the preparation of nanoemulgels, at concentrations up to 3% [10], it was concluded in previous studies that its incorporation up to 1% would produce the most suitable gel consistency, while maintaining its stability [38]. Therefore, a xanthan gum concentration between 0.25 and 1% was selected.

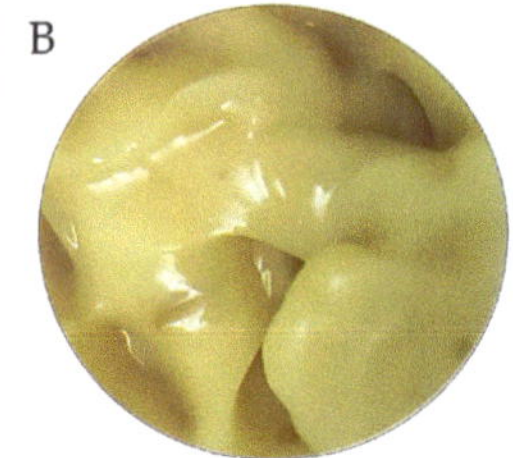

Figure 2. Macroscopic morphological profile of LN-F2, as an example of a lemon-peel-extract-based nanoemulgel formulation. (**A**) LN-F2 swatch, demonstrating the spreadability of the formulation; (**B**) LN-F2 inside a container, demonstration the formulation homogeneity.

In an initial macroscopic evaluation, all formulations demonstrated satisfactory spreadability (Figure 2A), homogeneity, and no indications of phase separation (Figure 2B), suggesting preliminary stability in the formulation [39]. Furthermore, there were no notable distinctions observed among the three tested nanoemulgels, each formulated with varying concentrations (0.25 to 1.00%) of xanthan gum, used as a gelling agent and stabilizer, to create the polymeric base gel. Consequently, based on a first analysis, all formulations appeared to possess physical and morphological characteristics suitable for the intended application.

3.3. LE and LN Formulations' Physicochemical Analysis

3.3.1. LE and LN Formulations' FTIR Analysis

The FTIR spectra of LE and LN formulations, as well as of the individual formulation components (Figure 3), were acquired to characterize their chemical structure and the interaction between compounds within formulations.

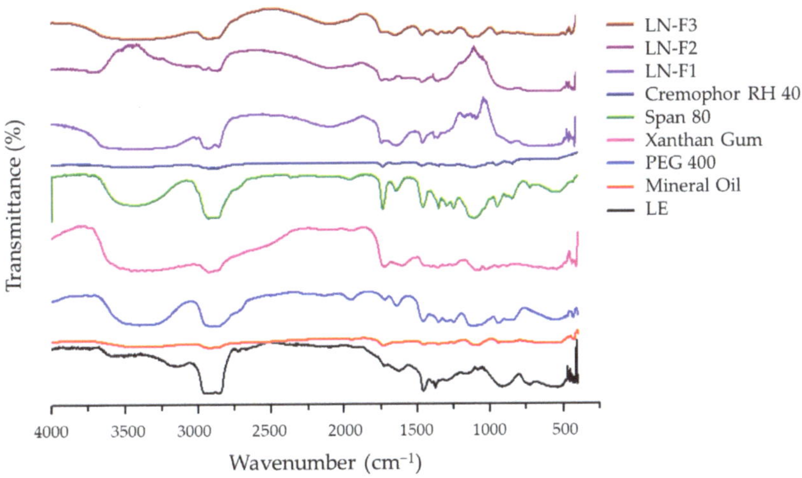

Figure 3. FTIR spectra of the individual components of the tested formulations (mineral oil, PEG 400, xanhan gum, Span 80, and Cremophor RH 40), the lemon peel extract (LE), and the corresponding nanoemulgel formulations (LN-F1, LN-F2, and LN-F3).

Regarding LE, the results indicated that its main component is naringinin, as evidenced by the reported peaks at 3285.79, 3117.36, 3035.80, 1629.76, 1602.09, 1519.88, 1498.10, and 1463.39 cm^{-1} [40]. Moreover, when comparing the spectra of the nanoemulgels (LN-F1, LN-F2, and LN-F3) with LE, changes in peak intensity and position of various components were observed, corroborating the assembly of the nanoemulgel through different chemical interactions [41]. In detail, the individual analyzed components were LE, Span 80, Cremophor RH 40, mineral oil, and PEG 400. The characteristic peaks of these components resemble those described in the literature, suggesting highly purified samples suitable for nanoemulgel formulations [42]. Analyzing the LE spectrum, the peak at 2750 cm^{-1} related to the -OH group of narginin is shifted or fused in the IR spectra of the nanoemulgels, indicating an interaction between the ingredients and the extract and suggesting successful extract incorporation into the nanoemulgels [43]. On the other hand, the FTIR spectra of LN-F1, LN-F2, and LN-F3 showed the characteristic peaks of Span 80, PEG 400, and xanthan gum. However, most of the peaks were diffused, especially those related to the hydroxyl group, asymmetric and symmetric aliphatic C-H stretching, and C=O ester bond. The intensity of the peak for alkyl substituted ether in the Span 80 spectrum decreased upon incorporation into the nanoemulgel formulations. Additionally, the stretching vibration of the ester functional group (C-O-C) was observed between 1172 and 1102 cm^{-1}. Furthermore, the intensity of LE characteristic peaks also decreased and shifted to a lower frequency after its formulation into the nanoemulgel (2915 and 2752 cm^{-1} for the asymmetric and symmetric stretching vibrations of -CH2, 1450.3 cm^{-1} for C=C stretching vibration, and 722 cm^{-1} for the C-H bend), with a minor shift noticed in the peak for C=C stretching vibration (1467.9 cm^{-1}).

It is important to note that the LE peaks within the nanoemulgel formulations did not significantly shift, and no new peaks were observed. Overall, these findings indicate that the ingredients used to formulate the nanoemulgel did not significatively affect the chemical structure of LE, suggesting that the extract and its potential bioactivities were preserved after inclusion in the nanoemulgels [44].

3.3.2. LN Formulations' Organoleptic Properties, Phase Separation, and Rheology Analysis

In order to assess whether the LN formulations possessed adequate organoleptic and physical–chemical characteristics for hand sanitization applications, their color, texture, stability in terms of phase separation under extreme conditions, and rheological behavior were analyzed.

In terms of color and texture, all samples presented a light-yellow color, were homogenous, demonstrated reasonable consistency, and no signs of primary phase separation were observed. However, all formulations underwent centrifugation stress to test for the occurrence of creaming, cracking, or coalescence to analyze their stability under extreme conditions. The results indicated that only LN-F2 maintained its morphology after centrifugation, showing no visible signs of phase separation. However, LN-F1 and LN-F3 exhibited phase separation after being submitted to this assessment. These results might be attributed to the quantities of xanthan gum and dH$_2$O used to formulate the polymeric gel base (phase B), specifically 0.025 g of xanthan gum and 5.075 g of dH$_2$O for LN-F1, and 0.100 g of xanthan gum and 5.000 g of dH$_2$O for LN-F3, leading to unstable o/w chemical interactions between the emulsion components and eventually resulting in phase separation [45].

Furthermore, for an ideal topical application, the formulation must be easily spreadable and non-dripping in nature. Regarding these results (Figure 4), all LN samples exhibited a Newtonian shear-thinning flow behavior, indicating a decrease in viscosity with increasing shear rate. Moreover, it was observed that the nanoemulgels' viscosity increased with higher xanthan gum concentrations. However, when the formulation is subjected to a shear force, its network structure breaks down, leading to a gradual decrease in viscosity. This shear-thinning property might be suitable for the target application, as it allows easier removal of the nanoemulgel from the container and facilitates its application on skin [46].

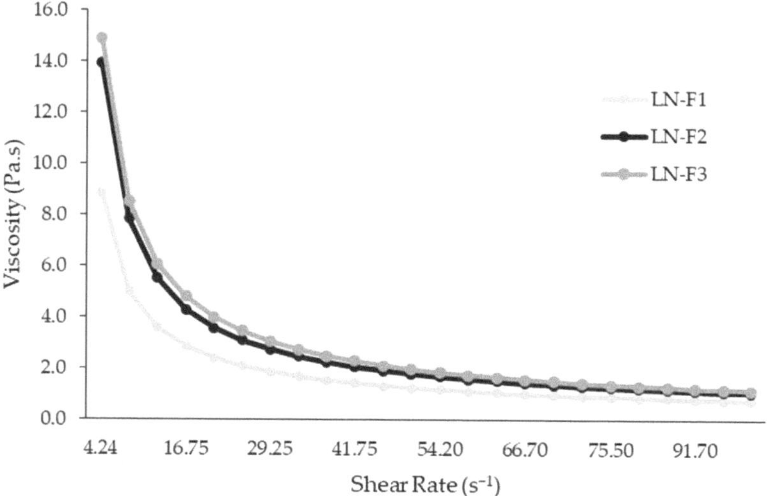

Figure 4. Rheogram of the three developed nanoemulgel formulations (LN-F1, LN-F2, and LN-F3). The plot is presented as an average of duplicates.

3.3.3. LN Formulations' Particle Size, Polydispersity Index, Zeta Potential, and pH Assessment

All LN formulations were characterized in terms of particle size (PS), polydispersity index (PDI), zeta potential, and pH (Table 2).

Table 2. Particle size, zeta potential, polydispersity index (PDI), and pH values of the developed nanoemulgel formulations (LN-F1, LN-F2, and LN-F3).

Formulation	Particle Size ± SD (nm)	Zeta Potential ± SD (mV)	PDI ± SD	pH ± SD
LN-F1	123.30 ± 0.18	−15.90 ± 1.97	0.422 ± 0.01	5.51 ± 0.24
LN-F2	152.80 ± 1.78	−43.60 ± 2.64	0.43 ± 0.06	5.59 ± 0.07
LN-F3	176.90 ± 4.36	−56.80 ± 3.77	0.46 ± 0.19	5.57 ± 0.73

As reported by Eid et al. [47], a slight increase in PS may occur due to the increment in viscosity, which is evident in the obtained results and aligns with the data obtained from the rheology assessment. Therefore, LN-F3, being the formulation with higher viscosity, also presented a higher PS value (176.90 nm). Moreover, the mean droplet size is below 200 nm in all formulations, consistent with previous results reported by Eid et al. [47]. The PDI is a valuable parameter to evaluate the stability of the nanoemulgel formulation, representing the distribution of a population's size within a given sample [48]. Since all LN formulations presented a PDI value lower than 0.5, it indicates a narrow and uniform particle size distribution within all samples [49]. On the other hand, the zeta potential is a relevant factor to assess the stability of nanosuspensions. According to the literature, acquiring a zeta potential of approximately ± 30 mV is typically ideal for the formulation of a stable solution. At this charge level, droplets exhibit repulsion, preventing coalescence and ensuring stability [50]. In this way, LN-F2 and LN-F3, showing a zeta potential greater than −30 mV, can be considered stable formulations, while LN-F1 presented a zeta potential value of −15.9 mV. Extremely positive or negative zeta potential values cause larger repulsive forces, whereas repulsion between particles with similar electric charges prevents aggregation of the particles, ensuring easy redispersion [51]. The negative surface charge of the nanoemulsion droplet might be explained by the mixture of Span 80 with Cremophor RH 40 ions, forming hydrogen bonds between the mixed surfactants and water molecules in the boundary layer of the o/w emulsion, in accordance with already reported results [52]. Regarding pH values, they were similar for all formulations (5.51–5.59).

3.4. In Vitro Release of the LE and LN Formulations' Polyphenolic Content

Citrus limon peel is highly rich in phenolic compounds, which have been studied for their antioxidant and antimicrobial capacities [8,53]. Therefore, the release profile of phenolic content from LE and LN formulations was analyzed. As shown in Figure 5, all the LN formulations released higher amounts of phenolic compounds than LE, emphasizing the benefits of using this carrier for delivering bioactive compounds to the skin. Regarding the LN formulations, LN-F1 released a superior amount of phenolic compounds, followed by LN-F2 and LN-F3. Moreover, a rapid release of lemon polyphenols from the LN was observed, possibly due to increased surface area of the fabricated nanodroplets that permeated easily through the dialysis membranes. Furthermore, since the LN droplet size was found to be smaller than 200 nm, it might enhance the solubility of phenolic compounds, as reported previously [54]. Also, as corroborated by the rheology results, as the concentration of xanthan gum increases, the viscosity rises while the release capacity of the formulation decreases. In this way, the higher viscosity of the LN formulation, which is also related to lower water content, may result in slower diffusion of the phenolic compounds, which is in agreement with previous studies [55]. In addition, the in vitro release profiles were treated with different mathematical models. Thus, the release pattern of phenolic compounds from all LN formulations and from LE followed the Higuchi equation, as indicated by the highest correlation coefficient (R^2). The Higuchi model, in conjunction with the

Korsmeyers–Peppas model, suggested a non-Fickian diffusion release from the gel matrix, in accordance with other published results for the same delivery system [56]. Therefore, the advantages of a delivery system through a nanoemulgel include good adhesion to the skin surface and high delivery capacity leading to a larger concentration gradient of bioactive compounds [57].

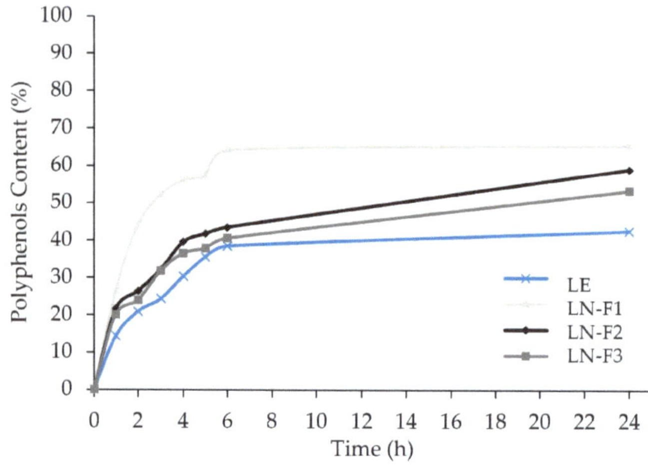

Figure 5. In vitro polyphenolic content release profile of the lemon-extract-based nanoemulgel formulations (LN-F1, LN-F2, and LN-F3) and the free lemon extract (LE).

3.5. LN Optimal Formulation Selection and Characterization

Considering all the abovementioned results, LN-F2 was selected as the optimal formulation due to improved physicochemical characteristics and favorable in vitro release capacity. Therefore, LN-F1 and LN-F3 were excluded from additional analysis, while LN-F2 was further characterized and evaluated for potential bioactivities.

3.5.1. Phenolic Content Analysis

To corroborate the previous analysis, the optimal selected formulation (LN-F2) was assessed for the identification and quantification of phenolic compounds. Therefore, according to the obained results (Table 3 and Figure 6) 12 phenolic compounds were identified within this formulation, with narirutin (195.000 ± 0.670 µg/g), naringenin (160.000 ± 0.803 µg/g), hesperidin (74.000 ± 0.410 µg/g), chlorogenic acid (9.530 ± 0.510 µg/g), diosmine (4.830 ± 0.030 µg/g), and coumaric acid (2.540 ± 0.010 µg/g) being the ones in major quantities. Although in lower concentrations, ellagic and ferulic acids, rutin, hesperetin, and methyl gallate were also identified. The concentration of each phenolic compound can vary depending on the lemon variety. Nonetheless, the obtained concentrations are in accordance with the values described in the literature for lemon extracts [53,58]. Furthermore, several of the identified compounds, such as naringenin, hesperidin, ellagic acid, and coumaric acid, are commonly found in *Citrus* species and have been researched for their antioxidant, anti-inflammatory, antimicrobial, and antiviral activities [31,59–61]. Therefore, besides LN-F2 being capable of maintaining the integrity of polyphenolic compounds after formulation, it presented promising results regarding the identified and quantified compounds due to their potential bioactivities.

3.5.2. TEM Characterization

A TEM characterization was carried out to analyze the morphology of LN-F2. The micrograph (Figure 7) revealed that droplets were spherical and corroborated that the average droplet size was less than 200 nm, being within the nanosize range [47]. Therefore, these results validate that the developed formulation can be defined as a nanoemulsion. Moreover, the TEM micrograph results align with the aforementioned size determination values.

Table 3. Identification and quantification (µg/g) of the phenolic compounds present in the selected nanoemulgel formulation (LN-F2).

Phenolic Compound	Calibration Curve Equation	R^2	LOD (ng/mL)	LOQ (ng/mL)	Rt (min)	Concentration ± SD (µg/g)
Narirutin	y = −0.002538650 + 0.0256495x	0.9998	10.25	20.01	10.84	195.000 ± 0.670
Naringenin	y = −0.000761480 + 0.0310687x	0.9996	12.31	23.57	15.02	160.320 ± 0.803
Hesperidin	y = 0.000321812 + 0.00008838x	0.9997	15.41	30.19	11.10	74.000 ± 0.410
Chlorogenic acid	y = −0.000162510 + 0.0082195x	0.9993	21.84	30.84	8.13	9.530 ± 0.510
Diosmin	y = −0.040602700 + 0.1132100x	0.9997	15.10	30.62	11.03	4.830 ± 0.030
Coumaric acid	y = −0.000916311 + 0.0643206x	0.9995	15.03	30.58	9.75	2.540 ± 0.010
Rutin	y = −0.002097290 + 0.0613716x	0.9998	15.06	30.37	9.91	1.490 ± 0.009
Ellagic acid	y = −0.000728175 + 0.0480413x	0.9996	15.14	30.89	10.12	1.200 ± 0.002
Ferulic acid	y = −0.000938050 + 0.0189100x	0.9987	14.98	30.97	10.38	1.170 ± 0.006
Saponarin	y = 0.001245340 + 0.0568465x	0.9993	15.54	30.47	8.66	0.420 ± 0.004
Hesperetin	y = 0.000213181 + 0.0007883x	0.9996	15.41	30.12	15.56	0.210 ± 0.001
Methyl gallate	y = −0.001515260 + 0.0801915x	0.9998	15.78	40.21	7.65	0.120 ± 0.001

R^2, correlation coefficient; LOD, limit of detection; LOQ, limit of quantification; Rt, retention time.

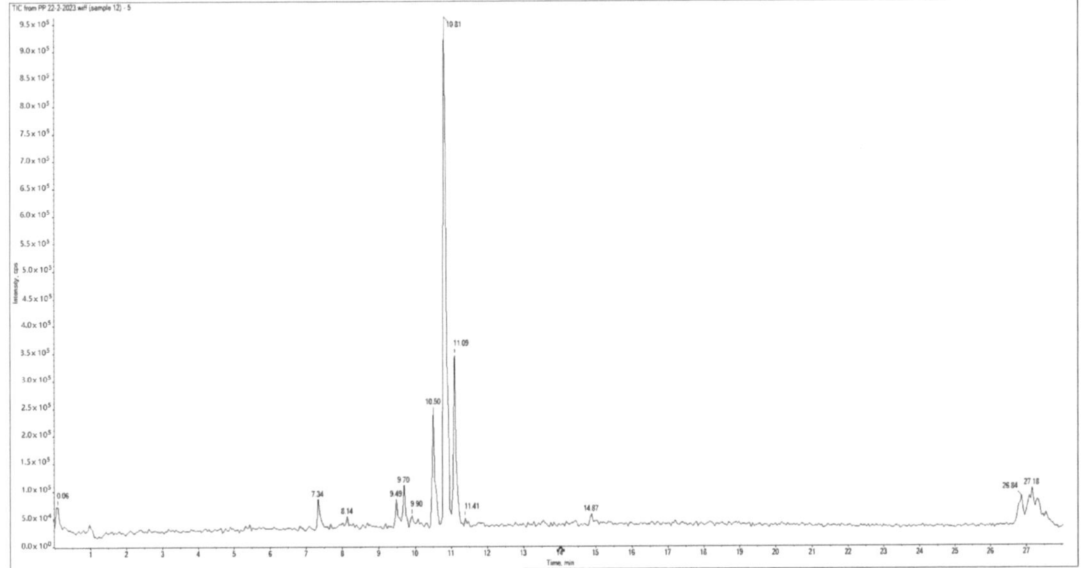

Figure 6. Chromatogram of the selected optimal nanoemulgel formulation (LN-F2).

3.5.3. Cytotoxicity Assessment on MA-104 Cells

To assess the potential cytotoxic effect of LE and LN-F2, MA-104 cells were exposed to several concentrations of each sample (ranging from 10 µg/100 mL to 1 mg /100 µL), and their morphology was continually evaluated by inverted light microscopy [23]. According to the results obtained through this test, no morphological alterations, such as loss of confluence, cell rounding and shrinking, cytoplasm granulation, and/or vacuolization, were observed in the tested cells for up to 72 h and at the concentrations of LE and LN-F2 that were applied. Furthermore, as mentioned by Diab et al. [59], who tested the effect of lemon extract (300, 200, and 500 µg/mL) on mouse splenocytes, no cytotoxic effects were reported, and even proliferative activity was observed in these cells after treatment.

Moreover, the results also suggest that the LN matrix did not have a cytotoxic effect on MA-104 under the tested conditions.

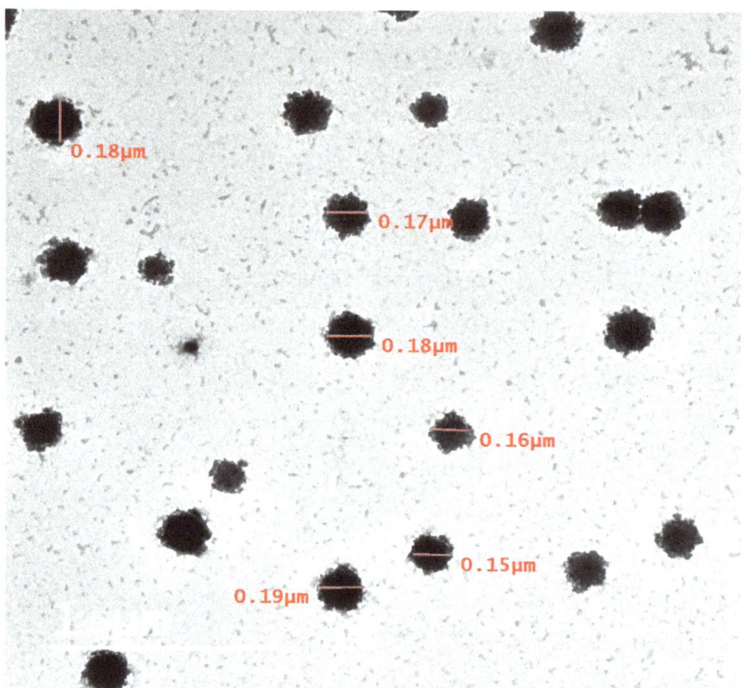

Figure 7. TEM micrograph of the selected optimal nanoemulgel formulation (LN-F2).

3.6. Bioactivity Analyses

3.6.1. LE and LN-F2 Antiviral Activity

After 30 min and 60 min of incubation, both LE and LN-F2 did not show any antiviral activity against rotavirus (SA-11). However, when tested against phiX 174 virus, after 30 min, both samples promote a final viral count reduction at concentrations of 10 mg/mL (Figure 8). Furthermore, LE at a concentration of 10 mg/mL showed a 22.58% virus reduction, while the LN-F2 containing 1 mg/mL or 10 mg/mL of LE showed a 3.23% and 24.19% reduction of phiX 174 infection, respectively. Therefore, these results suggest that, at the same LE concentration, the antiviral efficiency is similar for free LE and LE formulated into the nanoemulgel (LN-F2). Moreover, the results also suggest that the antiviral effectiveness of the LN-F2 is dose-dependent.

The results obtained for the phiX 174 bacteriophage are corroborated by previous results on this matter, since many studies have proved the activity of *Citrus* extracts against other types of viruses such as SARS-CoV-2 [62], HSV-1 [63], and Lassa virus [64]. These results might be related to the presence of bioactive compounds such as phenolic compounds (e.g., hesperidin or tangeritin) within the lemon extract, since they have been reported to have antiviral activity against different viruses [50,64]. One of the mechanisms for this bioactivity was described by Tang et al. [64], who stated that these types of extracts might be able to block the entry of the virus into the cell at the viral fusion step. Additionally, as already mentioned, it is important to consider that the extracts' bioactivities on certain pathogens might depend on the plant composition from where the extract is obtained. Furthermore, this composition varies according to several factors such as climatic and growing conditions. Nonetheless, to the best of our knowledge, this is the first report on the in vitro antiviral activity of a nanoemulgel formulation containing lemon peel extract against phiX 174.

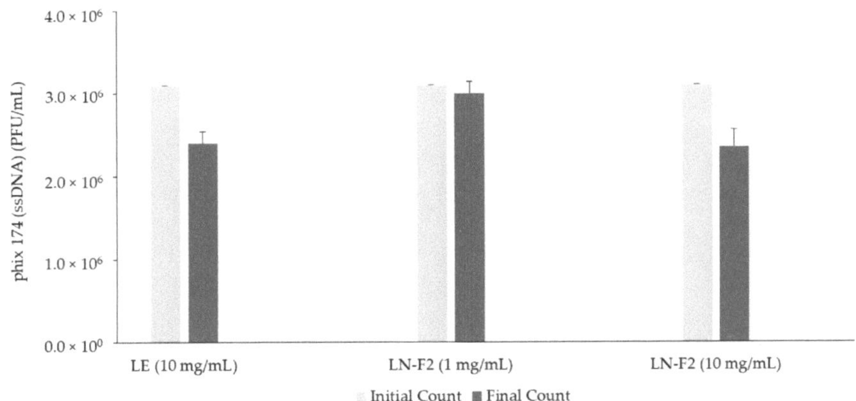

Figure 8. Viral counts of phiX 174 (ssDNA) before (initial) and after (final) exposure to 10 mg/mL of lemon extract (LN) and 1 mg/mL or 10 mg/mL of the selected optimal nanoemulgel formulation (LN-F2), for 30 min, at 37 °C. The results are presented in plaque-forming units (PFU)/mL and as mean ± SD.

3.6.2. LN-F2 In Vitro Antimicrobial Activity

To corroborate the previous antimicrobial results of LE and to assess if the extract retains this bioactivity after its formulation into the nanoemulgel matrix, the antimicrobial activity of LN-F2 was evaluated by well and disk diffusion assays. This test was carried out against two Gram-positive bacterial strains: *Staphylococcus aureus* and *Listeria monocytogenes*; three Gram-negative bacterial strains: *Escherichia coli*, *Salmonella enterica* serovar Typhimurium, *Pseudomonas aeruginosa*; and a yeast: *Candida albicans*. As shown in Figure 9, LN-F2 exhibited in vitro antimicrobial activity against all six pathogenic microorganisms by both well and disk diffusion methods, consistent with the MIC/MMC results previously obtained. Also, it was observed that the zones of inhibition (ZIs) achieved by using the well diffusion method were slightly larger than those obtained using the disk diffusion method. Moreover, it can be observed that, regarding both methods, the most affected microorganism was *E. coli*, with ZIs of 11 mm and 10 mm for the well and disk techniques, respectively. On the other hand, concerning only the well diffusion method, the effect against *Candida albicans* was the lowest, resulting in a ZI of 8 mm, while for the disk diffusion method the least affected microorganisms were *S.* Typhimurium, *P. aeruginosa*, *S. aureus*, and *C. albicans*, all of which presented a ZI of 8 mm. Overall, LN-F2 seems to exhibit similar effects against the tested microorganisms. Once again, this activity might be mainly related to the lemon peel extract components such as the major phenolic compounds found and natural organic acids such as ascorbic and citric acids, which have demonstrated great antimicrobial properties [65]. Therefore, several studies show that extracts obtained from the lemon peel exhibited antimicrobial properties against *Salmonella* Typhimurium, *Bacillus cereus*, *Listeria monocytogenes*, *E. coli*, and *Enterococcus faecalis* [66–68]. Moreover, xanthan gum has shown a synergistic effect and enhancement of the antioxidant and antimicrobial activity of the lemon peel extract against microbial pathogens. In addition, xanthan gum increases the thickness, stability, as well as moisture content of the gel due to the presence of –OH groups, which increase the formation of hydrogen bonds [65,69], aiding in overall extract stabilization and bioactive capacities. Although the antimicrobial capacity of lemon extracts has not been studied within a nanoemulgel formulation, a work carried out by Yabalak et al. [70] described that the MICs of *Citrus*-essential-oil-incorporating gelatine film solution against *E. coli* and *S. aureus* were about 10.1 and 9.1 mg/mL, respectively. Hence, as far as it is concerned, this is the first study where a nanoemulgel formulation based on *Citrus limon* peel extract showed antimicrobial activity against Gram-positive and Gram-negative bacteria and fungi.

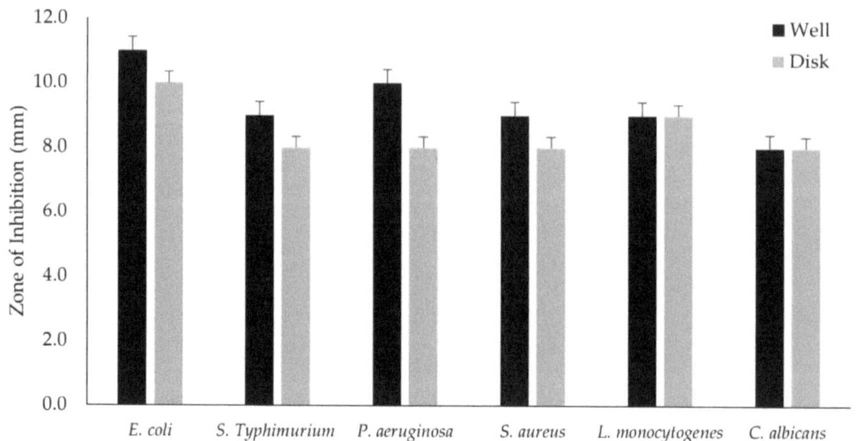

Figure 9. Zones of inhibition (mm) obtained by LN-F2 against *E. coli*, *S. enterica* serovar Typhimurium, *P. aeruginosa*, *S. aureus*, *L. monocytogenes*, and *C. albicans* by well and disk diffusion methods.

3.6.3. LN-F2 In Vivo Antimicrobial Activity

To assess the in vivo antimicrobial capacity of the formulated nanoemulgel, the hands of 20 volunteers were analyzed for the presence and quantity of bacteria and fungi, before and after the application of the LN-F2. Prior to the usage of LN-F2, the lowest total microbial count was 50 CFU/cm^2 for bacteria and 5 CFU/cm^2 for fungi, while the highest values for bacterial and fungal counts were 1500 CFU/cm^2 and 580 CFU/cm^2, respectively. Moreover, 3 out of 20 samples were free from any fungi before the application of the nanoemulgel (Table 4). After using LN-F2, the findings demonstrated that the formulated nanoemulgel has strong in vivo antimicrobial activity against the tested bacteria and fungus, promoting efficient removal of these microorganisms from the volunteers' hands (Figures 10 and 11). The average percentage of total bacterial counts removed (R %) from all samples was 95.11 ± 4.41%. Additionally, the average removal percentage of the total fungal counts was 97.01 ± 3.75. Therefore, it can be concluded that LN-F2, containing lemon peel extract, seems to have strong in vivo antimicrobial activity, which is in accordance with the previous analysis. Moreover, regarding the antifungal activity, these results are also corroborated by the findings reported in the literature that mention, for instance, the capacity of lemon extract to counteract *Alternaria* spp., *Curvularia* spp., *Fusarium* spp., *Trichophyton* spp., and *Geotrichum* spp. [71].

Table 4. Total bacterial and fungal counts in the microbial suspension samples collected from the volunteers' hands before and after the application of LN-F2.

Volunteer ID	Age	Gender	Total Bacterial Counts (CFU/cm^2)		R%	Total Fungal Counts (CFU/cm^2)		R%
			Before	After		Before	After	
1	39	M	1200	100	91.66	95	3	94.96
2	30	M	850	20	97.47	20	0	100.00
3	44	M	190	10	94.73	10	0	100.00
4	43	M	130	0	100.00	5	0	100.00
5	25	F	200	5	97.50	0	0	N/A
6	35	M	1100	120	89.09	100	9	91.00
7	40	M	500	30	94.00	150	0	100.00
8	41	M	90	0	100.00	12	0	100.00
9	39	F	420	30	92.85	190	4	97.89

Table 4. *Cont.*

Volunteer ID	Age	Gender	Total Bacterial Counts (CFU/cm^2)		R%	Total Fungal Counts (CFU/cm^2)		R%
			Before	After		Before	After	
10	26	M	1000	105	98.50	580	60	89.65
11	32	M	1500	250	83.33	120	10	91.66
12	38	F	750	45	94.00	35	0	100.00
13	48	M	60	0	100.00	0	0	N/A
14	67	M	250	10	96.00	18	0	100.00
15	65	M	690	55	92.02	90	5	94.44
16	44	M	500	40	92.00	150	10	93.33
17	30	M	990	30	96.96	80	3	96.25
18	39	M	250	20	92.00	30	0	100.00
19	43	M	120	0	100.00	0	0	N/A
20	30	M	50	0	100.00	20	0	100.00
Average ± SD	-	-	-	-	95.11 ± 4.41	-	-	97.01 ± 3.75

M, male; F, female; CFU, colony-forming unit; R%, removal percentage; N/A, not applicable.

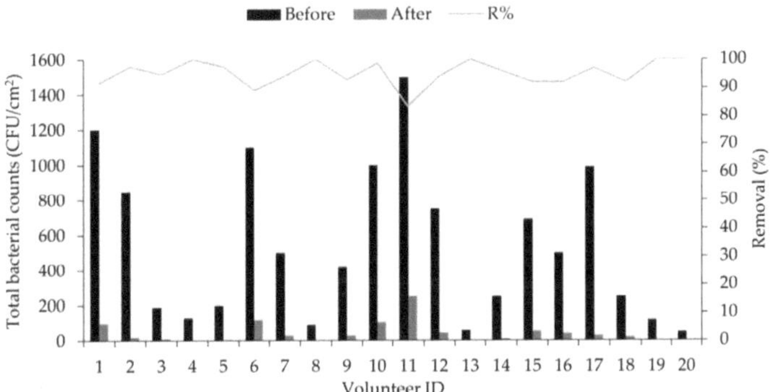

Figure 10. Total bacterial counts before and after the application of LN-F2 on volunteers' hands and the respective removal percentage (R%).

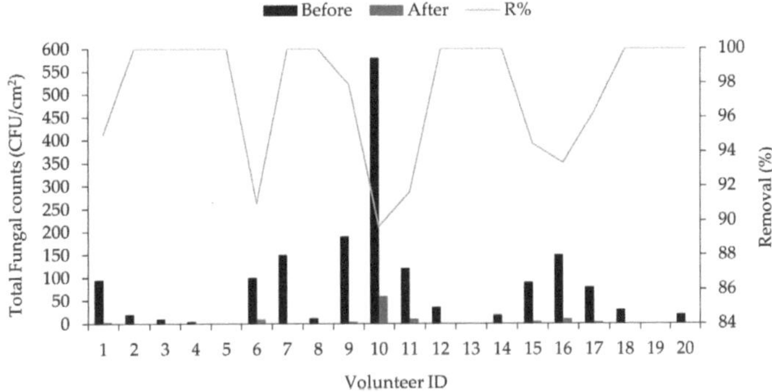

Figure 11. Total fungal counts before and after the application of LN-F2 on volunteers' hands and the respective removal percentage (R%).

Furthermore, following the completion of this assessment, there were no indications of irritation, dermal abrasion, trauma, or infection observed within or around the area of the formulation application. Additionally, the volunteers did not report any potential side effects after the application of the formulation.

4. Conclusions

The present study aimed to develop and characterize an alcohol-free lemon-peel-extract-based nanoemulgel for hand-sanitizing purposes. In conclusion, the lemon peel extract (LE) presented antimicrobial activity against various microorganisms, including *S. aureus*, *E. coli*, and *C. albicans*, with relatively low MIC values. Also, the correlation between MIC and MMC values further highlighted the extract's consistent antimicrobial activity across the different strains.

Moreover, three different nanoemulgels (LN-F1, LN-F2, and LN-F3) were developed using LE, with a focus on creating an optimal formulation for potential hand-sanitizing applications. The formulations exhibited satisfactory macroscopic characteristics, and the FTIR spectra indicated successful incorporation of LE without significant alterations to its chemical structure. In addition, the nanoemulsion's particle size, polydispersity index, zeta potential, and pH values met the criteria for stability. However, LN-F2 emerged as the optimal formulation based on stability under extreme conditions, rheological behavior, and effective in vitro release of phenolic compounds. Furthermore, the characterization of LN-F2 revealed the presence of bioactive compounds, the major ones being naringin, naringenin, hesperidin, and chlorogenic acid. LN-F2 also exhibited promising results in terms of safety, with no cytotoxic effects observed on MA-104 cells at the tested concentrations. In antiviral assays, both LE and LN-F2 did not affect rotavirus SA-11 but were able to induce the reduction of phiX 174 virus. Additionally, LN-F2 demonstrated in vitro antimicrobial activity against various pathogenic microorganisms, aligning with the LE MIC/MMC results. Also, the in vivo antimicrobial efficacy of LN-F2 was validated through the assessment of its efficacy against microorganisms present on the volunteers' hands, revealing substantial reductions in bacterial and fungal counts after application. Therefore, the strong antimicrobial activity of LN-F2, both in vitro and in vivo, suggests its potential as an effective hand-sanitizing formulation.

Overall, this work provides comprehensive insights into the antimicrobial and formulation aspects of lemon peel extract, highlighting its enhanced potential for applications in the fields of antimicrobial and antiviral formulations, particularly when incorporated into advanced delivery systems such as nanoemulgel matrices.

Author Contributions: Conceptualization, F.M.I., E.S.S. and E.S.E.H.; methodology, F.M.I., R.S.M., E.S.S., M.A.E.-L. and S.A.-E.; validation, F.M.I., R.S.M., E.S.S., M.A.E.-L. and S.A.-E.; formal analysis, F.M.I.; investigation, F.M.I., R.S.M., E.S.S., M.A.E.-L., S.I.S. and S.A.-E.; resources, F.M.I., R.S.M., E.S.S., M.A.E.-L. and S.A.-E.; data curation, F.M.I., R.S.M., E.S.S., M.A.E.-L., S.A.-E. and C.V.R.; writing—original draft preparation, F.M.I., R.S.M., E.S.S., M.A.E.-L. and S.A.-E.; writing—review and editing, F.M.I., C.V.R. and M.P.; visualization, F.M.I.; supervision, E.S.E.H.; project administration, E.S.E.H.; funding acquisition, E.S.E.H. All authors have read and agreed to the published version of the manuscript.

Funding: The development of the present work was supported by the MEDISMART project Mediterranean Citrus innovative soft processing solutions for S.M.A.R.T (Sustainable, Mediterranean, Agronomically evolved nutritionally enriched Traditional) products (reference number: 2019-SECTION2), and through the Science, Technology and Innovation Funding Authority (STDF), Egypt, and Portuguese national funds from Fundação para a Ciência e Tecnologia (FCT) (reference: DOI 10.54499/PRIMA/0014/2019)

Institutional Review Board Statement: The study was conducted in accordance with the Declaration of Helsinki and approved by the National Research Center's Ethics and Animal Care Committee (Ethic No. 20236).

Informed Consent Statement: Informed consent was obtained from all subjects involved in the study.

Data Availability Statement: The data generated and/or analyzed during the current study are available from the corresponding author on reasonable request.

Acknowledgments: The authors acknowledge the MEDISMART project (reference number: 2019-SECTION2), Science, Technology and Innovation Funding Authority (STDF), Egypt, and Fundação para a Ciência e Tecnologia (FCT) (reference: DOI 10.54499/PRIMA/0014/2019) for the financial assistance provided. Furthermore, the authors extend their appreciation for the scientific collaboration with CBQF within the framework of the UIDB/50016/2020 project.

Conflicts of Interest: The authors declare no conflicts of interest.

References

1. Golin, A.P.; Choi, D.; Ghahary, A. Hand sanitizers: A review of ingredients, mechanisms of action, modes of delivery, and efficacy against coronaviruses. *Am. J. Infect. Control* **2020**, *48*, 1062–1067. [CrossRef] [PubMed]
2. Saha, T.; Khadka, P.; Das, S.C. Alcohol-based hand sanitizer—composition, proper use and precautions. *Germs* **2021**, *11*, 408–417. [CrossRef] [PubMed]
3. Hamad Vuai, S.A.; Sahini, M.G.; Sule, K.S.; Ripanda, A.S.; Mwanga, H.M. A comparative in-vitro study on antimicrobial efficacy of on-market alcohol-based hand washing sanitizers towards combating microbes and its application in combating COVID-19 global outbreak. *Heliyon* **2022**, *8*, e11689. [CrossRef] [PubMed]
4. Ibrahim, F.M.; Mohammed, R.S.; Abdelsalam, E.; Ashour, W.E.; Magalhães, D.; Pintado, M.; El Habbasha, E.S. Egyptian Citrus Essential Oils Recovered from Lemon, Orange, and Mandarin Peels: Phytochemical and Biological Value. *Horticulturae* **2024**, *10*, 180. [CrossRef]
5. Vilas-Boas, A.A.; Magalhães, D.; Campos, D.A.; Porretta, S.; Dellapina, G.; Poli, G.; Istanbullu, Y.; Demir, S.; San Martín, Á.M.; García-Gómez, P.; et al. Innovative Processing Technologies to Develop a New Segment of Functional Citrus-Based Beverages: Current and Future Trends. *Foods* **2022**, *11*, 3859. [CrossRef] [PubMed]
6. Vilas Boas, A.; Gómez-García, R.; Campos, D.; Correia, M.; Pintado, M. Integrated Biorefinery Strategy for Orange Juice By-products Valorization: A Sustainable Protocol to Obtain Bioactive Compounds. In *Food Waste Conversion*; Methods and Protocols in Food Science; Humana: New York, NY, USA, 2023; pp. 113–124.
7. Shri Balakrishna, A.; Saradindu, G.; Giriraj, Y.; Kavita, S.; Sirsendu, G.; Sushil, J. Formulation, Evaluation and Antibacterial Efficiency of water-based herbal Hand Sanitizer Gel. *bioRxiv* **2018**, 373928. [CrossRef]
8. Rodrigues, C.V.; Pintado, M. Hesperidin from Orange Peel as a Promising Skincare Bioactive: An Overview. *Int. J. Mol. Sci.* **2024**, *25*, 1890. [CrossRef] [PubMed]
9. Liu, Y.; Benohoud, M.; Galani Yamdeu, J.H.; Gong, Y.Y.; Orfila, C. Green extraction of polyphenols from citrus peel by-products and their antifungal activity against Aspergillus flavus. *Food Chem. X* **2021**, *12*, 100144. [CrossRef]
10. Choudhury, H.; Gorain, B.; Pandey, M.; Chatterjee, L.A.; Sengupta, P.; Das, A.; Molugulu, N.; Kesharwani, P. Recent Update on Nanoemulgel as Topical Drug Delivery System. *J. Pharm. Sci.* **2017**, *106*, 1736–1751. [CrossRef]
11. Algahtani, M.S.; Ahmad, M.Z.; Ahmad, J. Nanoemulgel for Improved Topical Delivery of Retinyl Palmitate: Formulation Design and Stability Evaluation. *Nanomaterials* **2020**, *10*, 848. [CrossRef]
12. Bashir, M.; Ahmad, J.; Asif, M.; Khan, S.U.; Irfan, M.; Ibrahim, A.Y.; Asghar, S.; Khan, I.U.; Iqbal, M.S.; Haseeb, A.; et al. Nanoemulgel, an Innovative Carrier for Diflunisal Topical Delivery with Profound Anti-Inflammatory Effect: In vitro and in vivo Evaluation. *Int. J. Nanomed.* **2021**, *16*, 1457–1472. [CrossRef] [PubMed]
13. Morsy, M.A.; Abdel-Latif, R.G.; Nair, A.B.; Venugopala, K.N.; Ahmed, A.F.; Elsewedy, H.S.; Shehata, T.M. Preparation and evaluation of atorvastatin-loaded nanoemulgel on wound-healing efficacy. *Pharmaceutics* **2019**, *11*, 609. [CrossRef] [PubMed]
14. Asfour, M.H.; Mohsen, A.M. Formulation and evaluation of pH-sensitive rutin nanospheres against colon carcinoma using HCT-116 cell line. *J. Adv. Res.* **2018**, *9*, 17–26. [CrossRef]
15. Elhabak, M.; Ibrahim, S.; Abouelatta, S.M. Topical delivery of l-ascorbic acid spanlastics for stability enhancement and treatment of UVB induced damaged skin. *Drug Deliv.* **2021**, *28*, 445–453. [CrossRef]
16. Chellapa, P.; Eid, A.M.; Elmarzugi, N. Preparation and characterization of virgin coconut oil nanoemulgel. *J. Chem. Pharm. Res.* **2015**, *7*, 787–793.
17. Srivastava, M.; Kohli, K.; Ali, M. Formulation development of novel in situ nanoemulgel (NEG) of ketoprofen for the treatment of periodontitis. *Drug Deliv.* **2016**, *23*, 154–166. [CrossRef]
18. Sohail, M.; Naveed, A.; Abdul, R.; Khan, H.M.S.; Khan, H. An approach to enhanced stability: Formulation and characterization of Solanum lycopersicum derived lycopene based topical emulgel. *Saudi Pharm. J.* **2018**, *26*, 1170–1177. [CrossRef] [PubMed]
19. Mostafa, D.M.; Abd El-Alim, S.H.; Asfour, M.H.; Al-Okbi, S.Y.; Mohamed, D.A.; Hamed, T.E.-S.; Awad, G. Transdermal fennel essential oil nanoemulsions with promising hepatic dysfunction healing effect: In vitro and in vivo study. *Pharm. Dev. Technol.* **2019**, *24*, 729–738. [CrossRef]
20. Abd El-Alim, S.H.; Salama, A.; Darwish, A.B. Provesicular elastic carriers of Simvastatin for enhanced wound healing activity: An in-vitro/in-vivo study. *Int. J. Pharm.* **2020**, *585*, 119470. [CrossRef]
21. Zhang, X.; Liu, D.; Jin, T.Z.; Chen, W.; He, Q.; Zou, Z.; Zhao, H.; Ye, X.; Guo, M. Preparation and characterization of gellan gum-chitosan polyelectrolyte complex films with the incorporation of thyme essential oil nanoemulsion. *Food Hydrocoll.* **2021**, *114*, 106570. [CrossRef]
22. Ammar, N.M.; Hassan, H.A.; Mohammed, M.A.; Serag, A.; Abd El-Alim, S.H.; Elmotasem, H.; El Raey, M.; El Gendy, A.N.; Sobeh, M.; Abdel-Hamid, A.-H.Z. Metabolomic profiling to reveal the therapeutic potency of Posidonia oceanica nanoparticles in diabetic rats. *RSC Adv.* **2021**, *11*, 8398–8410. [CrossRef] [PubMed]

23. Wu, S.; Zeng, L.; Wang, C.; Yang, Y.; Zhou, W.; Li, F.; Tan, Z. Assessment of the cytotoxicity of ionic liquids on Spodoptera frugiperda 9 (Sf-9) cell lines via in vitro assays. *J. Hazard. Mater.* **2018**, *348*, 1–9. [CrossRef] [PubMed]

24. England, P.H. *Detection and Enumeration of Bacteria in Swabs and Other Environmental Samples*, 4th ed.; Volume National Infection Service, Food, Water and Environmental Microbiology Standard Method FNES4 (E1): London, UK, 2017; p. 22.

25. Aa, L.; Is, H.; Jh, D.; Jfr, L. Bacterial contamination of the hands of food handlers as indicator of hand washing efficacy in some convenient food industries. *Pak. J. Med. Sci.* **2014**, *30*, 755–758. [PubMed]

26. Rice, E.W.; Bridgewater, L.; Association, A.P.H. *Standard Methods for the Examination of Water and Wastewater*; American Public Health Association: Washington, DC, USA, 2012; Volume 10.

27. Elwakeel, K.Z.; El-Liethy, M.A.; Ahmed, M.S.; Ezzat, S.M.; Kamel, M.M. Facile synthesis of magnetic disinfectant immobilized with silver ions for water pathogenic microorganism's deactivation. *Environ. Sci. Pollut. Res. Int.* **2018**, *25*, 22797–22809. [CrossRef] [PubMed]

28. Magréault, S.; Jauréguy, F.; Carbonnelle, E.; Zahar, J.R. When and How to Use MIC in Clinical Practice? *Antibiotics* **2022**, *11*, 1748. [CrossRef] [PubMed]

29. Parvekar, P.; Palaskar, J.; Metgud, S.; Maria, R.; Dutta, S. The minimum inhibitory concentration (MIC) and minimum bactericidal concentration (MBC) of silver nanoparticles against *Staphylococcus aureus. Biomater. Investig. Dent.* **2020**, *7*, 105–109. [CrossRef] [PubMed]

30. Otang, W.M.; Afolayan, A.J. Antimicrobial and antioxidant efficacy of *Citrus limon* L. peel extracts used for skin diseases by Xhosa tribe of Amathole District, Eastern Cape, South Africa. *S. Afr. J. Bot.* **2016**, *102*, 46–49. [CrossRef]

31. Henderson, A.H.; Fachrial, E.; Lister, I.N.E. Antimicrobial Activity of Lemon (*Citrus limon*) Peel Extract Against Escherichia coli. *Am. Sci. Res. J. Eng. Technol. Sci.* **2018**, *39*, 268–273.

32. John, S.; Monica, S.; Priyadarshini, S.; Sivaraj, C.; Arumugam, P. Antioxidant and antimicrobial activity of lemon peel. *Int. J. Pharm. Sci. Rev. Res.* **2017**, *46*, 115–118.

33. Saleem, M.; Durani, A.I.; Asari, A.; Ahmed, M.; Ahmad, M.; Yousaf, N.; Muddassar, M. Investigation of antioxidant and antibacterial effects of citrus fruits peels extracts using different extracting agents: Phytochemical analysis with in silico studies. *Heliyon* **2023**, *9*, e15433. [CrossRef]

34. Shanmugam, R.; Lakki Reddy Venkata, B.R.; Geetha, R.V. Broad spectrum antibacterial silver nanoparticle green synthesis: Characterization, and mechanism of action. In *Green Synthesis, Characterization and Applications of Nanoparticles*; Elsevier: Amsterdam, The Netherlands, 2019; pp. 429–444.

35. Karaman, R.; Jubeh, B.; Breijyeh, Z. Resistance of Gram-Positive Bacteria to Current Antibacterial Agents and Overcoming Approaches. *Molecules* **2020**, *25*, 2888. [CrossRef] [PubMed]

36. Hamida, R.S.; Ali, M.A.; Goda, D.A.; Khalil, M.I.; Al-Zaban, M.I. Novel Biogenic Silver Nanoparticle-Induced Reactive Oxygen Species Inhibit the Biofilm Formation and Virulence Activities of Methicillin-Resistant Staphylococcus aureus (MRSA) Strain. *Front. Bioeng. Biotechnol.* **2020**, *8*, 433. [CrossRef] [PubMed]

37. Qiu, C.; Zhao, M.; McClements, D. Improving the stability of wheat protein-stabilized emulsions: Effect of pectin and xanthan gum addition. *Food Hydrocoll.* **2015**, *43*, 377–387. [CrossRef]

38. Mulia, K.; Ramadhan, R.M.A.; Krisanti, E.A.J.M.W.C. Formulation and characterization of nanoemulgel mangosteen extract in virgin coconut oil for topical formulation. *MATEC Web. Conf.* **2018**, *156*, 01013. [CrossRef]

39. Matman, N.; Min Oo, Y.; Amnuaikit, T.; Somnuk, K. Continuous production of nanoemulsion for skincare product using a 3D-printed rotor-stator hydrodynamic cavitation reactor. *Ultrason. Sonochemistry* **2022**, *83*, 105926. [CrossRef] [PubMed]

40. Semalty, A.; Semalty, M.; Singh, D.; Rawat, M. Preparation and characterization of phospholipid complexes of naringenin for effective drug delivery. *J. Incl. Phenom. Macrocycl. Chem.* **2010**, *67*, 253–260. [CrossRef]

41. Thombare, N.; Mahto, A.; Singh, D.; Roy Chowdhury, A.; Ansari, M. Comparative FTIR Characterization of Various Natural Gums: A Criterion for Their Identification. *J. Polym. Environ.* **2023**, *31*, 3372–3380. [CrossRef]

42. Rehman, M.; Rasul, A.; Khan, M.; Hanif, M.; Naaem Aamir, M.; Khan, H.m.; Hameed, M.; Akram, M. Development of niosomal formulations loaded with cyclosporine A and evaluation of its compatibility. *Trop. J. Pharm. Res.* **2018**, *17*, 1457–1464. [CrossRef]

43. Kondiah, P.P.D.; Rants'o, T.A.; Mdanda, S.; Mohlomi, L.M.; Choonara, Y.E. A Poly (Caprolactone)-Cellulose Nanocomposite Hydrogel for Transdermal Delivery of Hydrocortisone in Treating Psoriasis Vulgaris. *Polymers* **2022**, *14*, 2633. [CrossRef]

44. Kanaze, F.; Kokkalou, E.; Niopas, I.; Georgarakis, M.; Stergiou, A.; Bikiaris, D. Thermalanalysis study of flavonoid solid dispersions having enhanced solubility. *J. Therm. Anal. Calorim.* **2006**, *83*, 283–290. [CrossRef]

45. Mushtaq, A.; Mohd Wani, S.; Malik, A.R.; Gull, A.; Ramniwas, S.; Ahmad Nayik, G.; Ercisli, S.; Alina Marc, R.; Ullah, R.; Bari, A. Recent insights into Nanoemulsions: Their preparation, properties and applications. *Food Chem. X* **2023**, *18*, 100684. [CrossRef] [PubMed]

46. Verdú-Soriano, J.; Casado-Díaz, A.; de Cristino-Espinar, M.; Luna-Morales, S.; Dios-Guerra, C.; Moreno-Moreno, P.; Dorado, G.; Quesada-Gómez, J.M.; Rodríguez-Mañas, L.; Lázaro-Martínez, J.L. Hard-to-Heal Wound Healing: Superiority of Hydrogel EHO-85 (Containing Olea europaea Leaf Extract) vs. a Standard Hydrogel. A Randomized Controlled Trial. *Gels* **2023**, *9*, 962. [CrossRef] [PubMed]

47. Eid, A.; Issa, L.; Al-kharouf, O.; Jaber, R.; Hreash, F. Development of Coriandrum sativum Oil Nanoemulgel and Evaluation of Its Antimicrobial and Anticancer Activity. *BioMed Res. Int.* **2021**, *2021*, 5247816. [CrossRef] [PubMed]

48. Chookiat, S.; Theansungnoen, T.; Kiattisin, K.; Intharuksa, A. Nanoemulsions Containing Mucuna pruriens (L.) DC. Seed Extract for Cosmetic Applications. *Cosmetics* **2024**, *11*, 29. [CrossRef]
49. Jusril, N.A.; Abu Bakar, S.I.; Khalil, K.A.; Md Saad, W.M.; Wen, N.K.; Adenan, M.I. Development and Optimization of Nanoemulsion from Ethanolic Extract of Centella asiatica (NanoSECA) Using D-Optimal Mixture Design to Improve Blood-Brain Barrier Permeability. *Evid.-Based Complement. Altern. Med.* **2022**, *2022*, 3483511. [CrossRef] [PubMed]
50. Ding, Z.; Jiang, Y.; Liu, X. Chapter 12—Nanoemulsions-Based Drug Delivery for Brain Tumors. In *Nanotechnology-Based Targeted Drug Delivery Systems for Brain Tumors*; Kesharwani, P., Gupta, U., Eds.; Academic Press: Cambridge, MA, USA, 2018; pp. 327–358.
51. Honary, S.; Zahir, F. Effect of Zeta Potential on the Properties of Nano-Drug Delivery Systems—A Review (Part 1). *Trop. J. Pharm. Res.* **2013**, *12*, 255–264. [CrossRef]
52. Sungpud, C.; Panpipat, W.; Chaijan, M.; Sae Yoon, A. Techno-biofunctionality of mangostin extract-loaded virgin coconut oil nanoemulsion and nanoemulgel. *PLoS ONE* **2020**, *15*, e0227979. [CrossRef] [PubMed]
53. Magalhães, D.; Vilas-Boas, A.A.; Teixeira, P.; Pintado, M. Functional Ingredients and Additives from Lemon by-Products and Their Applications in Food Preservation: A Review. *Foods* **2023**, *12*, 1095. [CrossRef]
54. Gadhave, D.; Tupe, S.; Tagalpallewar, A.; Gorain, B.; Choudhury, H.; Kokare, C. Nose-to-brain delivery of amisulpride-loaded lipid-based poloxamer-gellan gum nanoemulgel: In vitro and in vivo pharmacological studies. *Int. J. Pharm.* **2021**, *607*, 121050. [CrossRef] [PubMed]
55. Soliman, W.E.; Shehata, T.M.; Mohamed, M.E.; Younis, N.S.; Elsewedy, H.S. Enhancement of Curcumin Anti-Inflammatory Effect via Formulation into Myrrh Oil-Based Nanoemulgel. *Polymers* **2021**, *13*, 577. [CrossRef]
56. Sultan, M.H.; Javed, S.; Madkhali, O.A.; Alam, M.I.; Almoshari, Y.; Bakkari, M.A.; Sivadasan, D.; Salawi, A.; Jabeen, A.; Ahsan, W. Development and Optimization of Methylcellulose-Based Nanoemulgel Loaded with Nigella sativa Oil for Oral Health Management: Quadratic Model Approach. *Molecules* **2022**, *27*, 1796. [CrossRef] [PubMed]
57. Eid, A.M.; El-Enshasy, H.A.; Aziz, R.; Elmarzugi, N.A. Preparation, characterization and anti-inflammatory activity of Swietenia macrophylla nanoemulgel. *J. Nanomed. Nanotechnol.* **2014**, *5*, 1–10. [CrossRef]
58. Xi, W.; Lu, J.; Qun, J.; Jiao, B. Characterization of phenolic profile and antioxidant capacity of different fruit part from lemon (*Citrus limon* Burm.) cultivars. *J. Food Sci. Technol.* **2017**, *54*, 1108–1118. [CrossRef] [PubMed]
59. Diab, K.A. In Vitro Studies on Phytochemical Content, Antioxidant, Anticancer, Immunomodulatory, and Antigenotoxic Activities of Lemon, Grapefruit, and Mandarin Citrus Peels. *Asian Pac. J. Cancer Prev.* **2016**, *17*, 3559–3567. [PubMed]
60. Imran, M.; Basharat, D.; Khalid, S.; Aslam, M.; Syed, F.; Jabeen, S.; Kamran, H.; Muhammad, Z.; Shahid, M.Z.; Tufail, T.; et al. Citrus Peel Polyphenols: Recent Updates and Perspectives. *Int. J. Biosci.* **2020**, *16*, 53–70.
61. Shehata, M.G.; Awad, T.S.; Asker, D.; El Sohaimy, S.A.; Abd El-Aziz, N.M.; Youssef, M.M. Antioxidant and antimicrobial activities and UPLC-ESI-MS/MS polyphenolic profile of sweet orange peel extracts. *Curr. Res. Food Sci.* **2021**, *4*, 326–335. [CrossRef] [PubMed]
62. Saaty, A.H. Grapefruit Seed Extracts' Antibacterial and Antiviral Activity: Anti-Severe Acute Respiratory Syndrome Coronavirus 2 Impact. *Arch. Pharm. Pract.* **2022**, *13*, 69. [CrossRef]
63. Mejri, H.; Aidi Wannes, W.; Mahjoub, F.H.; Hammami, M.; Dussault, C.; Legault, J.; Saidani-Tounsi, M. Potential bio-functional properties of *Citrus aurantium* L. leaf: Chemical composition, antiviral activity on herpes simplex virus type-1, antiproliferative effects on human lung and colon cancer cells and oxidative protection. *Int. J. Environ. Health Res.* **2024**, *34*, 1113–1123. [CrossRef]
64. Tang, K.; He, S.; Zhang, X.; Guo, J.; Chen, Q.; Yan, F.; Banadyga, L.; Zhu, W.; Qiu, X.; Guo, Y. Tangeretin, an extract from Citrus peels, blocks cellular entry of arenaviruses that cause viral hemorrhagic fever. *Antivir. Res.* **2018**, *160*, 87–93. [CrossRef]
65. Meydanju, N.; Pirsa, S.; Farzi, J. Biodegradable film based on lemon peel powder containing xanthan gum and TiO$_2$–Ag nanoparticles: Investigation of physicochemical and antibacterial properties. *Polym. Test.* **2022**, *106*, 107445. [CrossRef]
66. Miyake, Y.; Hiramitsu, M. Isolation and extraction of antimicrobial substances against oral bacteria from lemon peel. *J. Food Sci. Technol.* **2011**, *48*, 635–639. [CrossRef] [PubMed]
67. Albak, F.; TekİN, A.R. Development of Functional Chocolate with Spices and Lemon Peel Powder by using Response Surface Method: Development of Functional Chocolate. *Akad. Gıda* **2014**, *12*, 19–25.
68. Çilingir, S.; Goksu, A.; Sabanci, S. Production of Pectin from Lemon Peel Powder Using Ohmic Heating-Assisted Extraction Process. *Food Bioprocess Technol.* **2021**, *14*, 1349–1360. [CrossRef]
69. Šafranko, S.; Šubarić, D.; Jerković, I.; Jokić, S. Citrus By-Products as a Valuable Source of Biologically Active Compounds with Promising Pharmaceutical, Biological and Biomedical Potential. *Pharmaceuticals* **2023**, *16*, 1081. [CrossRef]
70. Yabalak, E.; Erdoğan Eliuz, E.A.; Nazlı, M.D. Evaluation of Citrus reticulata essential oil: Chemical composition and antibacterial effectiveness incorporated gelatin on *E. coli* and *S. aureus*. *Int. J. Environ. Health Res.* **2022**, *32*, 1261–1270. [CrossRef]
71. Abirami, S.; Edwin Raj, B.; Soundarya, T.; Kannan, M.; Sugapriya, D.; Al-Dayan, N.; Ahmed Mohammed, A. Exploring antifungal activities of acetone extract of selected Indian medicinal plants against human dermal fungal pathogens. *Saudi J. Biol. Sci.* **2021**, *28*, 2180–2187. [CrossRef]

Article

Natural Extracts and Essential Oils as Ingredients in Cosmetics: Search for Potential Phytomarkers and Allergen Survey

Laura Rubio [1],*, Andrea Pita [1], Carmen Garcia-Jares [1,2] and Marta Lores [1]

[1] LIDSA, Department of Analytical Chemistry, Nutrition and Food Science, Universidade de Santiago de Compostela, E-15782 Santiago de Compostela, Spain; andrea.pita.diaz@rai.usc.es (A.P.); carmen.garcia.jares@usc.es (C.G.-J.); marta.lores@usc.es (M.L.)

[2] CRETUS, Department of Analytical Chemistry, Nutrition and Food Science, Universidade de Santiago de Compostela, E-15782 Santiago de Compostela, Spain

* Correspondence: laura.rubio.lareu@usc.es; Tel.: +34-881814379

Abstract: The increasing use of natural ingredients such as essential oils (EOs) and natural extracts (NEs) in cosmetics is an analytical and legislative challenge due to their complex composition, which includes recognized allergenic compounds. In this work, 17 EOs and NEs have been characterized by gas chromatography coupled to mass spectrometry (GC-MS) of dilutions of the original samples. Additionally, solid phase microextraction (SPME) was applied for the analysis of volatile components. The results obtained allowed the identification of more than 90 compounds, including 20 allergens, in the analyzed samples and the study of potential phytomarkers of the addition of EOs and ENs in cosmetics.

Keywords: natural extract; essential oil; cosmetics; preservatives; allergens; phytomarker; solid phase microextraction; gas chromatography; mass spectrometry

Citation: Rubio, L.; Pita, A.; Garcia-Jares, C.; Lores, M. Natural Extracts and Essential Oils as Ingredients in Cosmetics: Search for Potential Phytomarkers and Allergen Survey. *Cosmetics* **2024**, *11*, 84. https://doi.org/10.3390/cosmetics11030084

Academic Editor: Elisabetta Esposito

Received: 19 April 2024
Accepted: 13 May 2024
Published: 24 May 2024

1. Introduction

The widespread use of cosmetic and personal care products has raised some social concerns about the harmful effects that some ingredients used in the formulations may have on the consumer's health. Current trends are focused both on natural products as "greener" or "safer" than conventional ones [1] and also on the avoidance of certain chemical compounds in products, commonly called "free-from" cosmetics. However, it is true that all cosmetic formulations are complex mixtures of multiple components with very different origins, chemical natures, and functions.

Most cosmetic products can be easily degraded by microorganisms, and microbial contamination represents a major health risk for consumers [2]. Preservatives play a crucial antimicrobial role since their addition is mainly aimed at inhibiting this deterioration due to microbial growth, thus ensuring the stability of cosmetic formulations over time. Different preservation strategies are available to ensure the microbiological safety of cosmetic products [3]. Over the last few decades, the most common way was to use synthetic preservatives. According to the predominant functional group in their molecular structure, they can be classified into alcohols and derivatives, halogenated, organic acids, esters, and salts, among others [4].

Although the benefits and positive protective effects of preservatives are undeniable, since the main modes of exposure are inhalation or absorption through the skin, possible negative health effects have also been described, as excessive exposure can lead to irritation, contact allergy, or skin dermatitis [5]. For this reason, the safety of these chemical ingredients has been questioned, which leads to further restrictions in the regulations of application. In the EU framework, there is a limited number of permitted preservatives listed in Annex V of Regulation 1223 [6], where limitations, requirements, label warnings, and maximum permitted concentrations in ready-to-use products are established. The positive list of preservatives [7] includes more than 50 different substances or chemical families, and it is continuously undergoing legislative changes.

The negative public perception of traditional preservatives is prompting the cosmetic industry to search for alternative approaches to the preservation of cosmetic products, developing self-preserving cosmetics and even preservative-free products. In some of these formulations, traditional synthetic preservatives are replaced by antimicrobial substances, mainly of natural origin, which are generally considered multifunctional agents. Multifunctional additives are molecules providing more than one beneficial effect to the formulation or to the skin, e.g., glycols, glycerol ethers, essential oils, plant-based extracts, and fragrance ingredients. In this field, the demand for products containing natural extracts and mixtures of natural ingredients that are marketed with a preservative function is growing. So, by carefully selecting these ingredients, it is possible to reduce or eliminate the use of traditional preservatives. However, these multifunctional antimicrobial ingredients are not regulated for this function in Annex V. In this sense, a great number of natural extracts (NEs) and essential oils (EOs) are added to cosmetics with other traditional functions, e.g., as fragrance ingredients. In addition, due to their recognized antioxidant and antimicrobial properties [8–10], they can act in formulations in combination with chemical preservatives or alone as natural preservatives in preservative-free format cosmetics. Therefore, the use of these nature-based ingredients and their several functions in the production of cosmetics and related personal care products provides several advantages, such as improving the dermo-cosmetic and preservative properties, as well as the marketing image of the final product in view of current consumer demands. While synthetic preservatives are more affordable, the high prices of many NEs and EOs and their increasing demand make adulteration a frequent practice [11,12], meaning that there are other natural compounds and even other EOs that are cheaper and easier to acquire. Therefore, their authentication is crucially important for both consumers and companies.

Both NEs and EOs are complex mixtures of chemical substances whose major components are allergenic fragrances, and therefore, require analytical control. For many years, the Scientific Committee for Consumer Safety (SCCS) [13] has cataloged a total of 82 substances, including natural extracts and oils, as established human contact allergens. While only 26 allergenic fragrances in cosmetics were regulated by the EU [6], the new regulation [14] has extended the list of 26 regulated ingredients by a further 56 new allergens, making it mandatory to label a total of 82 substances identified as allergens in the coming years. The list comprises 28 natural extracts (NE_{28}) and 54 individual chemical (IC_{54}) compounds, making their analysis difficult in complex matrices such as cosmetics.

For now, cosmetics manufacturers are obliged to label only the presence of the 26 allergens in the finished product when their concentration is higher than 0.001% for leave-on cosmetics or higher than 0.01% in rinse-off products [6]. However, they often only consider for this calculation the allergens contributed by fragrance substances, as they have traditionally been the main and sometimes the only source of allergens in cosmetics. With the new trend towards the use of these multifunctional antimicrobial ingredients, this labeling will probably have to be revised upwards.

Over the last few years, a considerable number of articles have significantly contributed to improving the challenge of the analysis of traditional preservatives in cosmetics. New trends in the preparation and extraction procedures for cosmetic ingredients in general [15] have been sought in recent years, with inclinations towards simple, sustainable, and environmentally friendly methodologies. For preservatives, this includes procedures based on solid phase extraction (SPE), matrix solid phase dispersion (MSPD) [16], and additionally advanced microextraction techniques such as solid phase microextraction (SPME) and liquid–liquid microextraction (LLME), among others, combined with chromatographic techniques. Regarding the alternative options, different techniques have been applied for the analysis of essential oils and natural extracts, some of which are easy to handle, such as the measurement of physicochemical parameters. But considering that adulteration occurs at low concentration levels to avoid detection by common methods and the need to analyze their constituents, other analytical methods need to be improved to achieve these objectives. In this sense, adequate methodology for the analysis of these complex mixtures will allow the characterization of the new natural preservatives and the determination of the total allergen content. Gas chromatography coupled to mass spectrometry is the most commonly used analytical method for the analysis of allergenic fragrances, while liquid chromatography is used less frequently. Some published studies have been conducted for the analysis of certain individual chemical compounds listed [17–20]. Analytical techniques are crucial for the quality control of an ingredient or product, especially when the factors surrounding the natural raw materials, as in the case of essential oils and natural extracts, can affect their composition. In addition, cosmetics manufacturers must also guarantee their compliance with current regulations and ensure the absence of fraud among their products. Cosmetic companies have a significant challenge regarding analytical techniques to identify and quantify fragrances in the final cosmetic products. However, no analytical tools have been described to assess the NE_{28} and, consequently, the extended list of fragrance allergens. Thus, the aim of this work is to search for selective phytomarkers to determine the presence of NE_{28} in cosmetic formulations. In addition, an estimation of the real allergen levels in cosmetic products containing Nes and Eos would be possible by a prior assessment of the allergen content in those NE and EO pure original products, establishing groups based on the expected affinities due to a close botanical origin. To characterize the composition of the selected pure samples of Eos and Nes, direct injection of sample dilutions and solid-phase microextraction (SPME) were applied, followed by gas chromatography coupled to mass spectrometry (GC-MS) analysis. After the identification of allergens, potential phytomarker compounds were examined. The results obtained are useful to prevent and regulate fraud and thus contribute to improving the safety of users of cosmetics that include NEs and EOs in their formulation.

2. Materials and Methods

2.1. Chemicals, Reagents, and Materials

Ethyl acetate was supplied by Scharlab (Barcelona, Spain) and was of analytical grade. The SPME manual holders and fibers were supplied by Supelco (Bellefonte, PA, USA). The commercial fiber coating used throughout the present work was 65 μm polydimethylsiloxane/divinylbenzene (PDMS/DVB). Prior to first use, the fiber was conditioned as recommended by the manufacturer, inserting in the GC injector under helium flow at 250 °C for 30 min.

2.2. Samples

In the choice of samples, the 28 ENs highlighted as contact allergens by the SCCS of the European Commission [13] are used as references. Samples from two different origin groups were selected, allowing the search for frequently occurring substances exclusive to each group that act as markers. Fifteen pure EOs and two NEs (Jasmine Absolute and Rose Absolute) were commercially acquired, comprising eight obtained from flowers and nine from trees.

Table 1 lists the 17 samples divided into groups, indicating their CAS numbers and the solubility indicated by the safety data sheets (SDS).

The constituent substances of these complex samples of EOs and NEs include recognized allergens, which in turn have specific functions and properties that are transferred to the final product. Table 2 lists the individual allergenic substances included in the SDS of the purchased samples and details their use and function as ingredients in cosmetics [21,22]. The chemical formulas and structures of these compounds are included in Supplementary Table S1.

Table 1. Characteristics of the analyzed samples: sample code, CAS numbers and solubility indicated by the safety data sheets.

Code	CAS	Solubility [a]
Flowers		
Jasmine absolute	84776-64-7	H_2O (I), EtOH (S)
Rose absolute	90106-38-0	EtOH (S)
Geranium Egypt	90082-51-2	EtOH (S)
Lavender	90063-37-9	EtOH (S)
Lavandin super	91722-69-9	EtOH (S)
Verbena oil	Aromatic substances mixture	H_2O (I), EtOH (S)
Ylang Ylang Extra	83863-30-3	H_2O (I), EtOH (PS)
Ylang Ylang II	83863-30-3	H_2O (I), EtOH (PS) alcohol (PS)
Trees		
Cinnamon leaves	84649-98-9	H_2O (I), EtOH (S)
Cassia	84961-46-6	H_2O (I), EtOH (S)
Atlas cedar	92201-55-3	EtOH (S)
Cedar super	85085-41-2	H_2O (I), EtOH (S)
Clove buds	84961-50-2	EtOH (S)
Eucalyptus	84625-32-1	EtOH (S)
Laurel leaves	84603-73-6	H_2O (I), EtOH (S)
Perú oil super	8007-00-9	H_2O (I), EtOH (S)
Indian Sandalwood	84787-70-2	H_2O (I), EtOH (S)

[a] I: insoluble; S: soluble; PS: partially soluble.

Table 2. Compounds declared as allergens [13]: CAS numbers, their function or use in cosmetics, and samples in which their presence is detailed in the safety data sheets.

Compound	CAS	Cosmetic Function/Use	Declared in
Benzyl alcohol	100-51-6	Preservative/solvent	Jasmine absolute, rose absolute, Perú oil
Benzyl benzoate	120-51-4	Solvent	Jasmine absolute, cinnamon leaves, Perú oil, Ylang ylang II, Ylang ylang extra
Benzyl cinnamate	103-41-3	Masking agent	Perú oil
Benzyl salicylate	118-58-1	UV absorber	Ylang ylang II, Ylang ylang extra
Camphor	464-49-3	Denaturant/plasticiser	Lavandin super
Cinnamaldehyde	104-55-2	Denaturant	Cassia, cinnamon leaves
Cinnamyl alcohol	104-54-1	Masking agent	Cassia
Citral	5392-40-5	Masking agent	Jasmine absolute, rose absolute, geranium Egypt
Citronellol	106-22-9	Masking agent	Rose absolute, geranium Egypt, verbena oil
Coumarin	91-64-5	Masking agent	Cassia, lavandin super
Eugenol	97-53-0	Denaturant/tonic	Jasmine absolute, rose absolute, clove buds, laurel leaves, cinnamon leaves, Ylang ylang II, Ylang ylang extra
Farnesol	4602-84-0	Soothing/solvent/deodorant	Rose absolute, Ylang ylang II, Ylang ylang extra
Geraniol	106-24-1	Tonic	Rose absolute, geranium Egypt, lavender, lavandin super, Ylang ylang II, Ylang ylang extra, verbena oil
Isoeugenol	97-54-1	Masking agent	Ylang ylang II, Ylang ylang extra
Limonene	138-86-3	Solvent	Geranium Egypt, lavender, lavandin super, laurel leaves, eucalyptus, cinnamon leaves, verbena oil

Table 2. *Cont.*

Compound	CAS	Cosmetic Function/Use	Declared in
Linalool	78-70-6	Deodorant	Jasmine absolute, lavender, lavandin super, laurel leaves, cinnamon leaves, Ylang ylang II, Ylang ylang extra, verbena oil
Linalyl acetate	115-95-7	Masking agent	Lavender, lavandin super
Terpinen-4-ol	562-74-3	Denaturant/solvent	Lavender, laurel leaves
Terpinolene	586-62-9	Fragrance	Cinnamon leaves
α-Pinene	80-56-8	Antifoaming	Geranium Egypt, laurel leaves, eucalyptus, cinnamon leaves
α-Terpineol	98-55-5	Denaturant/solvent	Lavender, lavandin super
β-Caryophyllene	87-44-5	Skin continioning	Clove buds, geranium Egypt, lavender, lavandin super, Ylang ylang II, Ylang ylang extra, cinnamon leaves
β-Pinene	127-91-3	Perfuming	Geranium Egypt, laurel leaves, eucalyptus, cinnamon leaves

2.3. Sample Preparation

First, sample preparation was kept as simple as possible, including simple dilution. After preliminary tests, pure samples were diluted 1:100 in ethyl acetate for direct injection in the GC-MS.

For SPME, 10 μL of the sample was placed in a 10 mL vial. After sealing with an aluminum cap furnished with PTFE-faced septa, samples were left to equilibrate for 30 min, and then the SPME fiber (PDMS/DVB) was exposed to the headspace (HS-SPME) for 15 min at 25 °C. Once the exposure period had finished, the fiber was retracted into the needle of the holder syringe and immediately inserted into the GC injection port. The desorption time was set at 5 min, and the desorption temperature was kept at 260 °C. To avoid potential contamination and memory effects on the fiber, blanks were periodically run. Two replicates of each sample were processed for analysis.

2.4. GC-MS Analysis

The GC-MS analysis was performed using an Agilent 7890A chromatograph coupled to an Agilent 5975C inert mass spectra detector (MSD) with a triple-axis detector and an Agilent 7693 autosampler from Agilent Technologies (Palo Alto, CA, USA). Two chromatographic columns were used. The first column was a low-polarity ZB-Semivolatiles (30 m × 0.25 mm i.d., 0.25 μm film thickness) from Phenomenex (Torrance, CA, USA), operated with an oven temperature program that applies 50 °C (held 3 min) to 200 °C at 4 °C min^{-1}, and a final ramp to 290 °C at 20 °C min^{-1} (held 3 min) (total run time: 50 min); the second column was a polar J&W Scientific DB-WAX 128-7052 (50 m × 0.20 mm i.d., 0.2 μm film thickness) from Agilent Technologies, applying an oven program from 50 °C (held 3 min) to 240 °C at 4 °C min^{-1} (held 5.5 min), with a total run time of 56 min. Helium (purity 99.999%) was employed as carrier gas at a constant flow of 1.0 mL min^{-1} (first column) and 0.8 mL min^{-1} (second column), and the injection temperature was 270 °C and 240 °C, respectively. The injection volume was 1 μL for direct injection. The mass spectrometer detector (MSD) was operated in the electron impact (EI) ionization positive mode (+70 eV). The ion source temperature was 150 °C. The temperature of the transfer line was set at 290 °C and 240 °C for the first and second columns, respectively.

Full Scan (FS) acquisition mode was employed, monitoring mass/charge (m/z) fragments between 30 and 800. The tentative identification of the compounds was performed by comparison of the experimental MS spectra and those provided by the spectral library database (NIST MS Search 2.0).

3. Results and Discussion

The analysis of the diluted samples using the two columns, shows a large number of compounds present in each sample. The overall data were examined by comparing the identified compounds using both columns, evaluating their concordance with the information contained in the SDS, and, as a final objective, seeking to identify group phytomarkers (trees and flowers). All this, considering the extended list of 82 substances in the Cosmetics EU regulation and highlighted as contact allergens by SCCS [13].

Despite the widespread use of low-polarity columns in the analysis by GC-MS of EOs/NEs, the higher-polarity column allowed a better separation of sample components.

A minimum of 28 compounds in the clove buds and Indian sandalwood EOs and a maximum of 85 compounds in the laurel leaf EO were identified. The group of flower EOs is characterized by sharing more than 50 components common to all of them, while the group of tree EOs presents a more differentiated composition according to their origin.

The allergenic content among the EOs and NEs varies notably, ranging from apparently innocuous essential oils such as cedar super (with zero allergens present, Figure 1) to others such as Ylang ylang extra, which contains 14 allergens. The number of allergens varies according to the sample and is independent of the flower or tree origin. In this way, although more allergens were found in the flower EOs, some EOs from trees contained a large number of allergenic substances, as was the case for the cinnamon leaf EO (Figure 2).

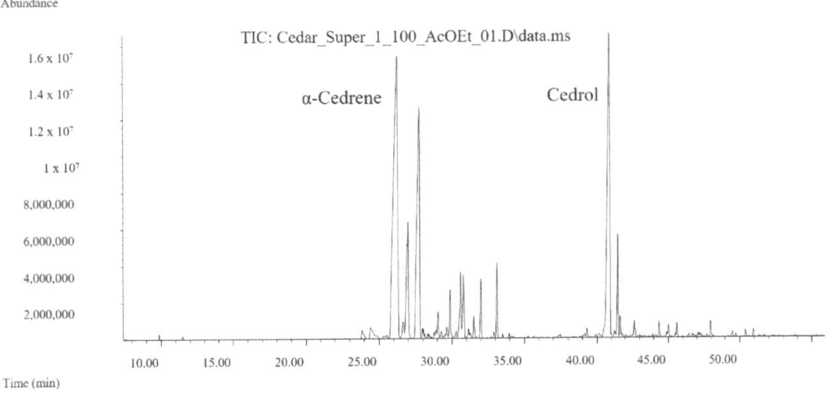

Figure 1. Example of a polar column chromatogram for the cedar super sample, indicating two major compounds.

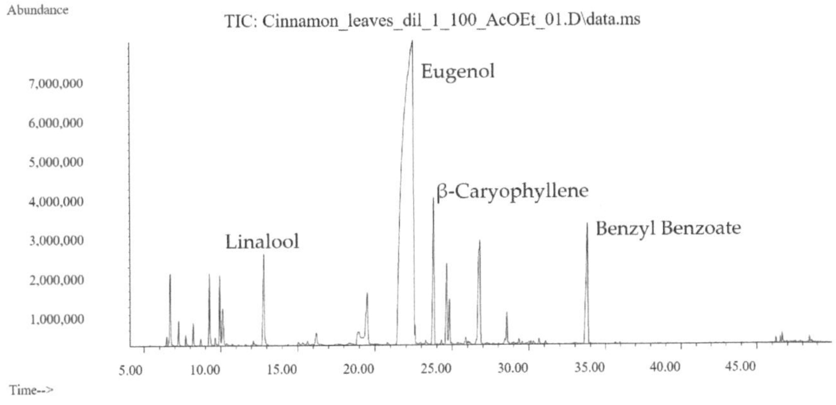

Figure 2. Example of a non-polar column chromatogram for the cinnamon leaves sample, indicating some fragrance allergens.

The use of phytomarkers for the presence of a certain EO can greatly facilitate the quality control of cosmetic products containing this EO in their ingredients. Some compounds were identified in a single sample and could therefore be used as unique markers for the presence of a certain EO in a cosmetic product with a label indicating the content of that EO in its composition, facilitating the detection of fraud in the case of its absence. Other compounds could act as origin group markers if they are present only in EOs extracted from trees or flowers.

A higher proportion of compounds unique to the species have been identified in samples from the flower group, so they are not useful as markers for the flower group. In addition, most of these compounds are not recognized allergens and therefore do not require follow-up monitoring. Therefore, it seems more interesting to look for group phytomarkers among the compounds whose control is necessary because they are recognized allergens [13]. Figure 3 depicts the occurrence of several allergens according to their origin (flowers or trees).

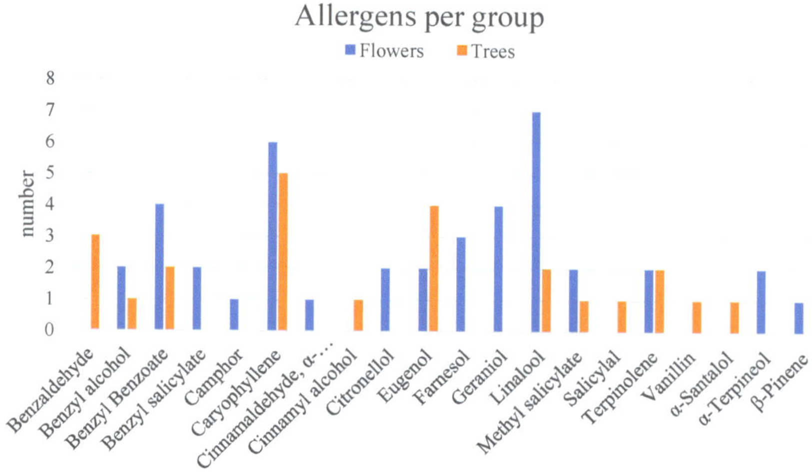

Figure 3. Distribution of allergens by the group of origin: flowers (blue) and trees (orange).

The figure shows the abundance of allergens in the analyzed EOs/NEs and also shows that most of the allergens found are ubiquitous substances that do not allow differentiation between the origin groups. The most common allergens are caryophyllene (appearing five times in the tree group and seven times in the flower group), while others are present only once (camphor and cinnamaldehyde in flowers; vanillin and α-santalol in trees). This fact, in addition to supporting the need to characterize the allergenic composition and regulate the use of EOs as natural preservatives due to their possible adverse effects, identifies those allergens with greater validity as sample or group markers.

Allergens found exclusively in the flower group were β-pinene, α-terpineol, geraniol, farnesol, citronellol, α-hexyl-cinnamaldehyde, benzyl salycilate, and camphor, while those exclusive to the tree group were α-santalol, vanillin, salycilal, cinnamyl alcohol, and benzaldehyde. Allergen markers found for specific NEs/EOs were salicylal and cinnamaldehyde in cassia oil, vanillin in Perú oil, α-santalol in Indian sandalwood, levomenthol in geranium Egypt oil, α-hexyl-cinnamaldehyde in jasmine absolute, and citral in rose absolute.

Table 3 shows the allergen composition of the samples and its agreement with the sample SDS. It is important to highlight the discordance between the SDS composition and the results found experimentally.

The fact that natural ingredients are complex mixtures of many chemical constituents, the precise nature of which is often unknown, presents problems regarding their allergic behavior [23]. As a result, their allergenic potency is also unclear. Several natural substances have caused contact allergies [24,25], and most contact fragrance allergens have been classified as moderate skin sensitizers [26]. The mechanisms of action of these substances include skin penetration and autoxidation (prehaptens). For example, in the case of contact allergy caused by the fragrance terpenes limonene and linalool (identified in the samples considered in this work), both are prehaptens, which means that pure fragrance chemicals with a low sensitizing power are modified outside the skin by air oxidation, resulting in a higher allergenic potential [27]. For both linalool and limonene, specific compounds formed during their oxidation have been identified as the main causes of sensitization.

Table 3. Allergens identified in the samples. Italics indicate agreement with the safety data sheets.

Compound	Identification by GC-MS Analysis
α-Terpineol	*Lavender, lavandin super,* laurel leaves, eucalyptus, cinnamon leaves, Perú oil, Ylan ylang extra, verbena oil
β-Caryophyllene	*Clove buds, geranium Egypt, lavender, lavandin super, Ylang ylang II, Ylang ylang extra, cinnamon leaves,* cassia, laurel leaves, eucalyptus, rose absolute, verbena oil
β-Pinene	*Cinnamon leaves,* lavender, lavandin super, Ylang ylang II, Ylang ylang extra, verbena oil
Benzyl alcohol	*Jasmine absolute, rose absolute, Perú oil,* Ylang ylang II, Ylang ylang extra
Benzyl benzoate	*Jasmine absolute, cinnamon leaves, Perú oil, Ylang ylang II, Ylang ylang extra,* verbena oil
Benzyl cinnamate	*Perú oil,* Ylang ylang extra
Benzyl salicylate	*Ylang ylang II, Ylang ylang extra,* jasmine absolute
Camphor	Lavender, *lavandin super*
Cinnamaldehyde	*Cassia*
Cinnamyl alcohol	*Cassia,* cinnamon leaves, Ylang ylang extra
Citral	*Rose absolute*
Citronellol	*Rose absolute, geranium Egypt,* eucalyptus
Eugenol	*Jasmine absolute, rose absolute, clove buds, laurel leaves, cinnamon leaves, Ylang ylang extra, cassia, Perú oil*
Farnesol	*Rose absolute, Ylang ylang II, Ylang ylang extra*
Geraniol	*Rose absolute, geranium Egypt, lavender, lavandin super, Ylang ylang II, Ylang ylang extra, verbena oil,* eucalyptus
Limonene	Atlas cedar, cassia
Linalool	*Jasmine absolute, lavender, lavandin super, laurel leaves, cinnamon leaves, Ylang ylang II, Ylang ylang extra, verbena oil,* clove buds, eucalyptus, geranium Egypt, rose absolute
Terpinen-4-ol	Eucalyptus, Perú oil, rose absolute
Terpinolene	*Cinnamon leaves,* laurel leaves, eucalyptus, lavender, lavandin super, Ylang ylang extra, geranium Egypt, verbena oil

Depending on the quality of the starting plant material and the extraction method used, the composition of these volatile natural complex mixtures can be different. These factors will determine their allergenic potential and are also important to ensure their safe use in personal care products. However, other papers listed specific substances studied in the present work as compounds identified in certain essential oils, such as linalool, α-terpineol, and camphor in lavender essential oil [28] or as citronellol, geraniol, citral, and farnesol in rose essential oil [29]. Regarding individual fragrances, eugenol is the main component in clove essential oil; geraniol also occurs naturally in geranium essential oil in lower concentrations, and linalool in cinnamon oil [29]. For cinnamon species, cinnamyl alcohol is commonly found [30].

As can be seen in Figure 4, samples generally contain more allergens than declared. The most pronounced difference in allergenic compounds identified experimentally compared to those reported in the SDS is found in the samples, namely Ylang ylang extra and verbena oil, with a difference of six and five, respectively. In other cases, the difference is smaller, as is the case with cinnamon leaves, laurel leaves, and lavender, in which one less compound is reported. In addition, there are some examples, like EOs of Atlas cedar or Indian sandalwood, whose data sheets do not mention any allergens and in whose samples L-α-terpineol and limonene, and α-santalol, respectively, were identified.

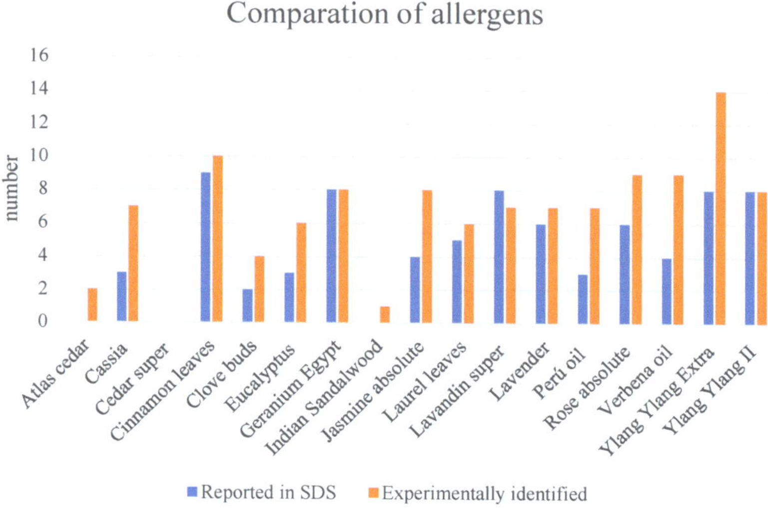

Figure 4. Number of allergens reported in the Eos/Nes corresponding SDS (blue) and identified in this work (orange).

On the contrary, four allergens (linalyl acetate, α-pinene, coumarin, and isoeugenol) are listed in the SDS, but their presence was not detected in the samples. In some cases, this may be because the CAS indicated by the SDS may actually correspond to a racemic mixture, but in others, it was not found in the sample's analysis.

Fewer compounds were identified using HS-SPME-GC-MS than by direct injection of the diluted samples, but still more compounds were identified than those listed in the SDS for each sample. In terms of the volatile allergenic substances present, cinnamon leaf EO stands out with 14 substances.

Regardless, as demonstrated throughout this study, individual samples of both essential oils and natural extracts contain numerous contact allergens. This poses an added problem, since research has previously demonstrated that exposure to a combination of two contact fragrance allergens has a synergistic effect compared to the expected response by testing with each substance alone [31]. Furthermore, a mixture of contact fragrance allergens has a higher potency with respect to the risk of sensitization compared to exposure to individual sensitizing fragrance substances [32]. Hence the importance of quality control of these innovative natural ingredients in order to ensure the safety of cosmetic and personal care products that incorporate such ingredients.

4. Conclusions

This work provides the basis for the identification of AEs and NEs in cosmetic products, applicable both to the quality control of raw materials and to the detection of possible fraud in procurement and/or labeling. The identification of the allergenic content of 17 samples of AEs and NEs revealed that all samples contained more allergenic substances than those declared in the corresponding safety data sheet. Although the list of samples to be considered could be extended, the results obtained support the creation of a list of specific plant markers for samples obtained from trees or flowers.

Supplementary Materials: The following supporting information can be downloaded at: https://www.mdpi.com/article/10.3390/cosmetics11030084/s1, Table S1: Compounds declared as allergens [13]: chemical formulas and structures.

Author Contributions: Conceptualization, L.R. and M.L.; methodology, L.R.; software, A.P. and L.R.; validation, L.R. and M.L.; formal analysis, L.R.; investigation, A.P. and L.R.; resources, C.G.-J. and M.L.; data curation, L.R. and M.L.; writing—original draft preparation, L.R.; writing—review and editing, C.G.-J. and M.L.; visualization, M.L.; supervision, C.G.-J. and M.L.; project administration, C.G.-J. and M.L.; funding acquisition, C.G.-J. and M.L. All authors have read and agreed to the published version of the manuscript.

Funding: This research was supported by the project ED431B 2023/04 (Galician Competitive Research Groups, Xunta de Galicia, Spain) and was based upon work from the Sample Preparation Study Group and Network, supported by the Division of Analytical Chemistry of the European Chemical Society. All these programs are co-funded by FEDER (EU).

Institutional Review Board Statement: Not applicable.

Informed Consent Statement: Not applicable.

Data Availability Statement: All the reported data are available in the manuscript.

Conflicts of Interest: The authors declare no conflicts of interest.

References

1. Klaschka, U. Natural personal care products-analysis of ingredient lists and legal situation. *Environ. Sci. Eur.* **2016**, *28*, 8. [CrossRef]
2. Lundov, M.D.; Moesby, L.; Zachariae, C.; Johansen, J.D. Contamination versus preservation of cosmetics: A review on legislation, usage, infections, and contact allergy. *Contact Dermat.* **2009**, *60*, 70–78. [CrossRef] [PubMed]
3. Halla, N.; Fernandes, I.P.; Heleno, S.A.; Costa, P.; Boucherit-Otmani, Z.; Boucherit, K.; Rodrigues, A.E.; Ferreira, I.C.F.R.; Barreiro, M.F. Cosmetics Preservation: A Review on Present Strategies. *Molecules* **2018**, *23*, 1571. [CrossRef] [PubMed]
4. Alvarez-Rivera, G.; Llompart, M.; Lores, M.; Garcia-Jares, C. Preservatives in cosmetics: Regulatory aspects and analytical methods. In *Analysis of Cosmetic Products*, 2nd ed.; Elsevier: Amsterdam, The Netherlands, 2018; pp. 175–224.
5. Schwensen, J.F.; White, I.R.; Thyssen, J.P.; Menné, T.; Johansen, J.D. Failures in risk assessment and risk management for cosmetic preservatives in Europe and the impact on public health. *Contact Dematitis* **2015**, *73*, 133–141. [CrossRef]
6. European Union. Regulation (EC) No 1223/2009 of the European Parliament and of the Council of 30 November 2009 on Cosmetic Products. *Eur. Off. J.* **2009**, *L342*, 59–209. Available online: https://eur-lex.europa.eu/legal-content/EN/TXT/HTML/?uri=OJ:L:2009:342:FULL&from=EN (accessed on 26 January 2024).
7. Lores, M.; Llompart, M.; Alvarez-Rivera, G.; Guerra, E.; Vila, M.; Celeiro, M.; Lamas, J.P.; Garcia-Jares, C. Positive lists of cosmetic ingredients: Analytical methodology for regulatory and safety controls—A review. *Anal. Chim. Acta* **2016**, *915*, 1–26. [CrossRef] [PubMed]
8. Antignac, E.; Nohynek, G.J.; Re, T.; Clouzeau, J.; Toutain, H. Safety of botanical ingredients in personal care products/cosmetics. *Food Chem. Toxicol.* **2011**, *49*, 324–341. [CrossRef]
9. Mahesh, S.K.; Fathima, J.; Veena, V.G. Cosmetic Potential of Natural Products: Industrial Applications. In *Natural Bio-Active Compounds*; Swamy, M.K., Akhtar, M.S., Eds.; Springer: Berlin/Heidelberg, Germany, 2019; Volume 2, pp. 215–250.
10. Teixeira, B.; Marques, A.; Ramos, C.; Neng, N.; Nogueira, J.; Saraiva, J.A.; Nunes, M.L. Chemical composition and antibacterial and antioxidant properties of commercial essential oils. *Ind. Crops. Prod.* **2013**, *43*, 587–595. [CrossRef]
11. Sell, C. Perfumery Materials of Natural Origin. In *The Chemistry of Fragrances*, 2nd ed.; RSC: London, UK, 2006.
12. Tiên, D.; Francis, H.; Sylvain, A.; Fernandez, X. Authenticity of essential oils. *TrAC Trends Anal. Chem.* **2014**, *66*, 146–157.
13. SCCS (Scientific Committee on Consumer Safety): Opinion on Fragrance Allergens in Cosmetic Products, 26–27 June 2012. Available online: https://ec.europa.eu/health/scientific_committees/consumer_safety/docs/sccs_o_102.pdf (accessed on 26 January 2024).
14. European Union. Commission Regulation (EU) 2023/1545 of 26 July 2023 amending Regulation (EC) No 1223/2009 of the European Parliament and of the Council as Regards Labelling of Fragrance Allergens in Cosmetic Products. *Eur. Off. J.* **2023**, *L188*, 1–23. Available online: https://eur-lex.europa.eu/legal-content/EN/TXT/HTML/?uri=CELEX:32023R1545 (accessed on 26 January 2024).
15. Celeiro, M.; Garcia-Jares, C.; Llompart, M.; Lores, M. Recent advances in sample preparation for cosmetics and personal care products analysis. *Molecules* **2021**, *26*, 4900. [CrossRef] [PubMed]
16. Sanchez-Prado, L.; Alvarez-Rivera, G.; Lamas, J.P.; Lores, M.; Garcia-Jares, C.; Llompart, M. Analysis of multi-class preservatives in leave-on and rinse-off cosmetics by matrix solid-phase dispersion. *Anal. Bioanal. Chem.* **2011**, *401*, 3293–3304. [CrossRef] [PubMed]

17. Belhassen, E.; Bressanello, D.; Merle, P.; Raynaud, E.; Bicchi, C.; Chaintreau, A.; Cordero, C. Routine quantification of 54 allergens in fragrances using comprehensive two-dimensional gas chromatography-quadrupole mass spectrometry with dual parallel secondary columns. Part I: Method development. *Flavour Fragr. J.* **2018**, *33*, 63–74. [CrossRef]
18. Rey, A.; Corbi, E.; Pérès, C.; David, N. Determination of suspected fragrance allergens extended list by 2DGC-MS in ready-to-inject samples. *J. Chromatogr. A* **2015**, *1404*, 95–103. [CrossRef] [PubMed]
19. Desmedt, B.; Canfyn, M.; Pype, M.; Baudewyns, S.; Hanot, V.; Courselle, P.; De Beer, J.O.; Rogiers, V.; De Paepe, K.; Deconinck, E. HS-GC-MS method for the analysis of fragrance allergens in complex cosmetic matrices. *Talanta* **2015**, *131*, 444–451. [CrossRef]
20. Tranchida, P.Q.; Maimone, M.; Franchina, F.A.; Bjerk, T.R.; Zini, C.A.; Purcaro, G.; Mondello, L. Four-stage (low-)flow modulation comprehensive gas chromatography-quadrupole mass spectrometry for the determination of recently-highlighted cosmetic allergens. *J. Chromatogr. A* **2016**, *1439*, 144–151. [CrossRef] [PubMed]
21. Cosmetic Ingredient Database. Available online: https://ec.europa.eu/growth/sectors/cosmetics/cosmetic-ingredient-database_en (accessed on 26 January 2024).
22. European Commision (EC). No 2006/257/CE: Commission Decision of 9 February 2006 Amending Decision 96/335/EC Establishing an Inventory and a Common Nomenclature of Ingredients Employed in Cosmetic Products. Available online: https://op.europa.eu/en/publication-detail/-/publication/db30de80-11f8-4358-b1d6-e38d6cf96625 (accessed on 26 January 2024).
23. Goossens, A. Les allergies de contact aux produits naturels des cosmétiques. *Rev. Fr. Allerg.* **2015**, *55*, 171–173. (In French) [CrossRef]
24. Corazza, M.; Borghi, A.; Gallo, R.; Schena, D.; Pigatto, P.; Lauriola, M.M.; Guarneri, F.; Stingeni, L.; Vincenzi, C.; Foti, C.; et al. Topical botanically derived products: Use, skin reactions, and usefulness of patch tests. A multicenter Italian study. *Contact Dermat.* **2014**, *70*, 90–97. [CrossRef]
25. Jack, A.R.; Norris, P.L.; Storrs, F.J. Allergic Contact Dermatitis to Plant Extracts in Cosmetics. *Semin. Cutan. Med. Surg.* **2013**, *32*, 140–146. [CrossRef]
26. Uter, W.; Johansen, J.D.; Börje, A.; Karlberg, A.-T.; Lidén, C.; Rastogi, S.; Roberts, D.; White, I.R. Categorization of fragrance contact allergens for prioritization of preventive measures: Clinical and experimental data and consideration of structure-activity relationships. *Contact Dermat.* **2013**, *69*, 196–230. [CrossRef]
27. Karlberg, A.T.; Börje, A.; Duus Johansen, J.; Lidén, C.; Rastogi, S.; Roberts, D.; Uter, W.; White, I.R. Activation of non-sensitizing or low-sensitizing fragrance substances into potent sensitizers—Prehaptens and prohaptens. *Contact Dermat.* **2013**, *69*, 323–334. [CrossRef] [PubMed]
28. Sarkic, A.; Stappen, I. Essential oils and their single compounds in cosmetics-A critical review. *Cosmetics* **2018**, *5*, 11. [CrossRef]
29. Verma, R.S.; Padalia, R.C.; Chauhan, A.; Singh, A.; Yadav, A.K. Volatile constituents of essential oil and rose water of damask rose (*Rosa damascena* Mill.) cultivars from North Indian hills. *Nat. Prod. Res.* **2011**, *25*, 1557–1584. [CrossRef]
30. Wang, R.; Wang, R.; Yang, B. Extraction of essential oils from five cinnamon leaves and identification of their volatile compound compositions. *Innov. Food. Sci. Emerg. Technol.* **2009**, *10*, 289–292. [CrossRef]
31. Johansen, J.D.; Skov, L.; Volund, A.; Andersen, K.; Menné, T. Allergens in combination have a synergistic effect on the elicitation response: A study of fragrance-sensitized individuals. *Br. J. Dermatol.* **1998**, *139*, 264–270. [CrossRef]
32. Bonefeld, C.M.; Nielsen, M.M.; Rubin, I.M.C.; Vennegaard, M.T.; Dabelsteen, S.; Gimenéz-Arnau, E.; Lepoittevin, J.-P.; Geisler, C.; Johansen, J.D. Enhanced sensitization and elicitation responses caused by mixtures of common fragrance allergens. *Contact Dermat.* **2011**, *65*, 336–342. [CrossRef] [PubMed]

Article

Extending the Physical Functionality of Bioactive Blends of *Astrocaryum* Pulp and Kernel Oils from Guyana

Laziz Bouzidi [1], Shaveshwar Deonarine [1], Navindra Soodoo [1], R. J. Neil Emery [2], Sanela Martic [3] and Suresh S. Narine [1,*]

1   Centre for Biomaterials Research, Departments of Physics & Astronomy and Chemistry, Trent University, 1600 West Bank Drive, Peterborough, ON K9L 0G2, Canada; lazizbouzidi@trentu.ca (L.B.); shaveshwardeonarine@trentu.ca (S.D.); navindrasoodoo@trentu.ca (N.S.)
2   Department of Biology, Trent University, 1600 West Bank Drive, Peterborough, ON K9L 0G2, Canada; nemery@trentu.ca
3   Department of Forensic Science, Trent University, 1600 West Bank Drive, Peterborough, ON K9L 0G2, Canada; sanelamartic@trentu.ca
*   Correspondence: sureshnarine@trentu.ca; Tel.: +1-705-748-1011; Fax: +1-705-750-2786

Abstract: Natural lipids with nutritional or therapeutic benefits that also provide desired texture, melting and organoleptic appeal (mouthfeel, skin feel) are difficult to procure for the food and cosmetics industries. Natural *Astrocaryum* pulp oil (AVP) and kernel fat (AVK) from Guyana were blended without further modification to study the potential of extending the physical functionality of the blends beyond that of crude AVK and AVP. An evaluation of non-lipid components by ESI-MS indicated twenty-four (24) bioactive molecules, mainly carotenoids (90%), polyphenols (9%) and sterols (1%) in AVP, indicating important health and therapeutic benefits. Only trace-to-negligible amounts of these compounds were detected in AVK. The thermal transition phase behavior, solid fat content (SFC), microstructure and textural properties of five AVP/AVL blends were used to construct phase diagrams of the AVK/AVP binary system. Binary phase diagrams constructed from the cooling and heating DSC thermograms of the mixtures and description of the liquidus line indicated a mixing behavior close to ideal with a tendency for order, with no phase separation. Melting onsets, solid fat content and measurements of solid-like texture all predictively increased with increasing AVK content. The descriptive decay parameters obtained for SFC, crystal size, hardness, firmness and spreadability were similar and predictive and indicate the way the binary system structure approaches that of a liquid or a functional solid. The bioactive content of the blends was accurately calculated; the work provides a blueprint for the blending of AVP and AVK to deliver targeted bioactive content, stability, spreadability, texture, melting profile, organoleptic appeal and solid content. SFCs at 20 °C ranged from 9.1% to 39.1%, melting onset from −17.5 °C to 27.8 °C, hardness from 0.1 N to 3.5 N and spreadability from 3.3 N·s to 147.1 N·s; indicating a useful dynamic range of physical properties suitable for bioactive oils to bioactive butters.

Keywords: *Astrocaryum*; Guyana; cosmetics; blending; thermal behavior; textural analysis; bioactive; solid fat content

Citation: Bouzidi, L.; Deonarine, S.; Soodoo, N.; Emery, R.J.N.; Martic, S.; Narine, S.S. Extending the Physical Functionality of Bioactive Blends of *Astrocaryum* Pulp and Kernel Oils from Guyana. *Cosmetics* **2024**, *11*, 107. https://doi.org/10.3390/cosmetics11040107

Academic Editor: Vasil Georgiev

Received: 20 May 2024
Revised: 11 June 2024
Accepted: 18 June 2024
Published: 26 June 2024

1. Introduction

The demand for natural, bioactive and environmentally friendly products is rapidly growing, driven by rising public interest in sustainability, concerns regarding adverse health effects of synthetic materials, and environmental damages caused by the production of conventional chemically produced materials [1]. Alternatives to traditional means of production including green and ecological approaches and the use of naturally sourced materials are correspondingly being increasingly explored. The pharmaceutical, cosmetic and food sectors, in particular, are actively seeking alternatives to chemically synthesized ingredients, particularly lipid-based materials. Natural lipids, including those from so-called exotic sources, are pursued as healthier alternatives. The goal is to tailor physical functionality in products while retaining bioactive and green characteristics. The cosmetic industry is actively researching the replacement of potentially harmful conventional cosmetics with products based on natural materials [2]. The food sector is also investing significant resources to achieve healthier and more functional products

Lipids extracted from oleaginous exotic plants, such as the oils and fats from tropical fruits and seeds, potentially offer attributes such as enhanced nutrition and active therapeutic benefits [3], but must be researched to ensure that they can also provide the physical functionality required in cosmetic and food products. Tropical oleaginous species are gaining increasing industrial and research interest particularly because of their high lipid content and the substantial and beneficial bioactive components of their fruits [3]. *Astrocaryum vulgare* (AV, awara in Guyana and tucumã do Amazonas in Brazil) is a species of the *Arecaceae* family native to the Amazonian region and Guiana shield that deserves special attention [4]. The AV fruit's fibrous mesocarp (pulp) contains carbohydrates (7–20%), fibers (11–30%) and lipids (25–50%) and relatively small amounts of proteins (3–10%) and minerals (2–3%) [5–7]. Their lipids possess distinctive bioactive and nutritional properties that can potentially enhance cosmetic, food and pharmaceutical products [8–10].

AV pulp (AVP) has a yield of 34 $w/w\%$ oil and AV kernel (AVK) has a yield of 27 $w/w\%$ fat, on a dry basis [11]. AVP produces an oil with a high concentration of unsaturated fatty acids (65–75%) [12], and AVK produces fats of high lauric acid content (45–50%) [13]. AVK lipid is a fat consisting of medium-chain triacylglycerols (TAGs), mainly *propane-1,2,3-triyl tridodecanoate* (LLL) (30.1%), *3-(tetradecanoyloxy)propane-1,2-diyl didodecanoate* (LLM) (26.10%), *3-(decanoyloxy)propane-1,2-diyl didodecanoate* (LLC$_{10}$) (12.0%), *3-(octanoyloxy)propane-1,2-diyl didodecanoate* (LLC$_8$) (10.5%) and *3-(dodecanoyloxy)propane-1,2-diyl ditetradecanoate* (MML) (10.0%) and AVP lipid is an oil composed of long-chain TAGs, mainly *2-(palmitoyloxy)propane-1,3-diyl dioleate* (POO) (43.0%), *propane-1,2,3-triyl trioleate* (OOO) (27.9%), *2-(oleoyloxy)propane-1,3-diyl dipalmitate* (POP) (17.3%), *3-(oleoyloxy)propane-1,2-diyl distearate* (SOO) (9.1%) and *propane-1,2,3-triyl tristearate* (SSO) (1.3%) [11].

AVP lipid extract contains high levels of bioactive compounds, particularly carotenoids (122.2–163.7 mg/100 g), sterols, phenolic compounds and α-tocopherol (5.3 mg/100 g) [14,15]. These compounds reportedly impart significant beneficial bioactivity. The bioactive components found in AVP have been demonstrated to impart antimicrobial, antifungal [16], anti-inflammatory [14], antiproliferative [17], cytoprotective [18], skin-protective [8] and antioxidant activity [15,19,20]. Non-lipid bioactive compounds have not been reported in AVK lipid extractions.

AVK lipid extract contains medium-chain fatty acids, making it especially suitable for delivering unique nutrition, flavor and texture in various foods [8,10]. Its moisturizing and emollient capacity can be exploited to produce functional cosmetic products such as lotions and creams [21–23], and it can potentially be safely incorporated in drug formulations and pharmaceutical delivery systems [24,25].

AVK lipid extract can be considered a butter. It has melting and SFC profiles closely resembling those of cocoa butter (CB), the staple lipid ingredient in many food, cosmetics and pharmaceutical formulations. The unique chemical composition [26] of CB results in the development of a crystal structure and microstructure in its solid phase that delivers a preferred physical functionality related to stability, spreadability, texture and mouthfeel [27–29].

AVK lipid extract has a melting point close to CB but higher than the other most-used commercial CB alternatives such as coconut oil (CO), palm kernel oil (PKO) and shea butter (SB) [11,30,31]. CO, PKO and SB are first fractionated, and the higher-melting fractions are used to make CB alternatives (called CB equivalents or CBEs) [32]. AVK fat's sharp melting profile suggests that it would be suitable as a CB replacement in confectionery materials, without the need to first fractionate.

Astrocaryum vulgare's biochemical functions place health-beneficial bioactive compounds in the pericarp but not in the kernel. This would seem to provide a biological design to attract animals that feed on the nutritious and bioactive pulp and distribute the kernel with its hard impenetrable protective "shell" so that the plant proliferates. However, AVP lipid extract is a liquid oil, even at 0 °C. Therefore, there is an opportunity to utilize AVK butter blended with AVP liquid to potentially create a bioactive butter with tailored physical properties in the solid state at useful temperatures. This could potentially lead to a range of natural materials for the food, cosmetic and pharmaceutical industries that impart both desired physical functionality and bioactive content.

Compared to other means of modification such as fractionation, enzymatic or chemical interesterification and hydrogenation, direct blending of fats and oils is the most cost-effective, least destructive and environmentally friendly approach [33]. Direct blending is a straightforward process that involves mechanically mixing fats and oils in a molten state [3]. The incorporation of selected unsaturated TAGs into systems with a high content of saturated TAGs can yield fats with a softer texture, lower melting points and improved plasticity [34]. Similar lauric-rich fats such as palm kernel oil and coconut oil are commonly blended with common unsaturated vegetable oils to make cocoa butter substitutes (CBSs) [35]. Despite completely different TAG compositions, the blends can be directed to exhibit melting points and polymorphism similar to CB such as those formulated by Biswas et al. [36] and Hashem et al. [37] with palm kernel oils and various highly unsaturated vegetable oils such as olive oil and palm oil.

The blends of AVK and AVP lipid extracts studied were formulated with the aim to mimic important fats such as cocoa butter (CB) and shea butter (SB) in terms of physical and functional properties and for suitability as ingredients in food, cosmetics and pharmaceutical applications. Five mixtures were studied (x_{AVP} = 0.2, 0.4, 0.5 and 1.0). The bioactive components of the blends were assessed using the relative content determined from the molar concentration of each mixture, and from information derived from electrospray ionization–mass spectrometry (ESI-MS) analysis of the AVP lipid extracts.

The thermal transition behavior, solid fat content (SFC) versus temperature from 5 °C to above the melting point (40 °C) and microstructure of the blends were used to describe the phase behavior of the AVK/AVP binary system. The thermal transition behavior of the mixtures was determined by differential scanning calorimetry (DSC). The liquidus line was established using the offset of melting from the heating thermograms and modeled using the Hildebrand theory and the Bragg–William (BW) approximation [38,39]. The SFC versus temperature curve of the blends was determined by wide-line pulsed nuclear magnetic resonance (p-NMR) measurements and the kinetics of crystallization modeled with the Gompertz model [31,40]. Textural attributes including hardness, firmness, spreadability and adhesiveness were determined using a texture analyzer. The microstructure of the blends was determined using polarized light microscopy (PLM) and related to the thermal transition behavior, SFC and texture. The structure and physical properties of blends were utilized to assess their equivalence with established industrial fats and how closely they mimic physical properties and functions.

2. Materials and Methods

2.1. Materials

2.1.1. Sample Collection, Processing and Handling

Sample AV fruits were handpicked in May 2023 from forests of Guyana; Administrative Regions 1 and 2. Collection, processing and handling of the fruits as well as oil extraction from AVP and AVK were described in detail in our previous publication [11]. In brief, the pulps were separated from the seeds. The kernels were obtained by splitting the seeds with a knife and hammer. Both fresh pulps and kernels were sun-dried under ambient conditions, pulverized to small particles (~10 μm), then stored in sealed Ziploc® bags at room temperature (25 °C). The oils were extracted using mechanical pressing. AVP oil (AVP) was extracted without heating the expeller and AVK fat (AVK) was extracted after it was preheated to 50 °C for 15 min. The temperature during extraction did not exceed 50 °C for AVP and was between 50 and 80 °C for AVK. It is therefore expected that no loss or alteration of the lipid occurred in both extracts, but some non-lipid components may have been lost in AVK. The oils were immediately stored in glass jars and maintained at 4 °C until the time of analysis.

2.1.2. Preparation of the Blends

Five mixtures of 50 g each were prepared at AVK/AVP molar ratios (x) of 1.0, 0.8, 0.6, 0.5 and 0.4 from melted AVK and AVP (Table 1) according to the average molecular mass of AVK (687.29 g/mol) and AVP (867.99 g/mole) determined from the fatty acid profile of the lipids. The mixtures were blended in conical flasks using a magnetic stirrer (1000 revolutions/min) at a constant temperature of 40 °C for one hour.

Table 1. Molar ratios and masses of AVK fat-AVP oil mixtures.

Mixture	x_{AVP}	Molar Ratio	Mass AVK (g)	Mass AVP (g)	Average Molecular Mass (g)
AVK	0.0_{AVP}	1.0:0	50	0	687.27
AVK$_{0.8}$AVP$_{0.2}$	0.2_{AVP}	0.8:0.2	38	12	723.41
AVK$_{0.6}$AVP$_{0.4}$	0.4_{AVP}	0.6:0.4	26	24	759.56
AVK$_{0.5}$AVP$_{0.5}$	0.5_{AVP}	0.5:0.5	22	28	777.63
AVK$_{0.4}$AVP$_{0.6}$	0.6_{AVP}	0.4:0.6	18	32	795.70
AVP	1.0_{AVP}	0:1.0	0	50	867.99

2.2. Methods

2.2.1. Electrospray Ionization–Mass Spectrometry (ESI-MS)

Electrospray ionization–mass spectrometry (ESI-MS) was performed on a Thermo QExactive Orbitrap mass spectrometer (Thermo Fisher Scientific, San Jose, CA, USA) in the positive ion mode. Samples were injected using a syringe infusion pump (Harvard Apparatus, Holliston, MA, USA) and analyzed using an instrument resolution of 17,500. ESI source temperature was 320 °C and scan rate 12.5 scan/s. Methanol (HPLC grade; VWR, Mississauga, ON, Canada) was used as the source of hydrogen ions, and chloroform (HPLC grade; VWR, Mississauga ON, Canada) was used to dissolve the lipid analyte. Samples (1 ppm wt./v) were dissolved in a 70:30 (v/v) chloroform/methanol solution. The MS data were processed using the Qual Browser tool of Thermo Xcalibur software version 3.1 (Thermo Scientific, San Jose, CA, USA). NIST, Lipid Maps®, MetaboQuest and XCMS online databases were used to identify the molecular ions.

2.2.2. Total Phenolic Content (UV-Vis Spectrophotometry)

Total phenolic content (TPC) was determined using Folin–Ciocalteu's method [41,42]. An amount of 10 μg of sample oil was dissolved in 1 mL deionized water. The solution (1.0 mL) was then mixed with 2 mL of 20% sodium carbonate (Na_2CO_3) (MilliporeSigma Canada Ltd., Oakville, ON, Canada) in deionized water and 0.5 mL 1 N Folin–Ciocalteu's phenol reagent (MilliporeSigma Canada Ltd., Oakville, ON, Canada). After incubation at room temperature (20 ± 1 °C) for 1 h, the absorbance at 725 nm against a reagent blank was measured using a UV-Vis spectrophotometer (Cary 60, Agilent Technologies, Santa Clara, CA, USA). The measurements were performed in triplicate. Calibration was performed using gallic acid (HPLC grade; VWR, Mississauga ON, Canada).

2.2.3. Estimation of Total Carotenoid Content (UV-Vis Spectrophotometry)

Total carotenoid content (TCC) was estimated according to an established method [43,44] using a Cary 60 UV-Vis spectrophotometer (Agilent Technologies, Santa Clara, CA, USA). The measurements were performed on samples of 0.5 g dissolved in 50 mL of hexane. The instrument was operated in scanning mode between 200 nm and 800 nm. The baseline was determined using hexane (HPLC grade; VWR, Mississauga ON, Canada). TCC (mg/L) was calculated using a published equation [45].

2.2.4. Differential Scanning Calorimetry (DSC)

The thermal transition behavior of the samples was determined using a Q200 DSC equipped with a refrigerated cooling system (TA Instruments, Newcastle, DE, USA) and calibrated with pure indium. Samples (4–6 mg) in hermetic aluminum pans were equilibrated at 60 °C for 5 min to erase thermal history, then cooled at 5 °C/min down to −80 °C where it was held isothermally for 5 min to record the crystallization behavior and then heated back to 60 °C at 10 °C/min to investigate the melting behavior. Thermal Advantage universal analysis software (TA Advantage v5.5.24) was used to analyze the DSC data including peak (T_p), onset (T_{On}) and offset (T_{Off}) temperatures and enthalpies (ΔH) of crystallization and melting. Non-resolved peaks were located using first and second derivatives of the heat flow versus temperature curve.

2.2.5. Solid Fat Content (SFC)

SFC was measured on a Minispec mq 20 pulsed NMR spectrometer (Bruker Ltd., Milton, ON, Canada) equipped with a combined high- and low-temperature probe and a BVT3000 temperature controller (Bruker Ltd.). Liquid nitrogen was used for tempering. The fully melted sample was pipetted into the bottom of an NMR tube, filling its bottom to 1 cm (~0.57 ± 0.05 mL), stored at different temperatures (5, 10, 15, 18, 20, 25, 30 and 40 ± 1 °C), and then measured. Bruker's Minispec V2.58 Rev12 and Minispec plus V1.1 Rev05 software were used to collect the SFC data. Uncertainties attached are calculated standard deviations of at least three runs.

2.2.6. Texture Analysis

The textural properties of the samples were evaluated at room temperature using a texture analyzer (TA-TX HD, Texture Technologies Corp, Trenton, NJ, USA). A 90° conical spreadability cone (Model HDP/SR, Stable Micro Systems, Ltd., Godalming, UK) and matching conical sample holder were used to measure firmness, work of shear (spreadability), stickiness and work of adhesion (adhesiveness). Samples were measured after 12 h at room temperature in the compression mode at a test speed of 2.0 mm/s. The force (N) necessary to deform a column sample of 17 mm height was used as a firmness value.

Hardness was measured with a 5 mm diameter ball probe (Model P/5S, Stable Micro Systems, Ltd.). Samples were prepared in 100 mL plastic cups, achieving a uniform height of 20 mm. Hardness was determined as the force (N) required to achieve a penetration depth of 10 mm.

2.2.7. Polarized Light Microscopy (PLM)

A Leica DM2500P polarized light microscope (Leica Microsystems, Wetzlar, Germany) fitted with a digital camera (Leica DFC420C) was used for microstructure studies. A Linkam LS 350 temperature-controlled stage (Linkam Scientific Instruments, Tadworth, Surrey, UK) fitted to the PLM was used to thermally condition the samples. A small droplet of the sample was carefully pressed between a preheated glass slide and coverslip, ensuring a uniform thin layer. The sample was equilibrated at 60 °C for 10 min to erase crystal memory, then cooled down to −65 °C at 1 °C/min. PLM images (50×) were taken at 1 min intervals.

2.3. Statistical Analysis

Pearson Product Moment Correlation was used to measure the linear correlation between variables. Differences between groups were checked using analysis of variance (ANOVA) at a significance level of $p = 0.05$. The coefficient of determination (R^2), Chi-square (χ^2) and residual sum of squares (RSS) were used to evaluate the performance of the models. The Durbin–Watson (DW) statistic was used to test for autocorrelation in the residuals. Standard error and standard deviation were used to assess the physical relevance of the models' parameters. The statistical analysis was performed using the Sigmastat module of Sigmaplot 12.5 software. Built-in and user-defined equations in the nonlinear regression wizard of Sigmaplot 12.5 software were used to fit the data.

3. Results and Discussion

3.1. Non-Lipid Bioactive Composition

The total carotenoid content (*TCC*) was estimated using Equation (1) [45].

$$TCC(\text{mg/L}) = \frac{A_{446} \times V \times 10^4}{2592 \times m} \tag{1}$$

where A_{446} = absorbance at 446 nm, 2592 = extinction coefficient for carotenoids, V = volume of hexane (mL) and m = mass of sample (g). This equation, although specific to beta carotene, is more suitably expressed as the total carotenoid content as it also accounts for other carotenoid compounds being absorbed at the same wavelength as beta carotene [46].

Determination of total phenolic content (TPC) was determined using the linear calibration curve obtained with gallic acid in the 0.5 to 10 mg/L range (Equation (2); R^2 = 0.995).

$$Abs_{745} = 0.126 GAE - 0.0576 \tag{2}$$

where Abs_{745} is the absorbance, 745 nm, and GAE is the gallic acid equivalent in mg/L.

TCC and *TPC* results are provided in Table 2, together with ESI-MS results.

Twenty-four (24) bioactive compounds, falling into three classes—carotenoids, polyphenols and sterols—were identified in AVP. ESI-MS analysis of AVK indicated less than 0.5% of total bioactive compounds. Spectrophotometry indicated only trace amounts of carotenoids and polyphenols. These findings support a biological design of *Astrocaryum vulgare*'s fruit adapted to optimum plant proliferation where animals are attracted by the nutrition and health-beneficial compounds preferentially stored in the pulp while the seed is protected in a hard impenetrable "shell". MS and spectrophotometry results for the bioactive components of AVK are provided in Table S1 for the reader's review.

Table 2. Bioactive compounds identified in *Astrocaryum vulgare* pulp oil (AVP) using ESI-MS in positive mode in combination with UV-Vis spectrophotometry for quantification of total carotenoids (*TCC*) and polyphenols (TPC).

Compounds	Chemical Formula	Theoretical *m/z*	Experimental *m/z*	Adduct	Relative Intensity (%)	Concentration (mg/L)	Literature Values (mg/L)	Refs
				Carotenoids				
All-trans-β-Carotene All-trans-α-carotene 13-cis-β-carotene 15-cis-β-carotene	$C_{40}H_{56}$	536.4382	536.4379	[M]	84.32	1928.65	1296 ± 157.4	[14]
All-trans-lutein Zeinoxanthin Cis-lutein	$C_{40}H_{56}O_2$	568.4280	568.5626	[M]	4.17	95.38	0.79 * 1.02 * 0.04 *	[47]
All-trans-β-cryptoxanthin All-trans-α-cryptoxanthin Zeaxanthin	$C_{40}H_{56}O$	552.4331	552.4586	[M]	1.31	29.60	1.64 * 1.30 * 0.16 *	[47]
Estimated Total Carotenoid Content (mg/L)						2054.0 ± 91.0	778–2081	[14,41,48]
				Polyphenols				
Chlorogenic acid	$C_{16}H_{18}O_9$	354.0951	354.2691	[M + Na]$^+$	1.84	0.22	8.30 ± 0.2 *	
Caffeic acid	$C_9H_8O_4$	180.0422	203.0990	[M + Na]$^+$	3.19	0.38	0.07 ± 0.02 *	[15]
Rutin (quercetin glucoside)	$C_{27}H_{30}O_{16}$	610.1534	633.5074	[M + Na]$^+$	2.55	0.30	7.34 ± 0.1 *	

Table 2. *Cont.*

Compounds	Chemical Formula	Theoretical *m/z*	Experimental *m/z*	Adduct	Relative Intensity (%)	Concentration (mg/L)	Literature Values (mg/L)	Refs
Catechin	$C_{15}H_{14}O_6$	290.0790	313.2376	$[M + Na]^+$	0.82	0.10	27.08 ± 0.04 *	
Myricetin	$C_{21}H_{20}O_{12}$	318.0376	341.1901	$[M + Na]^+$	0.41	0.05	2.01 ± 0.05 *	[15]
Kaempferol	$C_{15}H_{10}O_6$	286.0477	309.0997	$[M + Na]^+$	0.33	0.04	<0.07	
Total Phenolic Content (mg/L)						1.1 ± 0.1	117.42 ± 1.67 *	
Phytosterols								
β-Sitosterol	$C_{29}H_{50}O$	414.3862	437.2012	$[M + Na]^+$	0.28	-		
Cycloartenol Cycloeucalenol	$C_{30}H_{50}O$	426.3862	449.3476	$[M + Na]^+$	0.55	-		
Arundoin 24-methylenecycloartanol	$C_{31}H_{52}O$	440.4018	463.0541	$[M + Na]^+$	0.31	-		

*—Values reported for AVP pulp extract but not for AVP pulp oil.

3.1.1. Carotenoids

Spectrophotometry measurements of AVP oil revealed a high level of carotenoids (2053.67 ± 90.67 mg/L). *TCC* values reported in the literature for AVP are generally lower by 27 to 1276 mg/L [14,41]. *TCC* is attributable to several factors including maturation stage, climate, ecology, harvesting method, handling, storage and processing of the fruits as well as extraction temperature [49]. In the present case, AVP oil was extracted using cold pressing at mild temperatures (30–40 °C). The *TCC* of 686.89 mg/L reported for *Mauritia flexuosa* oil (known as ite in Guyana and buriti in Brazil) [50] and 603.39 mg/L reported for *Elais oleifera* (palm) oil [51] are a third of those measured in AVP in this study.

ESI-MS indicated that carotenes constitute 94% (1928.65 mg/L) of the *TCC*. This significant presence of carotenes gives the fruit and oil their characteristic orange-red color. The remaining 6% of carotenoids were identified as xanthophylls. Representative chemical structures of carotenoid compounds identified in AVP oil are shown in Figure 1.

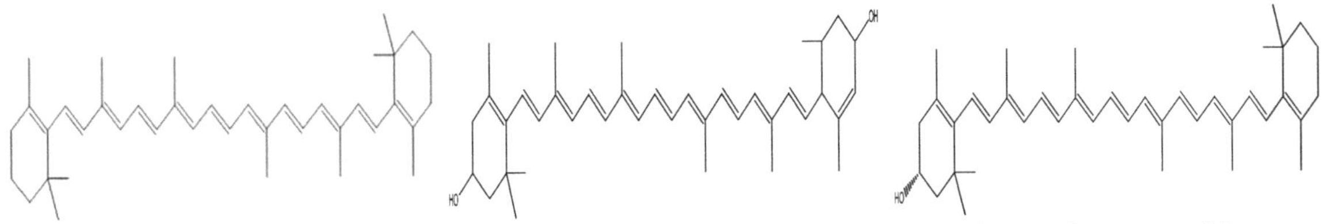

Figure 1. Chemical structures of representative carotenoid compounds identified in AVP oil.

The high level of β-carotene found in AVP makes it an important source of pro-vitamin A. In addition, AVP can potentially be used as a natural coloring agent in foods (e.g., cakes, crackers, candy) that can replace synthetic yellow-orange dyes such as tartrazine and sunset yellow, which have posed health and safety concerns for consumers [52]. Topical application of carotenoid-containing products offers protection against damage to the skin from sunlight, which reduces erythema (redness of the skin caused by inflammation). β-Carotene is known for its ability to protect against sunburns, making it a common ingredient in sunscreen formulations [53,54]. β-Carotene also helps protect the skin against oxidative damage, which manifests as wrinkles, by acting as a scavenger of singlet oxygen. The antioxidant properties of β-carotene also enhance the rate of skin regeneration, slowing the photoaging process [55].

Xanthophylls are a class of carotenoids that are oxygenated (e.g., lutein and cryptoxanthin, see Figure 1), which also offer skin benefits. They are known for their ability to increase skin hydration, particularly lutein, which penetrates intercellular lipids, reducing roughness of the skin.

3.1.2. Polyphenols

Six polyphenols were identified in AVP oil in small concentrations (TPC = 1.08 ± 0.1 mg/L). This is much lower than the TPC of AVP pulp extract (117.42 ± 1.67) reported in the literature [15], probably because most polyphenols are polar compounds and do not dissolve in non-polar lipids. The TPC of AVP is significantly lower compared to other plant oils of the Amazonian region such as *Caryocar brasiliense* (229.1 mg/L), *Attalea speciosa* (288.0 mg/L) and *Mauritia flexuosa* (309.9 mg/L) [56]. The amounts of polyphenols identified for AVP are lower than the minimum inhibitory concentration (MIC) against common microbes (e.g., *Escherichia coli* and *Staphylococcus aureus*), which ranges from 20 mg/L to 1024 mg/L [57–59]. This suggests that the antimicrobial activity of AVP against *E. coli*, *S. aureus*, *E. faecalis* and other microbes, reported by Rossato et al. [16], may be related to its β-carotene or the fatty acids present. β-Carotene has been shown to be effective against strains of *E. coli*, *S. aureus* at 100 mg/L [60]. Fatty acids, such as oleic acid, which makes up 65% of the fatty acid composition of AVP [12], can act as anionic surfactants that alter bacterial cell walls, which affects their ability to function at low pH levels [61].

3.1.3. Phytosterols

Phytosterols, compounds that closely resemble cholesterol in both structure and function, are found naturally. β-Sitosterol, cycloartenol, cycloeucalenol and arundoin 24-methylenecycloartanol were putatively identified in AVP oil by MS (Table 2). The same compounds were identified in AVP oil by Bony et al. [14] by gas chromatography, with a total content of phytosterols of 1497.2 ± 90.1 mg/L.

Phytosterols are often used as substitutes for lipids high in cholesterol to enhance cardiovascular health [62]. The antimicrobial [63] and anti-inflammatory activities [64] of phytosterols have been confirmed for AVP in in vivo studies [14]. Given its phytosterol content, AVP can potentially function as an edible oil that offers not only a distinct flavor but also health benefits from its anti-inflammatory properties.

3.2. Thermal Transition Behavior and Phase Diagram

The DSC cooling (5 °C/min) and heating (10 °C/min) thermograms of the AVK/AVP mixtures are displayed in Figure 2a,b, respectively. The corresponding thermal parameters and phase diagrams are presented in Figure 3a for the cooling cycle (T_{Con} and T_{C1-3}) and Figure 3b for the heating cycle (T_{Moff} and T_{M1-3}). The enthalpies of the individual endotherms represented in Figure 4 are used to estimate the relative content of the phases involved.

The DSC thermograms in Figure 2a,b indicate mainly one crystal phase for AVK (exotherm T_{C1} ~ 15 ± 1 °C and endotherm T_{M1} = 30 °C) and a minor phase at a very low temperature (T_{C3} ~ −40 °C and T_{M3} = −19 °C). Phase 1 accounts for ~96% of the total crystallization enthalpy (85 kJ/mol) and is related to the saturated medium-chain TAGs of the lipid. A slight shoulder (T_1^s at ~10 °C in Figure 2a and ~25 °C in Figure 2b) can be noted at the low-temperature side of T_1, which indicates a different small group of TAGs that melt at lower temperatures, namely OOO and POO found in AVK at 0.4% and 0.3%, respectively [11].

The AVP cooling thermogram shows a prolonged leading exotherm (T_1 in Figure 2a) over a relatively large window of temperature (~25 °C) followed by two crystallization events. The height of T_1 remained small but almost constant as it was cooled down to −30 °C, suggesting continuous nucleation and growth processes and the slow formation of small and probably disordered lamellar structures. According to its relative enthalpy, half of the material participated in this transformation. The lower-temperature exotherms were related to rapidly forming crystals from the nucleation of the low-melting groups of AVP TAGs, including those comprising mostly polyunsaturated fatty acids (PUFAs) such as linolenic acid. The phases observed during cooling were confirmed by successive melting events starting with a melt–recrystallization at ~−30 °C (Figure 2b).

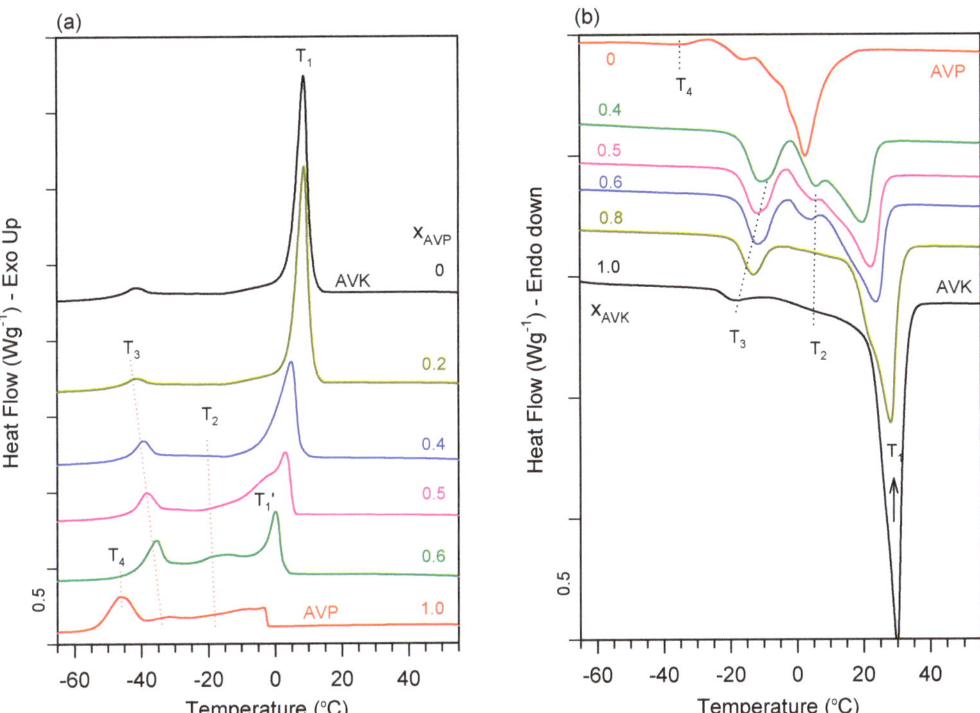

Figure 2. (**a**) DSC cooling (5 °C/min) and (**b**) heating (10 °C/min) thermograms of AVK/AVP mixtures.

As AVP was added, the total enthalpy of crystallization and melting both decreased linearly (Figure 3, $R^2 > 0.893$ and S.E.E < 6) at a rate of -31 ± 3 kJ/mol/mol. This indicates that the lattice energy of the fat steadily decreases at a constant rate as the unsaturated TAGs of AVP are mixed with the saturated TAGs of AVK.

With the increase in AVP concentration, the intensity of the leading peak (T_{C1} and T_{M1}), distinguishing the AVK solid, decreases, and its temperature shifts in a monotectic manner to low temperatures (★ in Figure 3a,b). T_{C1} and T_{M1} can be safely assigned to the crystallization and melting, respectively, of a phase made predominantly, if not exclusively, of AVK TAGs. The shoulder peak T_1^s becomes increasingly resolved and shifts to a lower temperature and appears to merge with the T_1 of AVP at concentrations higher than 0.50_{AVPO}. This suggests that AVK TAGs are increasingly taking part in the formation of the early AVP-AVK lamellae phase, which grows in an AVP-based crystal structure. The disappearance of the AVP lowest-temperature event (T_4 in crystallization in Figure 2a and melting and following recrystallization at −35 °C in Figure 2b) from the mixtures' transformation path ($X_{AVK} < 0.40$) and significant increase in the T_3 phase stability indicate that molecular organization improves even at very low temperatures and confirms a direct participation of AVK TAGs in the development of AVP-AVK crystals. The temperature of crystallization of T_3 (Figure 3b), and associated crystallization and melting enthalpies (Figure S1), increased linearly at a rate of 28 ± 3 kJ/mol/mol, indicating a gradual improvement in stability of the lowest-melting-temperature phase owing to the participation of AVK TAGs in its formation. The similar changes observed in T_2 also indicated a direct participation of AVK TAGs with AVP TAGs in the formation of the medium-temperature phase.

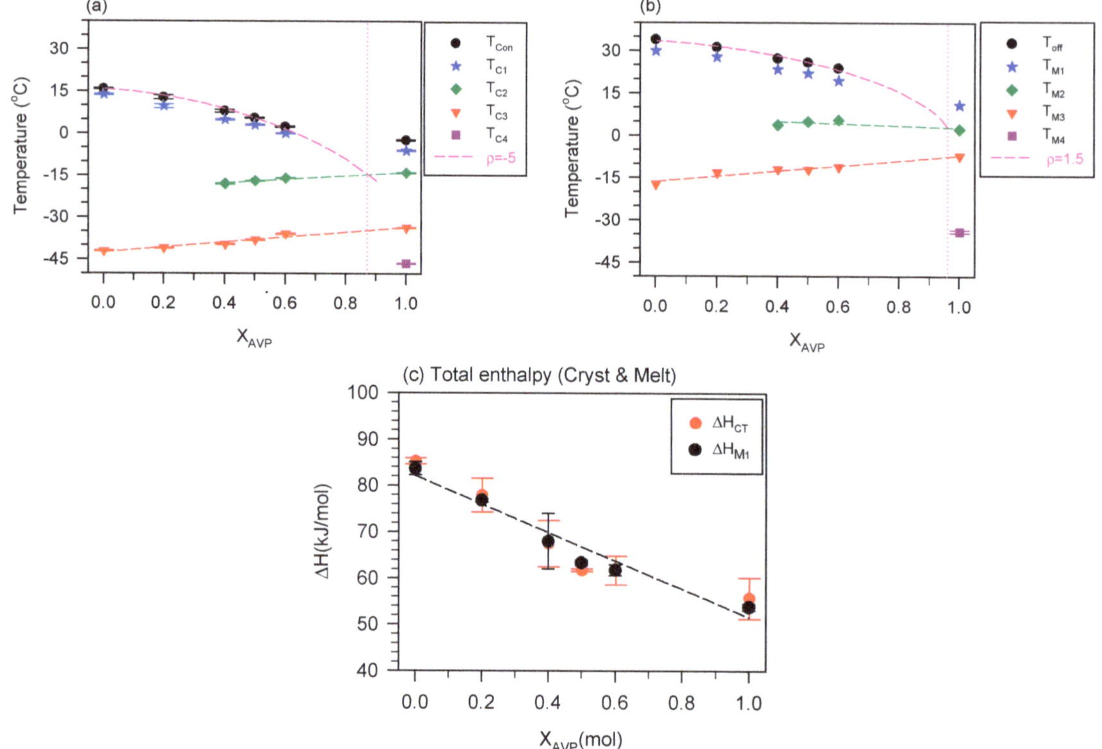

Figure 3. Thermal transition parameters and phase diagram $T(x_{AVP})$ obtained from DSC (**a**) cooling and (**b**) heating. Vertical dotted line is presumptive eutectic composition. Symbols in (**a**,**b**): •: onset temperature of crystallization (T_{Con}) or offset temperature of melting (T_{Moff}). ★, υ, θ and ν: crystallization and melting peak temperatures (T_{C1}, T_{C2}, T_{C3}, T_{C4}) and (T_{M1}, T_{M2}, T_{M3} T_{M4}), respectively. Lines passing through T_{Con} and T_{Moff} are simulations of the cooling and heating liquidus lines of the AVK/AVP binary system using the BW approximation (Equation (4), $\chi^2 = 1.13$ and 0.19, respectively). Dashed lines passing through T_{C2}, T_{C3}, T_{M2} and T_{M3} data are fit to straight lines ($R^2 > 0.996$). (**c**) Total enthalpy of crystallization and melting versus x_{AVP}. Dashed line passing through the enthalpy data are fit to a straight line ($R^2 > 0.893$, standard error of estimates < 6).

As AVP was added, T_{M1} decreased in an apparent monotectic manner but its FWHM remained constant at an estimated ~6 °C. This suggests that the crystals associated with T_{M1}, probably in the β' form, the main form shown by XRD for AVK (>90% reported in [65]), are essentially AVK TAGs. This AVK crystal phase seems to be relatively resilient to the influence of AVP components and remains well organized at the concentration levels considered here. The crystal nature of this phase and the phase associated with the shoulder peak were not determined for the mixtures, as such a task would involve investigation efforts involving XRD measurements, out of the scope of this work. However, if phase behavior studies of binary mixtures of pure lipid molecules such as those conducted by our research group on binary TAG systems [66,67], propanediol esters [68] and binary TAG/methyl esters systems [38,39,69] are of any indication, the most likely transformation path and phase content would be the common α to β' to β crystal forms. Note that more than one sub-form, particularly of β', may be present and that the β form may need a specific thermal treatment to be revealed.

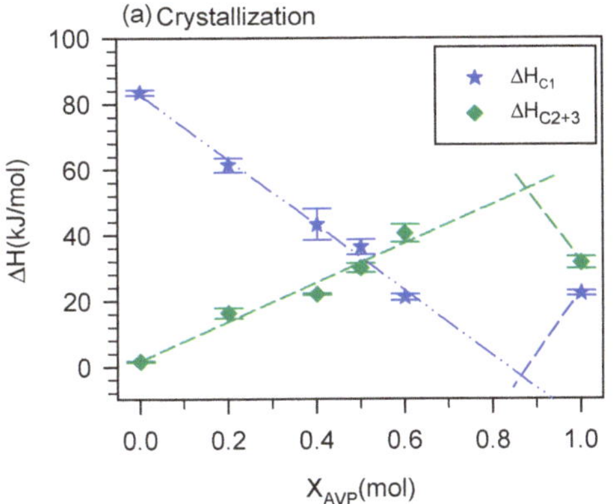
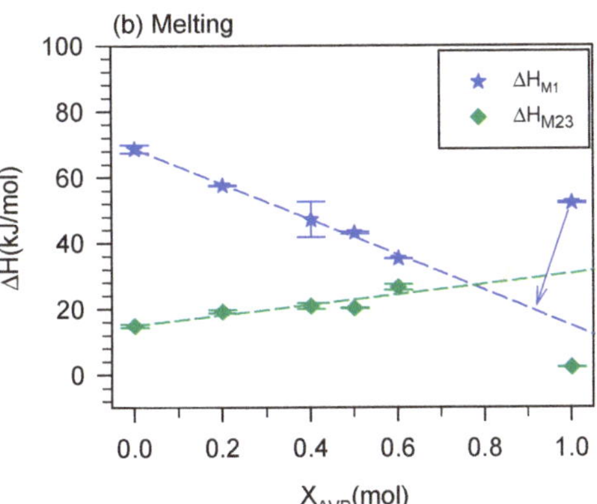

Figure 4. (**a**,**b**) Enthalpy of crystallization/melting of the leading peak ($\Delta H_{C/M1}$, ★) and combined trailing peaks ($\Delta H_{C/M2+3}$, υ). Dashed lines passing through the enthalpy data are fit to straight lines (R^2 = 0.816–0.995, S.E.E of 2–6). Solid drop arrow is a tentative location of the eutectic point using Tamman's diagram.

The AVK/AVP binary system can be related to the results obtained with pure TAGs in many aspects because of the composition homogeneity of each of these lipids and stark difference between the two ensembles [65]. Under these considerations, the binary system can be thought of on average as one made of two distinct "average TAG" molecules: a medium-chain lauric and myristic-acids TAG with a molecular mass of 687.27 g/mol on average for AVK (AVK primarily consisting of medium-chain saturated TAGs, such as LLL, LLM and MMM), and a long-chain oleic and palmitic-acids TAG with a molecular mass of 867.99 g/mol for AVP (predominantly consisting of long-chain unsaturated TAGs, including POO, PPO and OOO). This interpretation is corroborated by recent work on the mixing phase behavior of trilaurin and monounsaturated triacylglycerols based on palmitic and oleic fatty acids [70]. The phase diagrams of LLL/POP and LLL/PPO binary systems not only parallel the AVK/AVP binary system in terms of mixing behavior, they also show eutectic points at the very close level (X_{PPO} and X_{POP} = ~0.8). The small discrepancy in eutectic composition between these model systems and the AVK/AVP binary system (x_{AVP} = ~0.85) is attributable to the variety of medium-chain TAGs present in AVP. It is likely that the di-unsaturated TAGs and those comprising linoleic and linolenic fatty acids of AVK are responsible for pushing the eutectic point to higher oil concentrations and lower temperatures.

3.3. Thermodynamic Analysis of the Liquidus Line

Phase diagrams were constructed from both heating and cooling thermograms of the mixtures (cooling and heating phase diagrams in Figure 3a,b, respectively). Unlike DSC heating data, the phase diagram drawn from DSC cooling data does not represent the equilibrium state because of kinetic effects related to mass and energy transfer limitations. Cooling phase diagrams are, however, important as they can be directly contrasted with the pseudo-equilibrium phase diagrams from the DSC heating data. This allows us to gain insight into variations concerning concentrations of kinetic, mass and energy transfer effects. The cooling and heating liquidus lines of the AVP/AVK binary system were generated from the onset temperature of crystallization for the cooling data and offset temperature of melting (T_{Con} and T_{Moff} (•) in Figure 3a,b, respectively), as is typically done in the study of binary lipid mixtures [66–68,71–74]. T_{Moff} is particularly suitable for studying pseudo-equilibrium properties because it is determined by the most stable crystals. T_M (T_C) of the other peaks are used to represent the transition lines after correction for the transition widths of the pure components [75].

Two solidification lines constructed from T_2 and T_3 of the cooling and heating thermograms (υ and θ lines in Figure 3a,b) were very well described by slightly sloped straight lines ($R^2 > 0.951$). Line slopes are higher in the cooling phase diagram, which is attributable to greater kinetic effects during cooling. These lines are associated with the formation of α and maybe low-stability β' phases.

The liquidus line was calculated using the Hildebrand equation [76] coupled with the Bragg–William (BW) approximation for the non-ideality of mixing [77,78]. The result is included as dashed lines passing through the liquidus data Figure 3a,b for the cooling and heating phase diagrams, respectively. This model is commonly used to investigate the miscibility of lipid mixtures [39,66–68,79–82]. The deviation from an ideal behavior is described by a non-ideality of the mixing parameter, ρ (J/mol), defined as the difference in energy between mixed pairs (AB) and like pairs (AA and BB):

$$\rho = z\left(u_{AB} - \frac{u_{AA} + u_{BB}}{2}\right) \tag{3}$$

where z is the first coordination number, u_{AB}, u_{AA} and u_{BB}, the interaction energies for AB, AA and BB pairs, respectively. $\rho = 0$ indicates an ideal mixing, and a negative ρ reflects a tendency for order, whereas a positive ρ reflects a tendency of like molecules to cluster, which beyond some critical value leads to phase separation [74,76]

The branches of a liquidus line comprising a eutectic are described by the following equation [82,83]:

$$\ln X_{A(B)} + \frac{\rho\left(1 - X_{A(B)}\right)^2}{RT} = -\frac{\Delta H_{A(B)}}{R}\left(\frac{1}{T} - \frac{1}{T_{A(B)}}\right) \tag{4}$$

where R is the gas constant. X_A represents the mole fraction of A, ΔH_A and T_A are the molar heat of fusion and the melting point of component A. X_B, ΔH_B and T_B are those corresponding to component B. A or B are considered depending on whether the liquids segment composition is smaller or larger than the eutectic composition X_E.

Considering a monotectic phase diagram, the parameters of AVK (T_A and ΔH_A) were used to simulate the liquidus line. The standard method of least squares approach was used to obtain ρ and calculate the liquidus lines. The results for the cooling and heating liquids lines are shown in Table 3. The calculated p-values are comparable to published values for binary lipid systems [67,84].

Table 3. Parameters of the Bragg–William approximation (Equation (4)) used to simulate the liquidus line and corresponding values of the non-ideality of the mixing parameter, ρ. ΔH_A: enthalpy of melting and T_A: melting temperature.

Liquidus	Region	T_A (C)	T_A (K)	ΔH_A (kJ/mol)	ρ (kJ/mol)	χ^2
Cooling	$0 \leq X_A \leq 0.95_{AVPO}$	14.0	287.2 ± 0.5	76 ± 2	-5	1.13
Heating	$0 \leq X_A \leq 0.90_{AVPO}$	33.5	306.7 ± 0.5	83 ± 3	-1.5	0.19

The calculated liquidus lines perfectly matched the experimental data of AVK and the mixtures but did not pass through the AVP melting point, indicating a eutectic at a high AVP concentration. The T_2 solidification line is, therefore, a eutectic line, whose intersection with the calculated liquidus line indicates a presumptive eutectic point. The T_2 line of the cooling phase diagram intersected with the liquidus line at 0.85_{AVPO} and that of the heating phase diagram at 0.95_{AVP} (vertical dotted lines in Figure 3a,b). The eutectic of the AVK/AVP binary system is likely close to $\sim 0.95_{AVP}$, the point determined from the heating phase diagram, because quasi-equilibrium conditions are achieved during heating but not cooling.

Analyses of the enthalpy of crystallization and melting of the leading peak (T_1) as well as that of the combined trailing peaks T_2 and T_3 ($\Delta H_{C/M1}$, ★ and $\Delta H_{C/M2+3}$, υ in Figure 4a,b) suggest a Tamman's triangle shape. Tamman's diagram is an important outcome from the theory of phase diagrams [85,86] and is a characteristic of the eutectic effect and shows linear dependence of molar enthalpy on either side of the eutectic point [87]. The data of Figure 4a,b indicates a point, x_{AVP}, higher than the mixture with the maximum concentration studied here, i.e., 0.6_{AVP}, confirming the results from the cooling and heating phase diagrams of Figure 3a,b.

The presence of eutectic points close to one pure component is common in binary systems comprising pure components with significant differences in structure and melting temperature [88]. Given the high AVP content of the eutectic point, the AVK/AVP binary system may be considered monotectic for all intents and purposes.

The BW approach yielded small negative values for ρ for the simulation of both the cooling and heating liquidus lines (Table 3). The values of ρ obtained for the cooling and heating liquidus lines are rather small (-5 kJ/mol and -1.5 kJ/mol, respectively) and indicate a mixing behavior close to ideal with a tendency for order and exclude the tendency of like molecules to cluster and phase separation [82]. The result indicates that the formation of mixed pairs in the liquid state is not excluded. This result supports the conclusions from the analysis of the individual melting peaks in relation to their associated groups of TAGs. This gives credence to the conclusion that AVK molecules participate with AVP TAGs to increase the stability of the phase that AVP makes at low temperatures (T_{M2} and T_{M3} phases).

3.4. Solid Fat Content

$SFC(T)$ profiles of the AVK/AVP mixtures are reported in Figure 5a. Typical SFC values of CB from the literature [32,89,90] are included in Figure 5b for comparison purposes. When considered from high temperatures (the melt) to low temperatures (solidification), the $SFC(T)$ data of each mixture followed a growth trajectory that was very well described by the three-parameters Gompertz function (Equation (5)).

$$SFC(T) = SFC_0 \exp(-\exp(-\frac{T - T_0}{b})) \tag{5}$$

where SFC_0 is the maximum fraction of solid fat, which specifies the SFC magnitude, T_0 is the inflection point, which in the case of a growing $SFC(T)$ from zero (melt) to a maximum value (SFC_0) quantifies the induction temperature, and b is the maximum specific growth rate, which can be derived by calculating the first derivative at the inflection point.

The fit of the Gompertz function to the SFC versus temperature data of the mixtures above 5 °C yielded curves with excellent goodness of fit ($R^2 > 0.998$, standard error of estimates <0.94). The SFC measured at 5 °C is above the calculated maximum because of the unsaturated TAGs present, which start a new mechanism of crystallization that is described by a step segment. The presence of the segment that would represent this mechanism is obvious from the step increases in SFC for all the mixtures. It has not been measured because it is irrelevant to the present study. The model provided values of the parameters, which are significantly different for each material. The Gompertz SFC magnitude decreased with x_{AVP} following a straight line toward a theoretical value of zero for pure AVP (Figure 5c, $R^2 = 0.989$). The large slope obtained for the decrease ($-56 \pm 3\%$/mol) indicates a steep, gradual and continuous drop in overall SFC with increasing AVP. The parameter linked to the induction temperature of the mixtures ($T_{0\text{-SFC}}$; Figure 5d) also decreased with increasing AVP in an apparent monotectic manner. The rate of change of $T_{0\text{-SFC}}$ with AVP departs significantly from that of the DSC liquidus line. The rates of change of $T_0(x_{AVP})$ and $T_{Moff}(x_{AVP})$ data estimated by linear regression ($R^2 = 0.989$) were -7.6 ± 1.0 and -16.9 ± 0.7 °C/mol. This difference is because $T_{Moff}(x_{AVP})$ was obtained from samples crystallized at a fixed rate (5 °C/min), whereas the SFC was measured in samples quickly brought to temperature and left to crystallize isothermally.

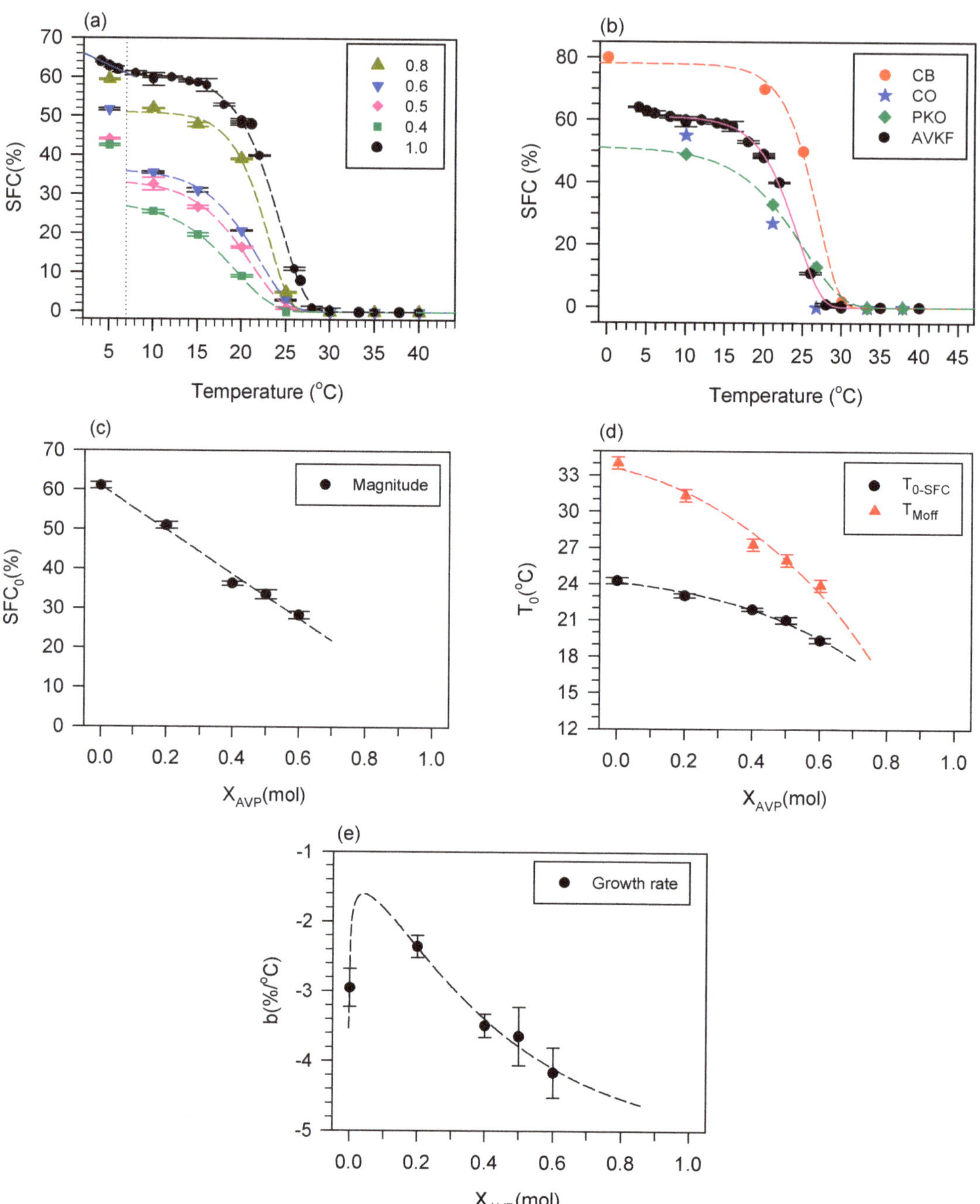

Figure 5. (**a**) $SFC(T)$ profiles of AVK and AVK/AVP mixtures. Dashed lines are fits of the Gompertz function (Equation (5), $R^2 > 0.998$) to the data. (**b**) SFC versus temperature of AVK (this work), cocoa butter (CB) [30], coconut oil (CO) [30], palm kernel oil (PKO) [30] and shea butter (SB) [91]. (**c–e**) Variation with AVP molar ration x_{AVP} of the Gompertz fit parameters: (**c**) SFC magnitude (SFC_0), (**d**) induction temperature (T_0) and (**e**) maximum SFC specific growth rate, b. Dashed line in (**c**) is a fit to a straight line ($R^2 = 0.989$) and dashed line in (**e**) is a fit to a Weibull function ($R^2 = 0.992$).

The specific growth rate of the mixtures started with a value for pure AVK lower than that of the 0.2_{AVP} mixture, then decreased in an exponential manner as the AVP concentration was increased, indicating the occurrence of a maximum between pure AVK and 0.20_{AVP} (Figure 5e). The fit of the $b(x_{AVP})$ data, excluding AVK, indicated that b of the mixtures with $x_{AVP} < 0.3$ is higher than that of AVK. The shape of the $b(x_{AVP})$ curve is reminiscent of a Weibull function with a prolonged tail. The fit of $b(x_{AVP})$ data to the Weibull function was excellent (dashed line in Figure 5e, $R^2 = 0.992$, standard error of estimates = 0.13) and yielded a maximum value at 0.04_{AVP} ($b = -1.6$). This suggests that AVP components have an accelerating effect on the SFC growth of the AVK/AVP mixtures that is exponentially overcome, bringing it to AVK levels at $x_{AVP}{\sim}0.3$. This is interesting as it shows that the crystallization of mixtures with $x_{AVP} < 0.3$ is faster than that of pure AVK.

As shown in Figure 5b, AVK has an SFC lower than that of CB and higher than that of the other most-used commercial fat alternatives such as coconut oil (CO) [30], palm kernel oil (PKO) [30] and shea butter (SB) [91].

The SFC values of AVK are overall higher than those of CO, PKO and SB at all temperatures above the melting point. Furthermore, its SFC is sharper in the melting range (20–30 °C). CO, PKO and SB are, in fact, too soft to make CB alternatives (called CB equivalents or CBEs) without being fractionized first [32]. AVK's sharp melting profile, high SFC in the hardness region (SFC > 70% for T < 25 °C) and intensive melting in the 27 to 30 °C range, which would bring about a cooling sensation in the mouth and flavor release, suggest that it would serve as an excellent confectionery fat. The zero SFC value at temperatures above 30 °C indicates the absence of a fat fraction, which may cause a waxy taste.

AVK is a lauric fat which, although probably not compatible with CB, would be a good CB substitute (CBS). Small corrections to the SFC profile of AVK may extend its compatibility with CB. The blending of AVK with the pulp oil of the same fruit achieved a range of materials with predictable SFC profiles. Overall, the SFC profiles of the mixtures and the Gompertz analysis indicate SFC phase diagrams with smooth boundaries at all temperatures starting just above the melt. The variation in the Gompertz modeling parameters is consistent with the monotectic nature of the liquidus line of the AVK/AVP binary system. These data indicate that it is possible to formulate AVK/AVP fats and, hence, lipid products that would have physical properties in a continuous spectrum of melting, hardness and plasticity. These fats would be suitable for varied applications ranging from hard coatings to soft fillings for foods, fillers, binders, lubricants, solubilizers, emulsifiers and emollients in a variety of cosmetics and pharmaceuticals delivery systems including tablets, capsules, suppositories, emulsions, ointments, creams and lotions. Furthermore, the AVK/AVP mixtures incorporate therapeutic attributes that are often absent in the conventional fats traditionally employed in these industries.

3.4.1. Implications for Food and Cosmetics Applications

Implications of the Mixtures' Melting Point in Foods and Cosmetics

Fats that have melting points (M_p) below 35 °C are highly valued as both cosmetic and confectionery fats because they melt at body temperature. This property allows them to provide a distinctive melting sensation either in the mouth or on the skin. Common examples of such fats include cocoa butter (CB) ($M_p = 23–33$ °C) [30,92], coconut oil (CO) ($M_p = 23–29$ °C) [30,92] and palm kernel oil (PKO) ($M_p = 27–29$ °C) [93]. The M_p of AVK and mixtures containing ≥ 0.5 mole fraction of AVP (T_{M1} in Figure 2b) are within M_p ranges of CB, CO and PKO, underscoring their potential for similar applications in the food and cosmetic industries.

Fat/oil phase separation is a notable issue in cosmetic products, and is a phenomenon that often appears as "sweating" characterized by the formation of droplets, in items such as balms and lipsticks, impacting their quality and appearance [94]. The mixing behavior as described by the phase diagram and BW model suggests that AVK/AVP mixtures are not likely to be affected by phase separation and may be considered for use in balms and lipsticks to mitigate issues associated with sweating.

Implications of the Mixtures' SFC in Cosmetics

The SFC significantly affects various qualities of cosmetic products, such as their visual attractiveness, temperature stability, ease of spreading and the sensation they leave on the skin. Low SFC levels prevent cosmetic fats from attaining the required flexibility, leading to an oily film on the skin. Conversely, overly high SFC levels may cause a product to impart a waxy texture. An ideally balanced SFC level, therefore, is key to improving a cosmetic's ease of application and its ability to moisturize the skin effectively, without any unwanted residue [95]. Ideally, fats for cosmetic purposes should have a moderate SFC (30–50%) at room temperature (20 °C), but transition rapidly to an SFC of 0% at body temperature (35 ± 2 °C). This ensures that the product is soft and spreads easily on the skin, but melts quickly on the skin, leaving a cooling sensation [95]. Common examples of cosmetics fats include raw cocoa butter, raw shea butter, mango butter and coconut oil, whose SFCs at 20 °C are 78% [96], 37% [91], 40% [97] and 27%, respectively [30].

The SFC of raw cocoa butter (CB) at 20 °C is higher by more than 28% than that of AVK and AVK/AVP mixtures. Because of this high SFC, raw CB is hard and brittle at room temperature, making it challenging to spread on the skin, and it leaves a waxy feel on the skin. CB is often mixed with other oils, such as mineral oil, to decrease its SFC, to be more suitable for cosmetic use. Notably, pure AVK, which has a lower SFC overall, would require less manipulation to achieve an SFC level suitable for cosmetic formulations.

The mixture containing a 0.2 mole fraction of AVP has an SFC at 20 °C (39%) comparable to shea butter (SB) [91] and mango butter (MB) [97], suggesting that it may be a viable substitute in creams and balms. The substitution would also offer therapeutic properties associated with the bioactive compounds present in AVP. The SFC profile of all AVK/AVP mixtures indicates rapid complete melting above 30 °C, a desirable characteristic in cosmetic fats as it would translate to the melting of the product in contact with the skin without leaving a waxy sensation.

The mixtures containing 0.4 and 0.5 mole fractions of AVP presented lower SFC values at 20 °C (20.7% and 16.7%, respectively) compared to SB, MB and CO and may be more suitable for cosmetic applications where fats with greater fluidity are desired, such as in the formulation of lotions and sunscreens. The mixture with 60% of AVP has an SFC of 9.1% at 20 °C, a relatively low level that may lead to oily residues when applied to the skin. Mixtures with such high AVP concentrations may not be suitable for use as cosmetic fats.

Implications of the Mixtures' SFC in Foods

The SFC plays an important role in the texture of various food products like chocolate, shortenings, margarines and butter [98]. The SFC's value at specific temperatures significantly influences the texture and mouthfeel of these products. For instance, margarine intended for use directly after refrigeration has an SFC of no more than 32% at 10 °C, and below 10% at room temperature (20–22 °C) to ensure optimal spreadability. Additionally, to avoid a waxy sensation in the mouth, the SFC of margarine must be under 3.5% at body temperature (35–37 °C) [99]. Figure 6 compares the SFC of the AVK/AVP mixtures to the recommended SFC of margarine at 10 °C, 20 °C and 35 °C.

The mixtures containing $\geq$ 0.5 mole fraction of AVP present SFC values at 10 °C, 20 °C and 35 °C that are considered too high for ideal textures in margarine. $AVP_{0.6}$ presents an SFC profile at 10 °C, 20 °C and 35 °C that meets the recommended SFC profile of good-quality margarines. The trends observed in SFC suggest that even higher concentrations can be used as substitutes for margarines and spreads, particularly those containing hydrogenated and partially hydrogenated fats and restricted in North America and Europe due to health concerns [100].

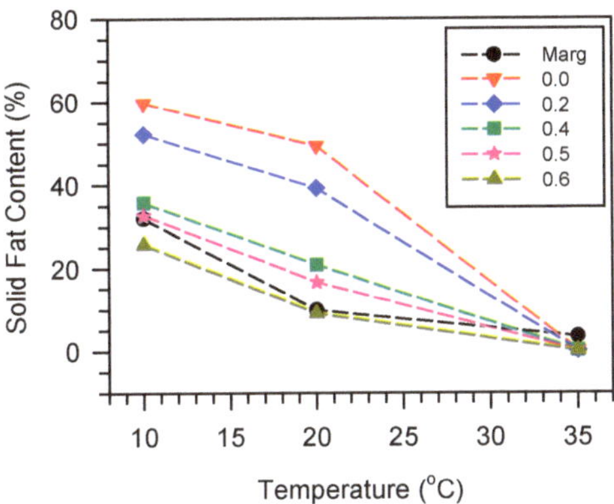

Figure 6. SFC of AVK/AVP mixtures (x_{AVP} = 0.0–0.6) compared to the recommended SFC for margarine (●, Marg).

Fats are also widely used in the food industry as filler shortenings in cookies or wafers. A good filler fat is formulated to provide structural support to the fragile (brittle) cookies or wafers, but also to have a steep melting profile, indicated by a rapid transition from a high SFC at room temperature to an SFC of approximately 0% at around 40 °C [31]. The *SFC(T)* curve of the 80% AVK and 20% AVP blend (Figure 5a) presents a relatively high SFC at 20 °C (39.1 ± 0.2%) and a steep decrease to 0% at 30 °C, indicating a very short plastic range and potential suitability as a filler.

The SFC profile of the AVK/AVP mixtures suggests that they are not optimal for use as shortenings in bakery products. Their SFC at 40 °C, which is 0%, would not achieve optimal baking results and proper development of dough structure, which require maintaining an SFC of at least 20% at 25 °C and no less than 5% at 40 °C [101].

3.5. Microstructure

Figure 7a shows PLM images of AVK and its mixtures captured while the sample was cooled from the melt at 1 °C/min. Figure 7b–e show the results of the analysis of the PLM including the induction temperature and evolution of the crystal size. The microstructure observed by PLM in AVKF and 0.8_{AVKF} began with the formation of small spots, which expanded radially to form spherulites typical of lauric-rich fats [102]. For the mixtures with higher AVP contents (0.4, 0.5 and 0.6), a significant change in the type of microstructure was observed. In these mixtures, small rods were first formed, then organized in spherical clusters.

The changes in microstructure between the mixtures can be related to crystal growth rate as depicted by the Gompertz parameter $b(x_{AVPO})$ (Equation (6)) $b(x_{AVPO})$ indicated that at low molar concentrations, AVP acts as a catalyst for crystallization up to a critical concentration ($\sim0.3_{AVP}$), at which time it starts to slow. This suggests that below a concentration threshold, the TAGs with long unsaturated fatty acid chains (e.g., OOO and POO) of AVPO act as a low-viscosity medium in which heat transfer and molecular movement of AVK components are improved compared to AVK in such a way that individual molecular organization and isotropic growth is facilitated. The rod-shaped microstructures observed at a higher-than-threshold concentration (x_{AVP} > 0.3) indicate that the AVP TAGs participate more actively in the formation of the early crystals, favoring a single active surface of growth. The clustering of the rods in increasingly smaller spheritic-like entities suggests a persisting but decreasing effect of AVK components on the overall microstructure development with a tendency to favor radial growth.

The phase behavior of trilaurin (LLL) and triolein (OOO) binary systems studied by Paluri et al. [103] mimics in many aspects the present AVK/AVP binary system. Although it differs in some aspects relating to thermal transformation, it shows striking similarities in terms of microstructure development. Particularly, the LLL/OOO mixture exhibited crystal morphologies like those observed in AVK/AVP mixtures. For instance, and similarly to their AVK/AVP counterparts, the $0.80_{LLL}/20_{OOO}$ mixture developed spherulitic-type crystals, and the $0.60_{LLL}/0.20_{OOO}$ mixture formed clusters of thin rods [103]. This is not surprising because of the close TAG composition of AVK and AVP to LLL and OOO. AVK is a lauric fat, comprising mostly lauric and myristic fatty acids, and AVP comprise long unsaturated fatty acid chains, mostly oleic acid. The dual effect of the long unsaturated fatty acid TAG on the development of the crystals of medium fatty acid TAGs is probably more prevalent, and its mechanisms warrant further investigations.

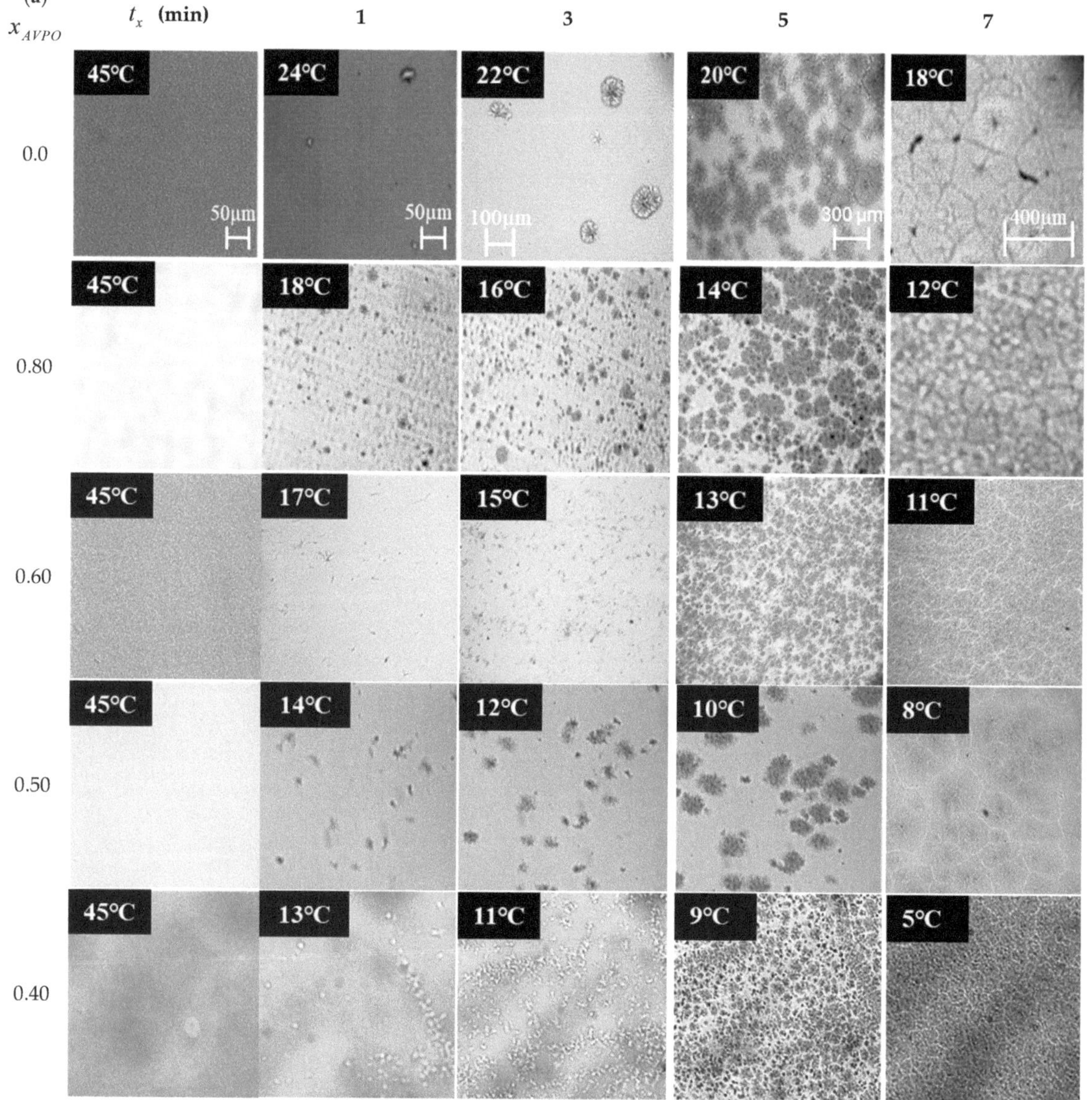

Figure 7. *Cont.*

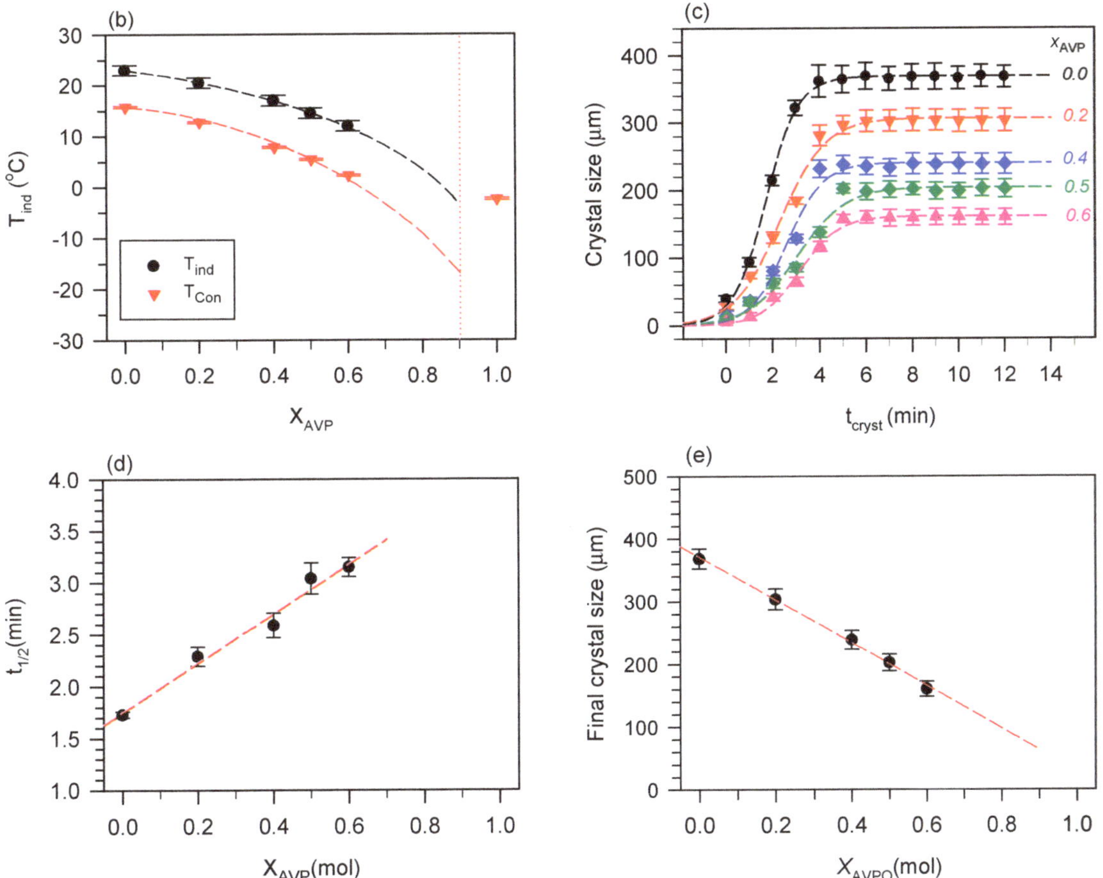

Figure 7. (**a**) PLM images ($\times 50$ magnification) of AVK/AVP mixture taken during cooling (1 °C/min) from the melt. t_x is the time (± 0.5 min) after induction temperature is reached. (**b**) Induction temperature versus AVP molar fraction (x_{AVP}) as determined by PLM (1 °C/min) compared to onset temperature of crystallization from DSC (5 °C/min). Dashed lines are simulation of the liquidus line using the Bragg–William approximation (Equation (4), $\chi^2 = 0.45$) and ρ is the calculated BW parameter. (**c**) Evolution of microstructure size with crystallization time as a function of mole fraction of AVP. Crystallization time is measured after induction time ($t_x = t - t_{ind}$). Dashed lines are simulation of the data with a 3-parameters sigmoid (Equation (6)) (**d**) $t_{1/2}$ at half maximum ϕ values from (**c**). (**e**) Final crystal size versus x_{AVP}. Dashed lines in (**d**,**e**) are fit to straight lines ($R^2 > 0.991$).

Figure 7b shows that as the AVP concentration was increased, the PLM induction temperature (T_{ind}) decreased in a monotectic manner, like the liquidus lines T_{C1} and T_{M1}. The T_{ind} versus AVP mole fraction curve lays with the same curvature between the T_{C1} and T_{M1} liquidus lines and is also very well described by the BW approximation with a non-ideality of the mixing parameter, ρ, of -2 kJ/mol. This value is between those obtained for the simulation of the cooling and heating liquidus lines (-5 and -1.5 kJ/mol, respectively). The mixing behavior indicated by microscopy is above that of the DSC cooling experiments because of differences in sample mass and shape as well as cooling rate (1 °C/min instead of 5 °C/min) and is significantly close to the quasi-equilibrium state indicated by the DSC heating experiments.

The evolution of crystal size with crystallization time (Figure 7c) follows an S-shape reminiscent of sigmoid functions, which were developed to model general growth under probabilistic assumptions [66]. A three-parameters sigmoid function (Equation (6)) described very well the crystal size versus crystallization time as measured after induction time ($t_x = t - t_{ind}$).

$$\phi(t_x) = \frac{\phi_0}{1 + e^{-\left(\frac{t_x - t_{1/2}}{b}\right)}} \tag{6}$$

The return value (ϕ axis) is in the range of 0 at $t_x = 0$ to the maximum ϕ_0 at the end of crystallization. $t_{1/2}$ is the inflexion point of the curve and defines the half-maximum y value. The slope of the curve is characterized by $1/b$ at its midpoint, and this represents the steepness of the curve. Parameter b was not significantly different for the mixtures (0.86 ± 0.14), indicating a similar homogeneous crystal development. The half-maximum ϕ values were achieved at increasingly longer times. Figure 7d shows that $t_{1/2}$ linearly increased with x_{AVP} ($R^2 = 0.995$) at a rate of 2.4 ± 0.2 min/mol. The final microstructure size versus AVP concentration decreased linearly (Figure 7e, $R^2 = 0.997$) at a rate of 3.42 μm/%mol. The size, shape and distribution of crystals are critical for achieving the desired consistency and consumer acceptance [104]. Products with smaller crystals tend to be firmer, whereas those containing crystals ranging from 30 to 200 μm often exhibit a sandy mouthfeel [105]. The crystals of AVKF/AVPO mixtures as crystallized here (cooling from the melt at 1 °C/min) did not present the sandy textures suggested by their size (between 160 μm to 370 μm). This is probably because these microstructures are made of small rod-shaped crystallites, which may be the origin of the smooth feel in the mouth.

3.6. Instrumental Textural Analysis

The principle of the test procedure was to measure the curves that appear during compression with a fixed force (shear) and when the male cone is retracted (adhesive curve) (Figure 8). The textural parameters of the samples were obtained from the shear and adhesive curves. Their peak heights indicate firmness and stickiness, respectively, and peak areas indicate the work of shear, which indicates spreadability, and the work of adhesion, which indicates adhesiveness, respectively.

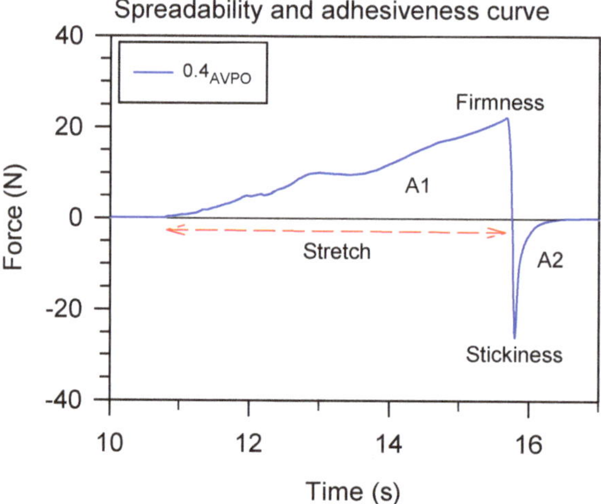

Figure 8. Shear and adhesive curves obtained during fixed force compression and drawing back of the cone spreadability rig for the 0.4_{AVP} mixture. Peak height during compression and drawing back: firmness and stickiness, respectively. Peak area A1 and A2: work of shear and adhesion, respectively.

Measurements of each sample were repeated three times, and the average value and the coefficient of variation were calculated. The stretch values of the curves were not significantly different, indicating similar adhesion to the cone, which justifies quantitative comparison between samples. The results obtained for the AVK/AVP mixtures are presented in Figure 9 for (a) firmness, (b) spreadability, (c) stickiness (adhesion force) and (d) adhesiveness (work of adhesion). The hardness results obtained with the 5 mm diameter ball probe are plotted in Figure 9a to allow for easy visual comparison with the firmness results.

Pearson Product Moment Correlation results (correlation coefficients R^2 and p-values) between spreadability, hardness, firmness, stickiness, work of adhesion, SFC and melting point are presented in Table 4.

Table 4. Pearson Product Moment Correlation results (correlation coefficients R and p-values) between crystal size, spreadability, hardness, firmness, stickiness, work of adhesion, SFC and melting point.

	Spreadability		Hardness		Firmness		Stickiness		Work of Adhesion		SFC		Melting Point		Crystal Point	
	R	$p<$	R	$p<$	R	$p<$	R	$p<$	R	$p<$	R	$p<$	R	$p<$	R	$p<$
Crystal size	0.864	0.06	0.890	0.04	0.879	0.05	0.462	0.43	0.703	0.19	0.991	0.001	0.997	0.0001	0.996	0.0002
Spreadability			0.9998	0.0001	0.999	0.00002	0.815	0.09	0.270	0.66	0.884	0.05	0.854	0.06	0.820	0.09
Hardness					0.999	7×10^{-5}	0.779	0.12	0.326	0.59	0.910	0.03	0.883	0.05	0.850	0.07
Firmness							0.793	0.11	0.304	0.62	0.900	0.04	0.871	0.05	0.838	0.08
Stickiness									0.304	0.62	0.465	0.43	0.427	0.47	0.384	0.52
Work of adhesion											0.688	0.20	0.728	0.16	0.762	0.13
SFC													0.996	0.0003	0.986	0.002

Hardness, firmness and spreadability determined by the cone geometry decrease in a similar exponential fashion toward zero (Figure 9a,b, $R^2 > 0.938$). The decay parameters obtained for hardness, firmness and spreadability are all close (steepness b = 6.4 ± 0.2, 6.9 ± 0.4 and 6.0 ± 0.8, respectively), indicating that the higher the AVP concentration, the more the mixture's structure approaches that of a liquid.

Stickiness versus x_{AVP} data from the cone spreadability rig (Figure 9d) are very well described by a three-parameters sigmoid function (Equation (6), $R^2 = 0.999$) and standard error of estimate (1.2). The calculated inflexion point of the curve is 0.37_{AVPO}, and the slope at the midpoint is 11 N/mol. The fundamental mechanisms that cause stickiness depend on the nature of the forces involved, adhesive, possibly in combination with cohesive, forces, as well as factors such as viscosity and viscoelasticity [106]. Differences in particle shape and size, liquid channels and external factors like temperature and moisture also contribute to stickiness. The growth model under probabilistic assumptions is a generalized characterization of the way in which an extent to which these forces and parameters contribute to the changes in stickiness with the increase in the oil concentration.

The work of adhesion versus AVP concentration showed a peak-like curve (Figure 9d) reminiscent of the maximum specific growth rate determined for the SFC (Figure 5e). Although the fit with a Weibull function yielded inconclusive parameters, the presumptive location of its peak and critical concentration (value at which the same adhesiveness is recorded) clearly matched. It is, therefore, safe to assume that the main factors affecting stickiness in the AVK/AVP binary system are those related to crystal growth and the resulting microstructure.

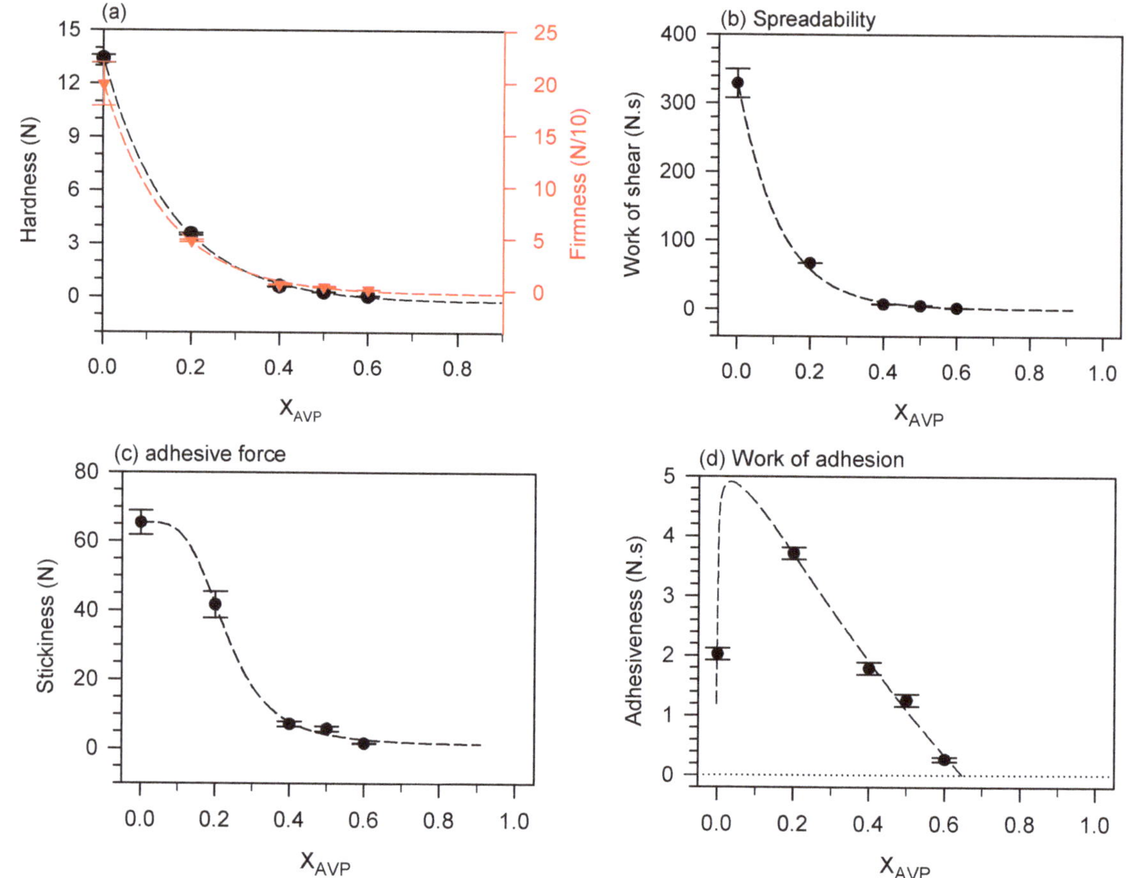

Figure 9. Instrumental textural properties of AVP/AVK blends—(**a**) hardness (●) and firmness (▼), (**b**) spreadability (work of shear). Dashed lines in (**a,b**) are fits to an exponential decay function, $R^2 > 0.996$. (**c**) Stickiness (●). Dashed line is a fit of a 3-parameters sigmoid function to the data (Equation (6), $R^2 = 0.999$ and standard error of estimates = 1.2), and (**d**) adhesiveness (work of adhesion) of AVK/AVP mixtures (●). Dashed line is a guide for the eye.

Correlation between the Textural Parameters and SFC, Melting Point and Crystal Size

Pearson Product Moment Correlation results indicated a higher correlation coefficient and lower *p*-value between firmness and spreadability than between firmness and hardness (Table 4). The relationships between firmness and spreadability and hardness, shown in Figure 10a, indicate that the discrepancy is mainly from the mixtures with high AVK amounts (AVK and 0.2_{AVP}). The comparison of all the mixtures indicated no correlation, for example between firmness and stickiness or spreadability. However, a closer look at the correlation graphs (see, for example, the case of firmness and stickiness in Figure 10b) indicates two strong linear correlations that can be delimited at ~0.3_{AVP}, one for the AVK-rich side and another for the AVP-rich side of the binary system. This, again, can be related to the stark differences in the crystal and SFC rates of growth discussed above. These data suggest that the properties related to the crystal growth of the present fat/oil mixtures are related to textural properties in two separate ways depending on the shape, size and probably the polymorphism of the phases present. The textural properties of fat are, in fact, the result of the three-dimensional network of fat crystals associated with a continuous oil phase [31,107–109]. The strength of crystal networks depends not only on the amount of solids present (% SFC) but also on the polymorphic behavior and the crystal sizes formed [110]. In the case of the AVK/AVP binary system, these two ways are dictated by the effects of the unsaturated long-chain TAGs of the oil on the crystallization of the medium-chain TAGs of the fat. The critical concentration at which the crystallization mechanisms change was shown to be 0.3_{AVP}.

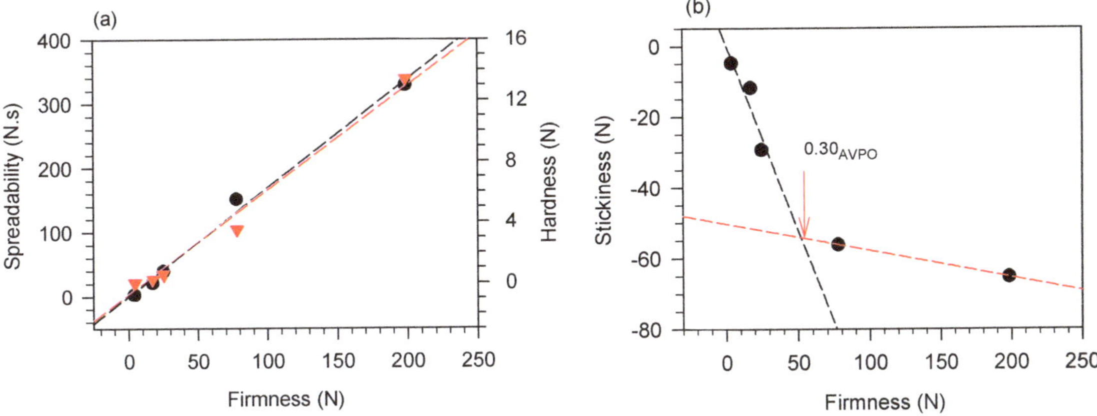

Figure 10. Relationship between (**a**) firmness and spreadability (•) firmness and hardness (▼), and (**b**) firmness and stickiness. Dashed lines in (**a**,**b**) are fits of the data to straight lines ($R^2 = 0.998$).

Stickiness and the work of adhesion did not correlate in any meaningful way with most of the other textural parameters (r < 0.5 and $p > 0.4$) for SFC, while melting and crystallization points were better at ~0.7, with $p > 0.2$ indicating a weak correlation, if any at all (Table 4). Considering the shapes of the relationships and S.E.E, a conclusion was drawn that for all the relationships listed in Table 4, the values of the Pearson correlation r < 0.9 do not indicate correlated parameters. The correlations in these cases are weak at best.

The analysis of the relationships between the parameters indicates that the SFC is strongly correlated with the melting and crystallization points (r > 0.986, $p < 0.002$), which is confirmed by an S.E.E of 0.4. It is also strongly correlated with crystal size (r = 0.991, $p < 0.001$ and standard error of estimate of 0.4). The correlation analysis depicts a picture in which, apart from the above few exceptions, the role of the crystallization parameters in the relationships between the structural properties, as determined with the TA-TX machine, are complex and need further investigations to be clarified. One approach is to consider single or group mixtures with similar crystallization processes, as suggested by Figure 10b.

Viscosity and viscoelastic properties that affect textural properties through interfacial bonds and energy dissipation at a molecular level, in a manner not yet clarified [106,111,112], may be considered as factors for an interpretation of the textural results.

4. Conclusions

The studied mixtures possess a large range of physicochemical and textural properties including good spreadability, ductility and plasticity. AVK/AVP mixtures can be developed for direct consumption or used as fat mimetics in both hard and soft confectionary applications depending on the fat/oil ratio. The mixtures demonstrated good mixing compatibility, as evidenced by their phase diagrams, and effectively mimicked the melting characteristics and SFC profiles of commonly used fats like cocoa butter, coconut oil and palm kernel oil. Importantly, this study has revealed the possibility of formulating functional fats for industrial uses from AVK and AVP—a simple, environmentally friendly and relatively cost-effective solution.

AVK/AVP mixtures can be innovatively used in formulations in multiple industries, catering to the growing consumer demand for products that not only deliver desired functionality, but also offer health benefits. These developments could also promote sustainability in production processes by reducing dependence on traditional fats that have greater environmental impacts.

Supplementary Materials: The following supporting information can be downloaded at https://www.mdpi.com/article/10.3390/cosmetics11040107/s1, Figure S1: Associated crystallization and melting enthalpies of T3 vs XAVP; Table S1: Non-lipid bioactive compounds in AVK detected by ESI-MS.

Author Contributions: Conceptualization, L.B., S.D., N.S. and S.S.N.; Methodology, L.B., S.D., N.S. and S.S.N.; Software, L.B., S.D., N.S. and S.S.N.; Validation, L.B., S.D., N.S., R.J.N.E., S.M. and S.S.N.; Formal analysis, L.B., S.D., N.S. and S.S.N.; Investigation, L.B., S.D., N.S. and S.S.N.; Resources, S.S.N.; Data curation, L.B., S.D., N.S. and S.S.N.; Writing—original draft, L.B., S.D., N.S. and S.S.N.; Writing—review & editing, L.B., S.D., N.S., R.J.N.E., S.M. and S.S.N.; Visualization, L.B., S.D., N.S. and S.S.N.; Supervision, L.B., R.J.N.E., S.M. and S.S.N.; Project administration, S.S.N.; Funding acquisition, S.S.N. All authors have read and agreed to the published version of the manuscript.

Funding: This research received funding indirectly from CGX Energy Inc. and Frontera Energy Corporation. The funders were not involved in the study design, collection, analysis, interpretation of data, the writing of this article or the decision to submit it for publication.

Data Availability Statement: Data is contained within the article and Supplementary Material.

Acknowledgments: The Sustainable Guyana Program, a partnership between Trent University and the University of Guyana funded by CGX Energy Inc. and Frontera Energy Corporation, the Natural Sciences and Engineering Research Council of Canada (NSERC) and Trent University are acknowledged. Collaboration with the communities of Surama in Region 9, Guyana and of Cabora, Kamwatta and Waramuri in Region 1, Guyana is also acknowledged.

Conflicts of Interest: The authors declare no conflict of interest. The authors declare that the research was conducted in the absence of any commercial or financial relationships that could be construed as a potential conflict of interest. The funders had no role in the design of the study; in the collection, analyses or interpretation of data; in the writing of the manuscript; or in the decision to publish the results.

Abbreviations

ESI-MS	Electrospray ionization–mass spectrometry
TAGs	Triacylglycerols
LLC_8	3-(octanoyloxy)propane-1,2-diyl didodecanoate
LLC_{10}	3-(decanoyloxy)propane-1,2-diyl didodecanoate
LLL	propane-1,2,3-triyl tridodecanoate
LLM	3-(tetradecanoyloxy)propane-1,2-diyl didodecanoate
MML	3-(dodecanoyloxy)propane-1,2-diyl ditetradecanoate
POP	2-(oleoyloxy)propane-1,3-diyl dipalmitate
PPO	3-(oleoyloxy)propane-1,2-diyl dipalmitate
POO	2-(palmitoyloxy)propane-1,3-diyl dioleate
OOO	propane-1,2,3-triyl trioleate
SOO	3-(oleoyloxy)propane-1,2-diyl distearate
SSO	propane-1,2,3-triyl tristearate
UV-Vis	Ultraviolet–visible
TPC	Total phenolic content
TCC	Total carotenoid content
DSC	Differential scanning calorimetry
FWHM	Full width at half max
T_{On}	Onset temperature
T_{Off}	Offset temperature
SFC	Solid fat content
PLM	Polarized light microscopy

References

1. USDA. *Oilseeds: Worldwide Markets and Trade*; U.S Department of Agriculture: Washington, DC, USA, 2024; p. 39.
2. Dini, I.; Laneri, S. The New Challenge of Green Cosmetics: Natural Food Ingredients for Cosmetic Formulations. *Molecules* **2021**, *26*, 3921. [CrossRef] [PubMed]
3. Santos, M.T.d.; Morgavi, P.; Le Roux, G.A.C. Exploring amazonian fats and oils blends by computational predictions of solid fat content. *OCL* **2018**, *25*, D107. [CrossRef]
4. Kahn, F. The genus Astrocaryum (Arecaceae). *Rev. Peru. Biol.* **2008**, *15*, 31–48. [CrossRef]
5. Craveiro Holanda Malveira Maia, G.; da Silva Campos, M.; Barros-Monteiro, J.; Eduardo Lucas Castillo, J.; Soares Faleiros, M.; Souza de Aquino Sales, R.; Moraes Lopes Galeno, D.; Lira, E.; das Chagas do Amaral Souza, F.; Ortiz, C.; et al. Effects of *Astrocaryum aculeatum* Meyer (Tucumã) on Diet-Induced Dyslipidemic Rats. *J. Nutr. Metab.* **2014**, *2014*, 202367. [CrossRef] [PubMed]
6. Silva, M.B.; Perez, V.H.; Pereira, N.R.; Silveira, T.d.C.; da Silva, N.R.F.; de Andrade, C.M.; Sampaio, R.M. Drying kinetic of tucum fruits (*Astrocaryum aculeatum* Meyer): Physicochemical and functional properties characterization. *J. Food Sci. Technol.* **2018**, *55*, 1656–1666. [CrossRef] [PubMed]
7. Menezes, E.G.O.; Barbosa, J.R.; Pires, F.C.S.; Ferreira, M.C.R.; de Souza e Silva, A.P.; Siqueira, L.M.M.; de Carvalho Junior, R.N. Development of a new scale-up equation to obtain Tucumã-of-Pará (*Astrocaryum vulgare* Mart.) oil rich in carotenoids using supercritical CO_2 as solvent. *J. Supercrit. Fluids* **2022**, *181*, 105481. [CrossRef]
8. Teixeira, G.L.; Ibañez, E.; Block, J.M. Emerging Lipids from Arecaceae Palm Fruits in Brazil. *Molecules* **2022**, *27*, 4188. [CrossRef] [PubMed]
9. Falcao, A.d.O.; Speranza, P.; Ueta, T.; Martins, M. Antioxidant potential and modulatory effects of restructured lipids from the Amazonian palms on liver cells. *Food Technol. Biotechnol.* **2017**, *55*, 553–561.
10. Bezerra, C.V.; Rodrigues, A.; de Oliveira, P.D.; da Silva, D.A.; da Silva, L.H.M. Technological properties of amazonian oils and fats and their applications in the food industry. *Food Chem.* **2017**, *221*, 1466–1473. [CrossRef] [PubMed]
11. Deonarine, S.; Soodoo, N.; Bouzidi, L.; Narine, S.S. Oil Extraction and Natural Drying Kinetics of the Pulp and Seeds of Commercially Important Oleaginous Fruit from the Rainforests of Guyana. *Processes* **2023**, *11*, 3292. [CrossRef]
12. Rodrigues, A.M.d.C.; Darnet, S.; Silva, L.H.M.d. Fatty acid profiles and tocopherol contents of buriti (*Mauritia flexuosa*), patawa (*Oenocarpus bataua*), tucuma (*Astrocaryum vulgare*), mari (*Poraqueiba paraensis*) and inaja (*Maximiliana maripa*) fruits. *J. Braz. Chem. Soc.* **2010**, *21*, 2000–2004. [CrossRef]
13. Bora, P.S.; Narain, N.; Rocha, R.V.M.; De Oliveira Monteiro, A.C.; De Azevedo Moreira, R. Characterization of the oil and protein fraction of (*Astrocaryum vulgare* Mart.) fruit pulp and kernel. *Cienc. Y Tecnol. Aliment.* **2001**, *3*, 111–116. [CrossRef]
14. Bony, E.; Boudard, F.; Brat, P.; Dussossoy, E.; Portet, K.; Poucheret, P.; Giaimis, J.; Michel, A. Awara (*Astrocaryum vulgare* M.) pulp oil: Chemical characterization, and anti-inflammatory properties in a mice model of endotoxic shock and a rat model of pulmonary inflammation. *Fitoterapia* **2012**, *83*, 33–43. [CrossRef] [PubMed]
15. da Silva Sousa, H.M.; Leal, G.F.; da Silva Gualberto, L.; de Freitas, B.C.B.; Guarda, P.M.; Borges, S.V.; Morais, R.A.; de Souza Martins, G.A. Exploration of the chemical characteristics and bioactive and antioxidant potential of tucumã (*Astrocaryum vulgare*), peach palm (*Bactris gasipaes*), and bacupari (*Garcinia gardneriana*) native Brazilian fruits. *Biomass Convers. Biorefinery* **2023**, *13*, 1–14. [CrossRef]
16. Rossato, A.; da Silva Silveira, L.; Lopes, L.Q.S.; De Sousa Filho, W.P.; Schaffer, L.F.; Santos, R.C.V.; Sagrillo, M.R. Evaluation in vitro of antimicrobial activity of tucumã oil (*Astrocaryum vulgare*). *Arch. Biosci. Health* **2019**, *1*, 99–112. [CrossRef]
17. Nascimento, K.; Copetti, P.M.; Fernandes, A.; Klein, B.; Fogaça, A.; Zepka, L.Q.; Wagner, R.; Ourique, A.F.; Sagrillo, M.R.; da Silva, J.E.P. Phytochemical analysis and evaluation of the antioxidant and antiproliferative effects of Tucumã oil nanocapsules in breast adenocarcinoma cells (MCF-7). *Nat. Prod. Res.* **2021**, *35*, 2060–2065. [CrossRef]
18. Copetti, P.; Oliveira, P.; Vaucher, R.; Duarte, M.; Krause, L. Tucumã extracts decreases PML/RARA gene expression in NB4/APL cell line. *Arch. Biosci. Health* **2019**, *1*, 77–98. [CrossRef]
19. Guex, C.G.; Cassanego, G.B.; Dornelles, R.C.; Casoti, R.; Engelmann, A.M.; Somacal, S.; Maciel, R.M.; Duarte, T.; Borges, W.d.S.; Andrade, C.M.d.; et al. Tucumã (*Astrocaryum aculeatum*) extract: Phytochemical characterization, acute and subacute oral toxicity studies in Wistar rats. *Drug Chem. Toxicol.* **2022**, *45*, 810–821. [CrossRef] [PubMed]
20. Bonadiman, B.; Chaves, C.; Assmann, C.E.; Weis, G.C.C.; Alves, A.O.; Gindri, A.L.; Chaves, C.; da Cruz, I.B.M.; Zamoner, A.; Bagatini, M.D. Tucumã (*Astrocaryum aculeatum*) Prevents Oxidative and DNA Damage to Retinal Pigment Epithelium Cells. *J. Med. Food* **2021**, *24*, 1050–1057. [CrossRef] [PubMed]
21. Burlando, B.; Cornara, L. Revisiting Amazonian Plants for Skin Care and Disease. *Cosmetics* **2017**, *4*, 25. [CrossRef]
22. Mosquera Narvaez, L.E.; Ferreira, L.M.; Sanches, S.; Alesa Gyles, D.; Silva-Júnior, J.O.; Ribeiro Costa, R.M. A Review of Potential Use of Amazonian Oils in the Synthesis of Organogels for Cosmetic Application. *Molecules* **2022**, *27*, 2733. [CrossRef] [PubMed]
23. Ibiapina, A.; Gualberto, L.d.S.; Dias, B.B.; Freitas, B.C.B.; Martins, G.A.d.S.; Melo Filho, A.A. Essential and fixed oils from Amazonian fruits: Proprieties and applications. *Crit. Rev. Food Sci. Nutr.* **2022**, *62*, 8842–8854. [CrossRef] [PubMed]
24. Cardona Jaramillo, J.E.; Carrillo Bautista, M.P.; Alvarez Solano, O.A.; Achenie, L.E.K.; González Barrios, A.F. Impact of the Mode of Extraction on the Lipidomic Profile of Oils Obtained from Selected Amazonian Fruits. *Biomolecules* **2019**, *9*, 329. [CrossRef] [PubMed]

25. Rabasco Álvarez, A.M.; González Rodríguez, M.L. Lipids in pharmaceutical and cosmetic preparations. *Grasas Y Aceites* **2000**, *51*, 74–96. [CrossRef]
26. Shukla, V. Confectionery Lipids. *Bailey's Ind. Oil Fat Prod.* **2005**, *4*, 159–173. [CrossRef]
27. Lipp, M.; Simoneau, C.; Ulberth, F.; Anklam, E.; Crews, C.; Brereton, P.; de Greyt, W.; Schwack, W.; Wiedmaier, C. Composition of Genuine Cocoa Butter and Cocoa Butter Equivalents. *J. Food Compos. Anal.* **2001**, *14*, 399–408. [CrossRef]
28. Shukla, V.; Kragballe, K. Exotic butters as cosmetic lipids. *Inform* **1998**, *9*, 512–516.
29. Watanabe, S.; Yoshikawa, S.; Sato, K. Formation and properties of dark chocolate prepared using fat mixtures of cocoa butter and symmetric/asymmetric stearic-oleic mixed-acid triacylglycerols: Impact of molecular compound crystals. *Food Chem.* **2021**, *339*, 127808. [CrossRef] [PubMed]
30. Weiss, T.J. *Food Oils and Their Uses*; AVI Publishing Company: Westport, CT, USA, 1983.
31. Ghotra, B.S.; Dyal, S.D.; Narine, S.S. Lipid shortenings: A review. *Food Res. Int.* **2002**, *35*, 1015–1048. [CrossRef]
32. Torbica, A.; Jovanovic, O.; Pajin, B. The advantages of solid fat content determination in cocoa butter and cocoa butter equivalents by the Karlshamns method. *Eur. Food Res. Technol.* **2006**, *222*, 385–391. [CrossRef]
33. Nor Hayati, I.; Che Man, Y.B.; Tan, C.P.; Nor Aini, I. Physicochemical characteristics of soybean oil, palm kernel olein, and their binary blends. *Int. J. Food Sci. Technol.* **2009**, *44*, 152–161. [CrossRef]
34. Hashempour-Baltork, F.; Torbati, M.; Azadmard-Damirchi, S.; Savage, G.P. Vegetable oil blending: A review of physicochemical, nutritional and health effects. *Trends Food Sci. Technol.* **2016**, *57*, 52–58. [CrossRef]
35. Zhang, Z.; Song, J.; Lee, W.J.; Xie, X.; Wang, Y. Characterization of enzymatically interesterified palm oil-based fats and its potential application as cocoa butter substitute. *Food Chem.* **2020**, *318*, 126518. [CrossRef]
36. Biswas, N.; Cheow, Y.L.; Tan, C.P.; Siow, L.F. Blending of Palm Mid-Fraction, Refined Bleached Deodorized Palm Kernel Oil or Palm Stearin for Cocoa Butter Alternative. *J. Am. Oil Chem. Soc.* **2016**, *93*, 1415–1427. [CrossRef]
37. Hashem, A.; Abdul-fadl, M.; Arafat, S.; Aboulhoda, B. Producing cocoa butter substitutes by blending process of some vegetable oils. *J. Biol. Chem. Environ. Sci.* **2018**, *13*, 133–154.
38. Mohanan, A.; Darling, B.; Bouzidi, L.; Narine, S.S. Mitigating crystallization of saturated FAMES (fatty acid methyl esters) in biodiesel. 3. The binary phase behavior of 1,3-dioleoyl-2-palmitoyl glycerol—Methyl palmitate—A multi-length scale structural elucidation of mechanism responsible for inhibiting FAME crystallization. *Energy* **2015**, *86*, 500–513. [CrossRef]
39. Boodhoo, M.V.; Bouzidi, L.; Narine, S.S. The binary phase behavior of 1,3-dicaproyl-2-stearoyl-*sn*-glycerol and 1,2-dicaproyl-3-stearoyl-*sn*-glycerol. *Chem. Phys. Lipids* **2009**, *157*, 21–39. [CrossRef]
40. Foubert, I.; Dewettinck, K.; Vanrolleghem, P.A. Modelling of the crystallization kinetics of fats. *Trends Food Sci. Technol.* **2003**, *14*, 79–92. [CrossRef]
41. Sagrillo, M.R.; Garcia, L.F.M.; de Souza Filho, O.C.; Duarte, M.M.M.F.; Ribeiro, E.E.; Cadoná, F.C.; da Cruz, I.B.M. Tucumã fruit extracts (*Astrocaryum aculeatum* Meyer) decrease cytotoxic effects of hydrogen peroxide on human lymphocytes. *Food Chem.* **2015**, *173*, 741–748. [CrossRef] [PubMed]
42. Sánchez-Rangel, J.C.; Benavides, J.; Heredia, J.B.; Cisneros-Zevallos, L.; Jacobo-Velázquez, D.A. The Folin–Ciocalteu assay revisited: Improvement of its specificity for total phenolic content determination. *Anal. Methods* **2013**, *5*, 5990–5999. [CrossRef]
43. de Carvalho, L.M.J.; Gomes, P.B.; Godoy, R.L.d.O.; Pacheco, S.; do Monte, P.H.F.; de Carvalho, J.L.V.; Nutti, M.R.; Neves, A.C.L.; Vieira, A.C.R.A.; Ramos, S.R.R. Total carotenoid content, α-carotene and β-carotene, of landrace pumpkins (*Cucurbita moschata* Duch): A preliminary study. *Food Res. Int.* **2012**, *47*, 337–340. [CrossRef]
44. Siqueira, E.M.d.A.; Rosa, F.R.; Fustinoni, A.M.; de Sant'Ana, L.P.; Arruda, S.F. Brazilian Savanna Fruits Contain Higher Bioactive Compounds Content and Higher Antioxidant Activity Relative to the Conventional Red Delicious Apple. *PLoS ONE* **2013**, *8*, e72826. [CrossRef] [PubMed]
45. Rodriguez-Amaya, D.B.; Kimura, M. *HarvestPlus Handbook for Carotenoid Analysis*; International Food Policy Research Institute (IFPRI) Washington: Washington, DC, USA, 2004; Volume 2.
46. Tee, E.S.; Lim, C.-L. The analysis of carotenoids and retinoids: A review. *Food Chem.* **1991**, *41*, 147–193. [CrossRef]
47. de Rosso, V.V.; Mercadante, A.Z. Identification and Quantification of Carotenoids, By HPLC-PDA-MS/MS, from Amazonian Fruits. *J. Agric. Food Chem.* **2007**, *55*, 5062–5072. [CrossRef] [PubMed]
48. Ferreira, M.J.A.; Mota, M.F.S.; Mariano, R.G.B.; Freitas, S.P. Evaluation of liquid-liquid extraction to reducing the acidity index of the tucuma (*Astrocaryum vulgare* Mart.) pulp oil. *Sep. Purif. Technol.* **2021**, *257*, 117894. [CrossRef]
49. Rodriguez-Amaya, D.B. *A Guide to Carotenoid Analysis in Foods*; ILSI Press: Washington, DC, USA, 2001; Volume 71.
50. Freitas, M.L.F.; Chisté, R.C.; Polachini, T.C.; Sardella, L.A.C.Z.; Aranha, C.P.M.; Ribeiro, A.P.B.; Nicoletti, V.R. Quality characteristics and thermal behavior of buriti (*Mauritia flexuosa* L.) oil. *Grasas Y Aceites* **2017**, *68*, e220. [CrossRef]
51. Ferreira, C.D.; da Conceição, E.J.L.; Machado, B.A.S.; Hermes, V.S.; de Oliveira Rios, A.; Druzian, J.I.; Nunes, I.L. Physicochemical Characterization and Oxidative Stability of Microencapsulated Crude Palm Oil by Spray Drying. *Food Bioprocess Technol.* **2016**, *9*, 124–136. [CrossRef]
52. Amchova, P.; Kotolova, H.; Ruda-Kucerova, J. Health safety issues of synthetic food colorants. *Regul. Toxicol. Pharmacol.* **2015**, *73*, 914–922. [CrossRef] [PubMed]
53. Köpcke, W.; Krutmann, J. Protection from sunburn with beta-Carotene—A meta-analysis. *Photochem. Photobiol.* **2008**, *84*, 284–288. [CrossRef] [PubMed]

54. Eicker, J.; Kürten, V.; Wild, S.; Riss, G.; Goralczyk, R.; Krutmann, J.; Berneburg, M. Betacarotene supplementation protects from photoaging-associated mitochondrial DNA mutation. Photochem. *Photobiol. Sci.* **2003**, *2*, 655–659. [CrossRef] [PubMed]
55. Morganti, P.; Bruno, C.; Guarneri, F.; Cardillo, A.; Del Ciotto, P.; Valenzano, F. Role of topical and nutritional supplement to modify the oxidative stress. *Int. J. Cosmet Sci.* **2002**, *24*, 331–339. [CrossRef] [PubMed]
56. Ferreira, B.S.; De Almeida, C.G.; Faza, L.P.; De Almeida, A.; Diniz, C.G.; Silva, V.L.d.; Grazul, R.M.; Le Hyaric, M. Comparative Properties of Amazonian Oils Obtained by Different Extraction Methods. *Molecules* **2011**, *16*, 5875–5885. [CrossRef] [PubMed]
57. Chai, B.; Jiang, W.; Hu, M.; Wu, Y.; Si, H. In vitro synergistic interactions of Protocatechuic acid and Chlorogenic acid in combination with antibiotics against animal pathogens. *Synergy* **2019**, *9*, 100055. [CrossRef]
58. Kępa, M.; Miklasińska-Majdanik, M.; Wojtyczka, R.D.; Idzik, D.; Korzeniowski, K.; Smoleń-Dzirba, J.; Wąsik, T.J. Antimicrobial Potential of Caffeic Acid against Staphylococcus aureus Clinical Strains. *Biomed. Res. Int.* **2018**, *2018*, 7413504. [CrossRef] [PubMed]
59. Miklasińska-Majdanik, M.; Kępa, M.; Wąsik, T.J.; Zapletal-Pudełko, K.; Klim, M.; Wojtyczka, R.D. The Direction of the Antibacterial Effect of Rutin Hydrate and Amikacin. *Antibiotics* **2023**, *12*, 1469. [CrossRef] [PubMed]
60. Abdulhadi, S.Y.; Gergees, R.N.; Hasan, G.Q. Molecular identification, antioxidant efficacy of phenolic compounds, and antimicrobial activity of beta-carotene isolated from fruiting bodies of *Suillus* sp. *Karbala Int. J. Mod. Sci.* **2020**, *6*, 4. [CrossRef]
61. Duarte, S.; Rosalen, P.L.; Hayacibara, M.F.; Cury, J.A.; Bowen, W.H.; Marquis, R.E.; Rehder, V.L.; Sartoratto, A.; Ikegaki, M.; Koo, H. The influence of a novel propolis on mutans streptococci biofilms and caries development in rats. *Arch. Oral Biol.* **2006**, *51*, 15–22. [CrossRef] [PubMed]
62. Barkas, F.; Bathrellou, E.; Nomikos, T.; Panagiotakos, D.; Liberopoulos, E.; Kontogianni, M.D. Plant Sterols and Plant Stanols in Cholesterol Management and Cardiovascular Prevention. *Nutrients* **2023**, *15*, 2845. [CrossRef] [PubMed]
63. Burčová, Z.; Kreps, F.; Greifová, M.; Jablonský, M.; Ház, A.; Schmidt, Š.; Šurina, I. Antibacterial and antifungal activity of phytosterols and methyl dehydroabietate of Norway spruce bark extracts. *J. Biotechnol.* **2018**, *282*, 18–24. [CrossRef] [PubMed]
64. Shih, C.-C.; Lin, C.-H.; Lin, W.-L. Effects of Momordica charantia on insulin resistance and visceral obesity in mice on high-fat diet. *Diabetes Res. Clin. Pract.* **2008**, *81*, 134–143. [CrossRef] [PubMed]
65. Deonarine, S.; Soodoo, N.; Bouzidi, L.; Emery, R.J.N.; Martic, S.; Narine, S.S. Molecular, Crystalline, and Microstructures of Lipids from Astrocaryum Species in Guyana and Their Thermal and Flow Behavior. *Thermo* **2024**, *4*, 140–163. [CrossRef]
66. Boodhoo, M.V.; Kutek, T.; Filip, V.; Narine, S.S. The binary phase behavior of 1,3-dimyristoyl-2-stearoyl-sn-glycerol and 1,2-dimyristoyl-3-stearoyl-sn-glycerol. *Chem. Phys. Lipids* **2008**, *154*, 7–18. [CrossRef] [PubMed]
67. Boodhoo, M.V.; Bouzidi, L.; Narine, S.S. The binary phase behavior of 1, 3-dipalmitoyl-2-stearoyl-sn-glycerol and 1, 2-dipalmitoyl-3-stearoyl-sn-glycerol. *Chem. Phys. Lipids* **2009**, *160*, 11–32. [CrossRef] [PubMed]
68. Abes, M.; Bouzidi, L.; Narine, S.S. Crystallization and phase behavior of 1,3-propanediol esters II. 1,3-Propanediol distearate/1,3-propanediol dipalmitate (SS/PP) and 1,3-propanediol distearate/1,3-propanediol dimyristate (SS/MM) binary systems. *Chem. Phys. Lipids* **2007**, *150*, 89–108. [CrossRef] [PubMed]
69. Baker, M.; Bouzidi, L.; Narine, S.S. Mitigating crystallization of saturated FAMEs (fatty acid methyl esters) in biodiesel: 2. The phase behavior of 2-stearoyl diolein–methyl stearate binary system. *Energy* **2015**, *83*, 647–657. [CrossRef]
70. Macridachis, J.; Bayés-García, L.; Calvet, T. Mixing phase behavior of trilaurin and monounsaturated triacylglycerols based on palmitic and oleic fatty acids. *J. Therm. Anal. Calorim.* **2023**, *148*, 12987–13001. [CrossRef]
71. Höhne, G.; Hemminger, W.; Flammersheim, H.-J. (Eds.) *Differential Scanning Calorimetry*, 2nd ed.; Springer: Berlin/Heidelberg, Germany, 2003; p. 298.
72. Inoue, T.; Hisatsugu, Y.; Ishikawa, R.; Suzuki, M. Solid–liquid phase behavior of binary fatty acid mixtures: 2. Mixtures of oleic acid with lauric acid, myristic acid, and palmitic acid. *Chem. Phys. Lipids* **2004**, *127*, 161–173. [CrossRef] [PubMed]
73. MacNaughtan, W.; Farhat, I.; Himawan, C.; Starov, V.; Stapley, A. A differential scanning calorimetry study of the crystallization kinetics of tristearin-tripalmitin mixtures. *J. Am. Oil Chem. Soc.* **2006**, *83*, 1–9. [CrossRef]
74. Costa, M.C.; Rolemberg, M.P.; Boros, L.A.D.; Krähenbühl, M.A.; de Oliveira, M.G.; Meirelles, A.J.A. Solid−Liquid Equilibrium of Binary Fatty Acid Mixtures. *J. Chem. Eng. Data* **2006**, *52*, 30–36. [CrossRef]
75. Inoue, T.; Motoda, I.; Hiramatsu, N.; Suzuki, M.; Sato, K. Phase-behavior of binary mixture of palmitoleic acid (cis-9-hexadecenoic acid) and asclepic acid (cis-11-octadecenoic acid). *Chem. Phys. Lipids* **1993**, *66*, 209–214. [CrossRef]
76. Hildebrand, J.H. Solubility XII. Regular solutions. *J. Am. Chem. Soc.* **1929**, *51*, 66–80. [CrossRef]
77. Bragg, W.L.; Williams, E.J. The effect of thermal agitation on atomic arrangement in alloys. *Proc. R. Soc.* **1934**, *145*, 699–730.
78. Moore, W.J. *Physical Chemistry*, 4th ed.; Prentice-Hall: Englewood Cliffs, NJ, USA, 1972; pp. 229–278.
79. Inoue, T.; Hisatsugu, Y.; Yamamoto, R.; Suzuki, M. Solid-liquid phase behavior of binary fatty acid mixtures 1. Oleic acid/stearic acid and oleic acid/behenic acid mixtures. *Chem. Phys. Lipids* **2004**, *127*, 143–152. [CrossRef] [PubMed]
80. Abes, M.; Bouzidi, L.; Narine, S.S. Crystallization and phase behavior of fatty acid esters of 1,3 propanediol III: 1,3 propanediol dicaprylate/1,3 propanediol distearate (CC/SS) and 1,3 propanediol dicaprylate/1,3 propanediol dipalmitate (CC/PP) binary systems. *Chem. Phys. Lipids* **2008**, *151*, 110–124. [CrossRef] [PubMed]
81. Lee, A.G. Lipid Phase-Transitions and Phase-Diagrams.1. Lipid Phase-Transitions. *Biochim. Biophys. Acta* **1977**, *472*, 237–281. [CrossRef] [PubMed]
82. Lee, A.G. Lipid Phase-Transitions and Phase-Diagrams.2. Mixtures Involving Lipids. *Biochim. Biophys. Acta* **1977**, *472*, 285–344. [CrossRef] [PubMed]

83. Tenchov, B.G. Non-uniform lipid distribution in membranes. *Prog. Surf. Sci.* **1985**, *20*, 273–340. [CrossRef]
84. Nibu, Y.; Inoue, T. Miscibility of binary phospholipid mixtures under hydrated and non-hydrated conditions.2. Phosphatidylethanolamines with different acyl-chain lengths. *Chem. Phys. Lipids* **1995**, *76*, 159–169. [CrossRef]
85. Rycerz, L. Practical remarks concerning phase diagrams determination on the basis of differential scanning calorimetry measurements. *J. Therm. Anal. Calorim.* **2013**, *113*, 231–238. [CrossRef]
86. Findlay, A. *The Phase Rule and Its Applications*; Longmans, Green and Company: Selinsgrove, PA, USA, 1904; p. 361.
87. Matsuoka, M.; Ozawa, R. Determination of solid-liquid phase equilibria of binary organic systems by differential scanning calorimetry. *J. Cryst. Growth* **1989**, *96*, 596–604. [CrossRef]
88. Timms, R.E. Phase behaviour of fats and their mixtures. *Prog. Lipid Res.* **1984**, *23*, 1–38. [CrossRef] [PubMed]
89. Petersson, B.; Anjou, K.; Sandström, L. Pulsed NMR Method for Solid Fat Content Determination in Tempering Fats, Part I: Cocoa Butters and Equivalents. *Fette Seifen Anstrichm.* **1985**, *87*, 225–230. [CrossRef]
90. Nillson, J. Measuring solid fat content. *Manuf. Confect.* **1986**, *5*, 88–91.
91. Lovett, P.N. 5—Shea butter: Properties and processing for use in food. In *Specialty Oils and Fats in Food and Nutrition*; Talbot, G., Ed.; Woodhead Publishing: Sawston, UK, 2015; pp. 125–158. [CrossRef]
92. Ghazani, S.M.; Marangoni, A.G. Molecular Origins of Polymorphism in Cocoa Butter. *Annu. Rev. Food Sci. Technol.* **2021**, *12*, 567–590. [CrossRef] [PubMed]
93. Siew, W.L. Crystallisation and melting behaviour of palm kernel oil and related products by differential scanning calorimetry. *Eur. J. Lipid Sci. Technol.* **2001**, *103*, 729–734. [CrossRef]
94. Matsuda, H.; Yamaguchi, M.; Arima, H. Separation and crystallization of oleaginous constituents in cosmetics: Sweating and blooming. In *Crystallization Processes in Fats and Lipid Systems*; CRC Press: Boca Raton, FL, USA, 2001; pp. 499–518.
95. Pereira, E.; Ferreira, M.C.; Sampaio, K.A.; Grimaldi, R.; Meirelles, A.J.d.A.; Maximo, G.J. Physical properties of Amazonian fats and oils and their blends. *Food Chem.* **2019**, *278*, 208–215. [CrossRef] [PubMed]
96. Torbica, A.; Jambrec, D.; Tomić, J.; Pajin, B.; Petrović, J.; Kravić, S.; Lončarević, I. Solid Fat Content, Pre-Crystallization Conditions, and Sensory Quality of Chocolate with Addition of Cocoa Butter Analogues. *Int. J. Food Prop.* **2016**, *19*, 1029–1043. [CrossRef]
97. Jahurul, M.H.A.; Soon, Y.; Shaarani Sharifudin, M.; Hasmadi, M.; Mansoor, A.H.; Zaidul, I.S.M.; Lee, J.S.; Ali, M.E.; Ghafoor, K.; Zzaman, W.; et al. Bambangan (*Mangifera pajang*) kernel fat: A potential new source of cocoa butter alternative. *Int. J. Food Sci. Technol.* **2018**, *53*, 1689–1697. [CrossRef]
98. da Silva Lannes, S.C.; Ignácio, R.M. Structuring fat foods. In *Food Industry*; Muzzalupo, I., Ed.; IntechOpen: London, UK, 2013; pp. 65–91. [CrossRef]
99. Silva, T.J.; Barrera-Arellano, D.; Ribeiro, A.P.B. Margarines: Historical approach, technological aspects, nutritional profile, and global trends. *Food Res. Int.* **2021**, *147*, 110486. [CrossRef] [PubMed]
100. Bhandari, S.D.; Delmonte, P.; Honigfort, M.; Yan, W.; Dionisi, F.; Fleith, M.; Iassonova, D.; Bergeson, L.L. Regulatory Changes Affecting the Production and Use of Fats and Oils: Focus on Partially Hydrogenated Oils. *J. Am. Oil Chem. Soc.* **2020**, *97*, 797–815. [CrossRef]
101. Latip, R.A.; Lee, Y.-Y.; Tang, T.-K.; Phuah, E.-T.; Tan, C.-P.; Lai, O.-M. Physicochemical properties and crystallisation behaviour of bakery shortening produced from stearin fraction of palm-based diacyglycerol blended with various vegetable oils. *Food Chem.* **2013**, *141*, 3938–3946. [CrossRef] [PubMed]
102. Hondoh, H.; Ueno, S. Polymorphism of edible fat crystals. *Prog. Cryst. Growth Charact. Mater.* **2016**, *62*, 398–399. [CrossRef]
103. Paluri, S.; Heldman, D.R.; Maleky, F. Effects of Structural Attributes and Phase Ratio on Moisture Diffusion in Crystallized Lipids. *Cryst. Growth Des.* **2017**, *17*, 4661–4669. [CrossRef]
104. Chawla, P.; Deman, J.; Smith, A. Crystal morphology of shortenings and margarines. *Food Struct.* **1990**, *9*, 2.
105. Watanabe, T. The impact of grain boundary character distribution on fracture in polycrystals. *Mater. Sci. Eng. A* **1994**, *176*, 39–49. [CrossRef]
106. Adhikari, B.; Howes, T.; Bhandari, B.R.; Truong, V. Stickiness in Foods: A Review of Mechanisms and Test Methods. *Int. J. Food Prop.* **2001**, *4*, 1–33. [CrossRef]
107. Macias-Rodriguez, B. Rheology-Based Techniques. In *Advances in Oleogel Development, Characterization, and Nutritional Aspects*; Springer: Berlin/Heidelberg, Germany, 2024; pp. 471–496.
108. Deman, J.; Beers, A. Fat crystal networks: Structure and rheological properties. *J. Texture Stud.* **1987**, *18*, 303–318. [CrossRef]
109. Glibowski, P.; Zarzycki, P.; Krzepkowska, M. The Rheological and Instrumental Textural Properties of Selected Table Fats. *Int. J. Food Prop.* **2008**, *11*, 678–686. [CrossRef]
110. DeMan, L.; DeMan, J.; Blackman, B. Polymorphic behavior of some fully hydrogenated oils and their mixtures with liquid oil. *J. Am. Oil Chem. Soc.* **1989**, *66*, 1777–1780. [CrossRef]

111. Guillemenet, J.; Bistac, S.; Schultz, J. Relationship between polymer viscoelastic properties and adhesive behaviour. *Int. J. Adhes. Adhes.* **2002**, *22*, 1–5. [CrossRef]
112. Hoppu, P.; Grönroos, A.; Schantz, S.; Juppo, A.M. New processing technique for viscous amorphous materials and characterisation of their stickiness and deformability. *Eur. J. Pharm. Biopharm.* **2009**, *72*, 183–188. [CrossRef] [PubMed]

Article

Apigenin and Phloretin Combination for Skin Aging and Hyperpigmentation Regulation

Alfredo Martínez-Gutiérrez *, Javier Sendros, Teresa Noya and Mari Carmen González

R+D Department, Mesoestetic Pharma Group S.L, Calle Tecnologia 25, 08840 Barcelona, Spain; jsendros@mesoestetic.com (J.S.); tnoya@mesoestetic.com (T.N.); mcgonzalez@mesoestetic.com (M.C.G.)
* Correspondence: amartinez@mesoestetic.com

Abstract: Melasma is a pathology with multifactorial causes that results in hyperpigmentation of sun-exposed areas, particularly facial skin. New treatments targeting the different factors regulating this condition need to be effective with and have limited adverse effects. Here, we describe a novel combination of two natural compounds (apigenin and phloretin) that has synergistic effects regulating melanogenesis in vitro. Both compounds inhibit Wnt-stimulated melanogenesis and induce autophagy in melanocytes. Apigenin induces *DKK1*, a Wnt pathway inhibitor, and reduces *VEGF*, a melanogenesis and proangiogenic factor, in fibroblasts. Moreover, apigenin induces miR-675, a melanogenesis inhibitor miRNA that is reduced in melasma skin in melanocytes. Both compounds showed senomorphic effects by regulating extracellular-matrix-related genes in senescent fibroblasts. Topical application of the compounds also showed significant melanin reduction in a reconstructed human epidermis after 7 days. Thus, the combination of apigenin and phloretin shows promising results as an effective topical treatment of skin hyperpigmentation conditions.

Keywords: skin aging; skin pigmentation; hyperpigmentation; melasma; melanocytes; epigenetics; synergy

Citation: Martínez-Gutiérrez, A.; Sendros, J.; Noya, T.; González, M.C. Apigenin and Phloretin Combination for Skin Aging and Hyperpigmentation Regulation. *Cosmetics* **2024**, *11*, 128. https://doi.org/10.3390/cosmetics11040128

Academic Editor: Kazuhisa Maeda

Received: 30 May 2024
Revised: 4 July 2024
Accepted: 15 July 2024
Published: 26 July 2024

Correction Statement: This article has been republished with a minor change. The change does not affect the scientific content of the article and further details are available within the backmatter of the website version of this article.

1. Introduction

The melanin pigment plays a key role in the photoprotection of skin, the largest organ of the body. However, the abnormal and uneven accumulation of melanin occurs in certain skin conditions, giving rise to skin hyperpigmentation. Melasma is one of these conditions, mainly occurring in facial skin but also in other photo-exposed areas. Although the pathophysiology is not fully understood, several factors can contribute to melanin accumulation in melasma, including hormonal levels, exposure to ultraviolet damage, or genetic factors [1].

At the tissue level, many paracrine factors released by surrounding cells can directly stimulate melanogenesis. Fibroblasts and keratinocytes release promelanogenic factors such as Wnt1, SCF, ET-1, HGF, VEGF, bFGF, or α-MSH, which can directly bind to their respective membrane receptors to activate the melanogenesis process [2]. Interestingly, some of these factors can favor hyperpigmentation through indirect mechanisms. For example, VEGF not only stimulates melanogenesis but also promotes angiogenesis, increasing dermal vascularity, which is another characteristic feature of melasma pathogenesis. On the other hand, neighboring melanocyte cells can also release anti-melanogenic factors, such as IL-6, IL-1β, or DKK1. DKK1 is an antagonist of the Wnt pathway and is highly expressed in human palmoplantar dermal fibroblasts [3]. Thus, the release of this factor is a key approach to limit excessive melanin production [4].

Additionally, hyperpigmented areas in melasma skin share common features with photoaging, including solar elastosis, damaged basement membrane, increased oxidative stress and the aforementioned increase in dermal vascularity [5]. Both melasma and skin photoaging are also characterized by the accumulation of senescent fibroblasts in the dermis [6,7]. These cells have been previously described as contributors to skin hyperpigmentation. Thus, controlling senescent cells' activity is another interesting factor to be targeted by depigmenting strategies.

At the cellular level, melanocytes from melasma skin show increased intracellular melanin production. Many pathways are dysregulated in these melanocytes, which contribute to this abnormal accumulation, including impaired autophagy or reduced miR-675. Autophagy deficiency favors melanosome retention and the release of proinflammatory mediators in melanocytes [8], while miR-675 downregulates MITF, the main transcription factor that controls the expression of the enzymes involved in melanogenesis (TYR, TYRP1, and TYRP2) [1].

Melasma management is complex due to the chronicity of this condition. Current treatments have either low efficacy or significant side effects such as skin irritation, erythema or post-inflammatory hyperpigmentation [9]. In this context, natural compounds with a good safety profile offer a good alternative to these treatments for melasma and hyperpigmentation management. Both apigenin and phloretin are natural compounds with multiple biological activities. As topical compounds, apigenin and phloretin show UV-protecting, antiaging, and anti-inflammatory effects [10,11], among others. This pleiotropic regulation of different cellular processes by both compounds could result in the synergistic regulation of skin pigmentation.

Here, we describe the synergistic effects of apigenin and phloretin on reducing melanin production in melanocytes. Additionally, we shed more light on the depigmenting mechanism of both compounds by regulating key pathways involved in melasma pathogenesis. Finally, we show that the combination of compounds reduces melanin accumulation in a reconstructed human epidermal 3D model containing phototype VI melanocytes.

2. Materials and Methods

2.1. Cell Culture

Human primary adult melanocytes (Cellsystems GmbH, Troisdorf, Germany) (passage 4–5) were cultured in DermaLife Ma Medium supplemented with Zellshield (Minerva Biolabs GmbH, Berlin, Germany). Human dermal adult fibroblasts (Promocell, Heidelberg, Germany) were cultured in DMEM (Capricorn Scientific GmbH, Ebsdorfergrund, Germany) supplemented with 10% fetal bovine serum (FBS) (Fisher Scientific, Hampton, NH, USA) and Zellshield (Minerva Biolabs GmbH, Berlin, Germany). Both cell types were incubated at 37 °C in 5% CO_2. Unless specified, all reagents were obtained from Merck Life Science, Darmstadt, Germany.

2.2. Cell Viability Assay

To test the maximum viable dose of the compounds in each cell type, fibroblasts or melanocytes were treated with the compounds at different concentrations for 24 h. A sulforhodamine B (SRB) colorimetric assay was performed to assess cell viability of the compounds as previously described [12]. The highest dose at which the compounds do not exert cell toxicity was determined to decide the concentrations of compounds used to study the mechanism of action in the following assays.

2.3. Melanin Synthesis Assay for Synergy Evaluation

A synergistic effect exists when the effect of a combination of compounds is greater than the sum of their individual effects. Thus, to study the possible synergy between the two compounds, we compared the effect of adding both compounds separately but successively to the same cells (additive effect) with the effect of adding both compounds at the same time (synergy, or combination effect). For this experiment, melanocytes were treated in different conditions for 2 days:

- Control cells (nontreated);
- Stimulated cells: treated every 24 h with 2 mM l-tyrosine plus 0.5 mM IBMX;

-   Additive effect: treated the first 24 h with 2 mM l-tyrosine plus 0.5 mM IBMX plus 10 µM phloretin and treated the next 24 h with 2 mM l-tyrosine plus 0.5 mM IBMX plus 0.5 µM apigenin;
-   Combination effect: treated every 24 h with 2 mM l-tyrosine plus 0.5 mM IBMX plus 10 µM phloretin and 0.5 µM apigenin.

After treatment, cells were harvested by trypsinization, and melanin was extracted and quantified as previously described [13]. Melanin content was normalized to cell count number in each condition and is expressed as percentage of control cells.

2.4. Wnt-Induced Melanogenesis Quantification in Melanocytes

To test whether the compounds could inhibit Wnt-stimulated melanogenesis, melanocytes were treated for 24 h with 6-bromoindirubin-3′-oxime (BIO), a Wnt pathway activator [14], plus 0.5 µM apigenin or 10 µM phloretin. Cells were then harvested by trypsinization. RNA was extracted, and retrotranscription and qPCR reaction was performed as previously described [15]. The primers used are detailed in Table S1.

2.5. Melanogenesis-Related miRNA and Autophagy-Related Gene Expression in Melanocytes

To test if the compounds regulate miRNA or autophagy-related gene expression in melanocytes, cells were treated with 0.5 µM apigenin or 10 µM phloretin for 24 h. Cells were harvested by trypsinization. Autophagy-related gene expression was analyzed as described in Section 2.4. The primers used are detailed in Table S1.

For miRNA level quantification, a Mir-XTM miRNA First Strand Synthesis kit (Takara Bio Inc., Kusatsu, Japan) was used to generate cDNA from miRNAs. This kit contains the universal 3′ primer, and the 5′ and 3′ primers for U6 to be used as housekeeping gene. The 5′ primer for miR-675 corresponds to its sequence (5′-CTGTATGCCCTCACCGCTCA-3′). qPCR protocol included one step at 95 °C for 10 s, 40 cycles at 95 °C for 5 s, and 60 °C for 20 s.

2.6. Melanogenesis Paracrine Regulators' Gene Expression in Fibroblasts

To test if the compounds regulate the gene expression of melanogenesis paracrine regulators, human dermal fibroblasts were treated with 25 µM apigenin or 50 µM phloretin for 24 h. Cells were harvested by trypsinization, and gene expression was analyzed as described in Section 2.4. The primers used are detailed in Table S1.

2.7. Senomorphic Effect in Senescent Fibroblasts

To study the effect of the compounds on senescent fibroblasts, cellular senescence was first induced by repeated UVB damage. For this, human dermal fibroblasts were irradiated with 25 mJ/cm^2 of UVB light using a Bio-Link Crosslinker BLX-E365 device (Vilber Lourmat, Marne-la-Vallée, France), incubated for 48 h, irradiated again with 25 mJ/cm^2 of UVB light, and incubated for 72 h more. Cells were then treated with 25 µM apigenin or 50 µM phloretin for 48 h, and gene expression was evaluated as described in Section 2.4. The primers used are detailed in Table S1.

2.8. Melanin Accumulation Quantification in 3D Reconstructed Human Epidermis

To test the efficacy of the combination in a 3D skin model, we used SkinEthicTM RHPE/Reconstructed Human Pigmented Epidermis Phototype VI (Episkin, Lyon, France) as a topical treatment with 25 µM apigenin plus 50 µM phloretin once a day for 5 days and harvested on day 7 for MTT and histology analysis. For the MTT assay, tissues were washed with PBS and incubated with MTT 0.5 mg/mL for 3 h at 37 °C. Then, tissues were incubated in isopropanol 100% for 2 h at room temperature. Absorbance of the solution was quantified at 570 nm. For the histology analysis, tissues were washed with PBS, incubated in 4% paraformaldehyde for 12 h, and embedded in paraffin. We created 10 µm sections using a microtome, and Fontana–Masson staining (Abcam, Cambridge, UK) was used to detect the melanin in the samples. All images were obtained by optical microscopy (Evident Scientific, Waltham, MA, USA).

2.9. Statistical Analysis

The experiments were performed three times under the same conditions, and the average and standard deviation calculated for each condition was compared using Student's *t*-test (* $p < 0.05$, ** $p < 0.01$, *** $p < 0.001$).

3. Results

3.1. Cell Viability of the Compounds

Human epidermal melanocytes and human dermal fibroblasts were treated with the compounds at different concentrations for 24 h, as shown in Figure 1. The highest viable dose for apigenin in the melanocytes and fibroblasts was 0.5 µM and 25 µM, respectively. The highest viable dose for phloretin in the melanocytes and fibroblasts was 10 µM and 50 µM, respectively. These doses were selected to perform the following experiments where the mechanism of action and synergy of the compounds is studied. Accordingly, the observed outcome of the different assays was produced by the effect of the compounds on the regulation of cellular processes not associated with cell damage or death.

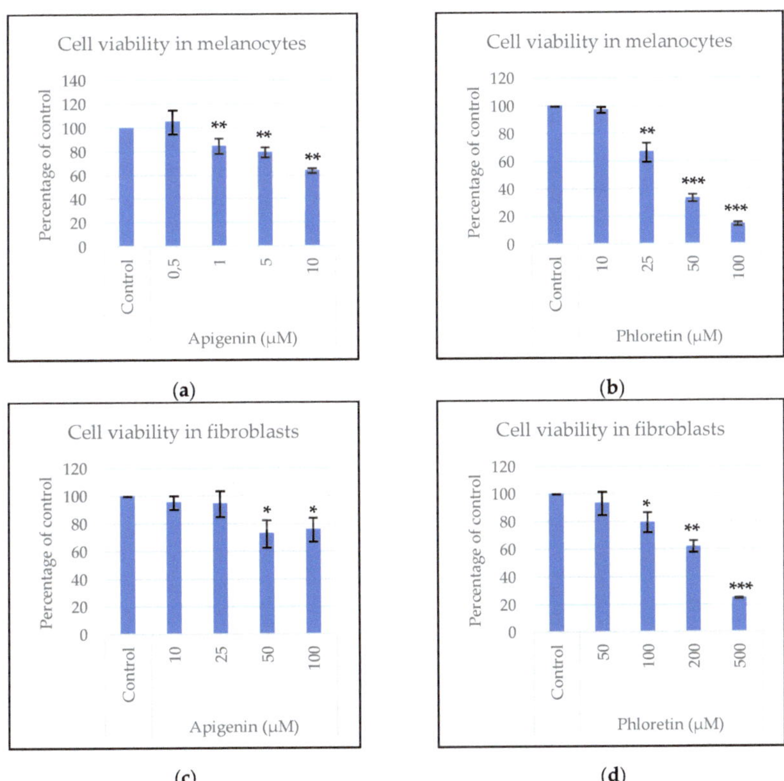

Figure 1. Cell viability of apigenin and phloretin in melanocytes (**a**,**b**) and fibroblasts (**c**,**d**) after 24 h. Statistical analysis: Significance of Student's *t*-test results is represented by * when $p < 0.05$, ** when $p < 0.01$, or *** when $p < 0.001$.

3.2. Synergistic Effect on Melanin Synthesis Inhibition

To determine if the apigenin and phloretin combination had a synergistic effect on melanin inhibition, melanocytes were stimulated and treated with the compounds separately and successively (additive effect) or treated with both compounds at the same time (combination effect). As shown in Figure 2, the stimulated melanocytes showed a significant increase in melanin accumulation compared to control cells. The additive incubation of the compounds significantly decreased melanin accumulation compared to that in the stimulated cells. Interestingly, both compounds incubated at the same time (combination) induced a larger decrease in melanin production than when the compounds were added separately (additive). This evidence proved that the compounds have a synergistic effect on melanin synthesis inhibition.

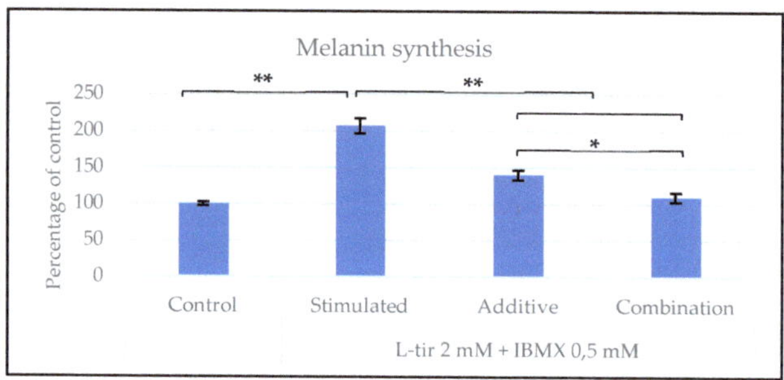

Figure 2. Melanin production after 48 h in nontreated cells (control), cells stimulated with 2 mM l-tyrosine + 0.5 mM IBMX (stimulated), cells stimulated with 2 mM l-tyrosine + 0.5 mM IBMX and treated for the first 24 h with 10 µM phloretin and the next 24 h with 0.5 µM apigenin (additive), and cells stimulated with 2 mM l-tyrosine + 0.5 mM IBMX and treated with both 10 µM phloretin and 0.5 µM apigenin at the same time (combination). Statistical analysis: significance of -Student's *t*-test results is represented by * when $p < 0.05$ or ** when $p < 0.01$.

3.3. Inhibition of Wnt-Stimulated Melanogenesis

As the Wnt pathway is upregulated in melasma skin [16], we studied whether the compounds could inhibit Wnt-stimulated melanogenesis. For this, we treated melanocytes with the Wnt activator BIO and the compounds for 24 h. The expression of melanogenesis-related genes was then quantified by qPCR (Figure 3). Apigenin reduced the expressions of *DCT* and *PAX3*, while phloretin reduced the expressions of *DCT*, *PAX3*, *PMEL17*, and *TYRP1*.

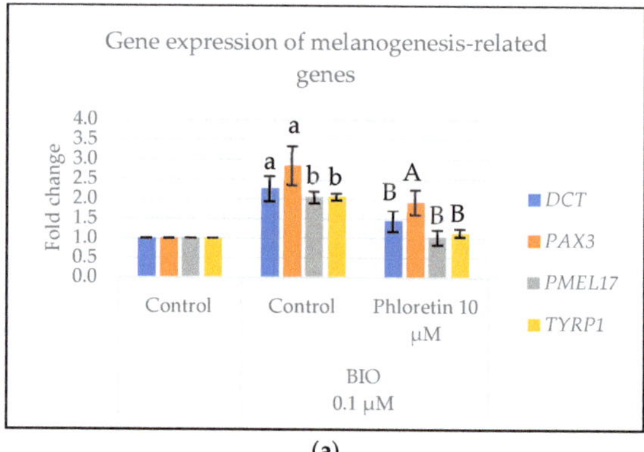

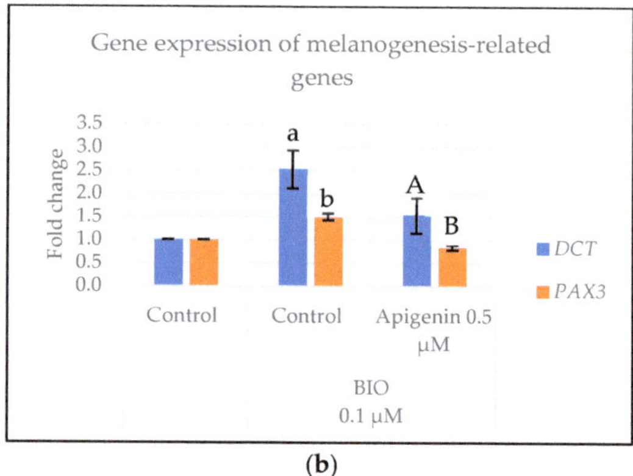

(**a**)   (**b**)

Figure 3. Inhibition of Wnt-activated melanogenesis in melanocytes treated for 24 h with phloretin (**a**) or apigenin (**b**). Statistical analysis: a, Student's *t*-test between control and BIO-stimulated control is $p < 0.05$; b, Student's *t*-test between control and BIO-stimulated control is $p < 0.01$; A, Student's *t*-test between BIO-stimulated control and BIO-stimulated cells treated with compound is $p < 0.05$; and B, Student's *t*-test between BIO-stimulated control and BIO-stimulated cells treated with compound is $p < 0.01$.

3.4. Regulation of miR-675 Levels and Autophagy-Related Gene Expression in Melanocytes

Other characteristic features of melanocytes from melasma skin are impaired autophagy and reduced miR-675 levels [1]. To test whether the compounds could regulate these processes, melanocytes were cultured for 24 h with the compounds, and the markers related to these pathways were quantified by qPCR. As shown in Figure 4, apigenin induced the levels of miR-675, while both apigenin and phloretin induced the expression of autophagy-related genes (*ATG5*, *ATG7*, and *LC3*).

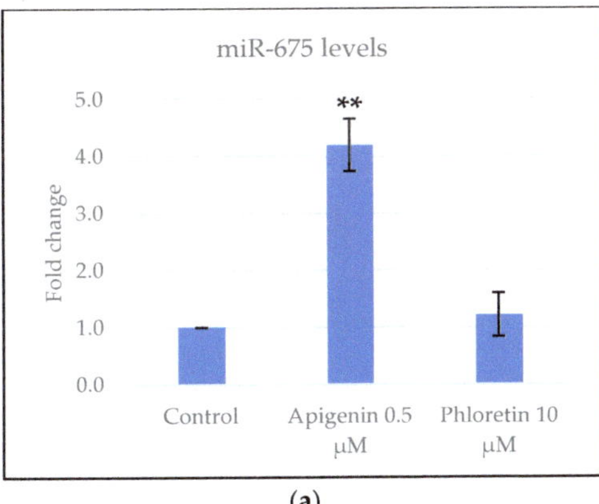
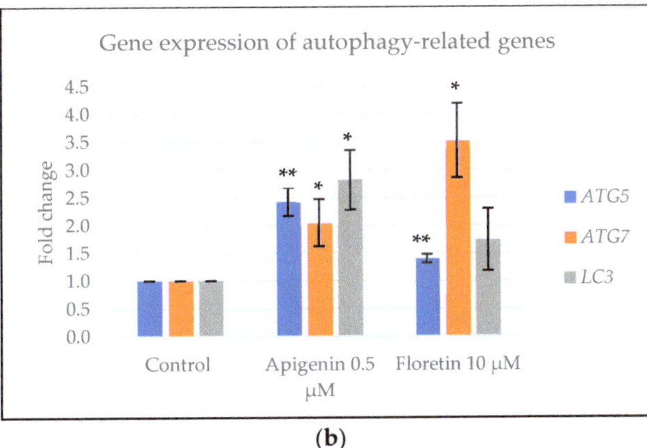

(a)   (b)

Figure 4. miR-675 (**a**) and autophagy-related gene expression (**b**) levels in melanocytes treated for 24 h with the compounds. Statistical analysis: significance of Student's *t*-test results is represented by * when $p < 0.05$ or ** when $p < 0.01$.

3.5. Regulation of Melanogenesis Paracrine Regulators in Fibroblasts

To test whether the compounds could regulate key paracrine regulators of melanogenesis [2], human dermal fibroblasts were treated for 24 h with the compounds, and the gene expressions of the paracrine regulators were quantified by qPCR. As shown in Figure 5, apigenin induced the expression of the Wnt pathway inhibitor *DKK1* and decreased the expression of the vascular growth factor *VEGF*, while phloretin did not affect the expression of these genes.

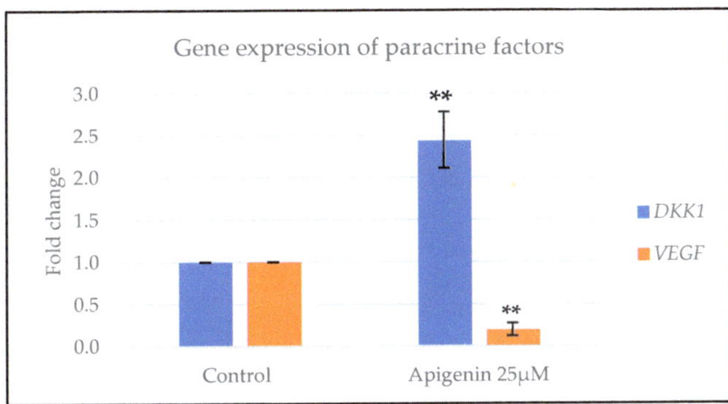

Figure 5. Gene expression of paracrine factors that regulate melanogenesis in fibroblasts treated with apigenin for 24 h. Statistical analysis: significance of Student's *t*-test results is represented by ** when $p < 0.01$.

3.6. Senomorphic Effect on Senescent Fibroblasts

As senescent fibroblasts have been described to influence skin pigmentation [6], we treated senescent human dermal fibroblasts with the compounds for 48 h and quantified the expression of extracellular matrix-related genes by qPCR to determine the regulation of both compounds in this process. As shown in Figure 6, *COL1A1* expression remained unaffected by the compounds. Apigenin significantly reduced *MMP1* expression and induced *TIMP1* expression, while phloretin significantly reduced *MMP1* expression and induced *ELN* and *TIMP1* expressions.

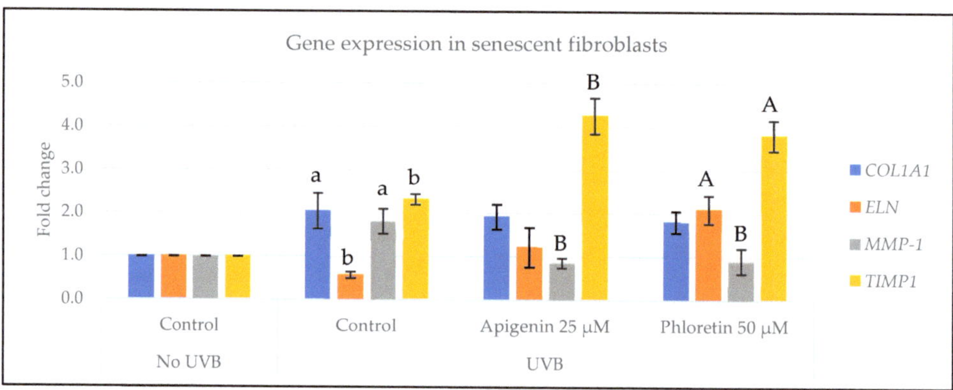

Figure 6. Gene expression of extracellular-matrix-related genes in senescent fibroblasts treated for 48 h with the compounds. Statistical analysis: a, Student's *t*-test between non-UVB control and UVB control is $p < 0.05$; b, Student's *t*-test between non-UVB control and UVB control is $p < 0.01$; A, Student's *t*-test between UVB control and UVB-treated with compound is $p < 0.05$; and B, Student's *t*-test between UVB control and UVB-treated with compound is $p < 0.01$.

3.7. Depigmenting Effect of the Combination in a 3D Epidermis Model

Finally, to validate the depigmenting efficacy of the compounds in a more complex skin model, a reconstructed human epidermis containing phototype VI was topically treated with the compounds daily for 5 days and melanin accumulation was evaluated by Fontana–Masson staining on day 7. As shown in Figure 7, the combination of apigenin and phloretin reduced the melanin content in the epidermal model after 7 days of treatment.

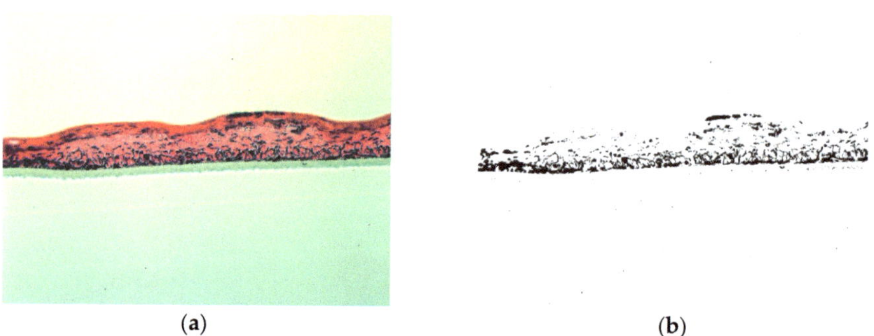

(**a**)   (**b**)

Figure 7. *Cont.*

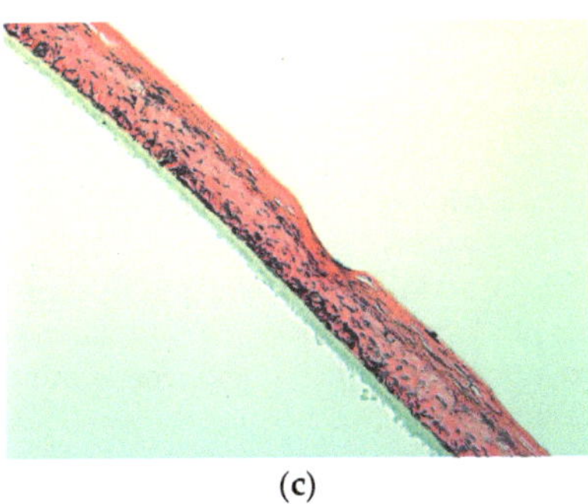
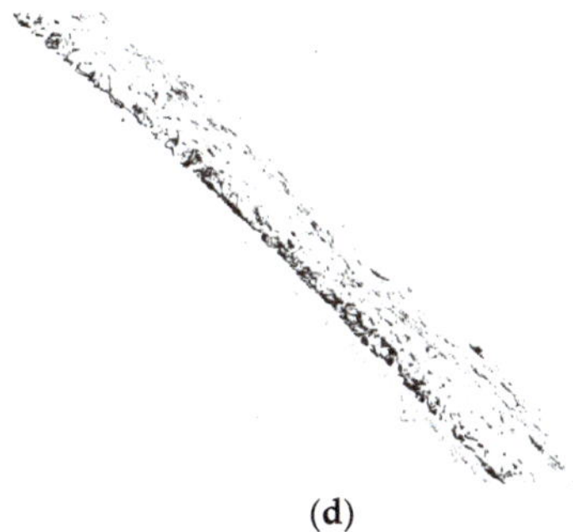

(**c**)   (**d**)

Figure 7. Representative optical microscopy images, 20× magnification, on day 7 from non-treated reconstructed human epidermis (**a**) and reconstructed human epidermis treated with 25 µM apigenin + 50 µM phloretin daily for 5 days (**c**). Fontana–Masson staining was performed to evidence melanin granules. Pictures showing the black pigment staining (**b**,**d**) are shown to better distinguish the melanin reduction after treatment.

4. Discussion

Skin hyperpigmentation, particularly melasma, has a complex pathophysiology involving different external and internal factors, such as ultraviolet radiation exposure, hormonal levels, or genetics [17]. Moreover, cells such as keratinocytes and fibroblasts release paracrine factors that result in hyperactivated melanocytes with increased melanogenic activity and impaired autophagy [1]. Here, we show that the combination of apigenin and phloretin is promising for the management of skin hyperpigmentation.

Phloretin was selected based on previous reports about its efficacy as a depigmenting compound, mainly by inhibiting tyrosinase activity [11,18]. Similarly, apigenin has been shown to exert many beneficial effects on skin, including the inhibition of tyrosinase [10]. Here, we demonstrate that the combination of both compounds is promising for decreasing the melanin production I then skin based on the synergistic effect of the compounds in melanogenesis inhibition (Figure 2).

Additionally, we better characterized the depigmenting mechanism of action of both compounds, beyond tyrosinase inhibition. We show that both compounds can inhibit Wnt-stimulated melanogenesis, which has not been reported before as far as we know. Interestingly, other models show that both apigenin and phloretin might have activating or inhibiting effects on the Wnt pathway depending on the cell type studied [19–26]. According to the Wnt regulatory mechanisms in previous research, apigenin and phloretin could be modulating the Wnt pathway at different levels, which could explain the observed synergistic effects of this combination. For example, these compounds might regulate glycogen synthase kinase 3 (GSK3) activity, or b-catenin levels, and/or transcriptional activity over specific melanogenesis genes such as TYR, TYRP1, or DCT.

Our results also show an upregulation of miR-675 by apigenin in melanocytes. This miRNA is reduced in melasma skin and targets MITF, a key transcription factor controlling the melanogenesis pathway [27]. Previous reports have shown that apigenin can influence other miRNAs in different cell models [28,29], but the effect of apigenin on miR-675 has been described for the first time in our research. The exact mechanism through which apigenin induces miR-675 levels might be via direct gene expression or miRNA processing regulation. Regarding autophagy activation, previous research has shown that both apigenin and phloretin can induce autophagy in different cell types, including keratinocytes, macrophages, and different cancer cell types [30–34]. Interestingly, one report showed that apigenin inhibited autophagy in normal keratinocytes but restored autophagy in UVB-damaged keratinocytes, indicating differential effects on the same cell type under different conditions [30]. Here, our results suggest that both apigenin and phloretin activate autophagy signaling in melanocytes. Interestingly, the autophagy pathway can be regulated through different targets, including mTOR, AMPK, or mitogen-activated protein kinase

(MAPK) signaling, among others [35]. Thus, apigenin and phloretin could also exert a synergistic depigmenting effect through the modulation of this pathway.

Up to this point, all the described mechanisms that regulate melanogenesis at the melanocyte level could explain the synergy observed by combining apigenin and phloretin. The modulation of Wnt pathway, along with autophagy and miR-675 levels, exerts a multistep regulation of the pigmentation process in melanocytes.

On the other hand, paracrine factors released by surrounding cells play a key role in melasma pathogenesis [2]. Here, we show that apigenin induces DKK1, an antagonist of the Wnt pathway with a proven role in melanogenesis inhibition [36]. No previous reports have described the relationship between apigenin and DKK1. On the other hand, previous research on the effect of apigenin on angiogenesis and VEGF is controversial, with results showing proangiogenic or antiangiogenic effects in different models [37–39]. Here, our results indicate that apigenin might have antiangiogenic properties by downregulating the levels of VEGF in human dermal fibroblasts. Regarding phloretin's effect on these paracrine factors, previous research shows that phloretin increases DKK1 in adipose tissue and upregulates or downregulates VEGF expression depending on the cell type [24,40–42]. Here, we observed that both DKK1 and VEGF gene expression remained unchanged in human dermal fibroblasts after phloretin treatment.

Senescent fibroblasts also contribute to increased pigmentation in aged skin [6]. Here, we showed that apigenin and phloretin exert senomorphic effects on senescent fibroblasts by regulating extracellular matrix proteins (Figure 6), which might have a beneficial effect on photoaged skin pigmentation. Interestingly, previous studies showed that apigenin induces collagen I synthesis in dermal fibroblasts while it does not affect MMP1 or TIMP1 expression [43]. According to our research, apigenin does not affect collagen I gene expression but downregulates and upregulates MMP1 and TIMP1 expression, respectively. The differences in these results might be explained by the fact that the previous study tested the effect of apigenin on normal human dermal fibroblasts, while we studied the effect of apigenin on senescent fibroblasts. This suggests different cell behavior in response to apigenin in the same cell type, being beneficial for skin functions in all the studied cases.

Regarding this effect on senescent cells by apigenin and phloretin, previous research already indicated that these compounds have senomorphic effects on senescent cells [44]. Perrott, K. et al. proved the positive effect of Apigenin in modulating the senescence-associated secretory phenotype (SASP) of senescent skin fibroblasts, mainly by regulating inflammatory markers [45]. Other reports identified the senescent-cell-modulating effects of phloretin [46]. Our research confirms the positive effect of apigenin and phloretin on senescent human dermal fibroblasts by regulating extracellular-matrix-related proteins.

Finally, the reduction in melanin accumulation in the reconstructed human epidermis model after 7 days of treatment with the compounds suggests that the combination of both compounds might have promising effects when applied in vivo to hyperpigmented skin. Moreover, this experiment sheds light on the safety of the compounds for topical application, as daily treatment of the 3D reconstructed epidermis for 7 days did not affect tissue viability. Furthermore, both compounds have been used for years in topical products for different chronic dermatological conditions [10,11], which is another proof of the long-term safety of both compounds.

The mechanisms of action of these compounds on melanogenesis inhibition, including Wnt pathway, autophagy, miR-675 levels, and paracrine factors, offer a complementary approach to targeting skin hyperpigmentation. Most natural compounds currently used in the cosmetic market target the activity of tyrosinase, the key and central enzyme involved in the synthesis of melanin [47], and only a few are focused on other key pathways such as inflammation, hormonal, or vascularization pathways [15,48].

Future experiments could include the evaluation of other paracrine factors such as HGF, SCF, or GDF15 on fibroblasts after treatment with apigenin and phloretin. Additionally, the clinical evaluation of a topical formulation including these ingredients is necessary to validate the efficacy of this combination of compounds on melasma skin. This further research would overcome the main limitations of this study. These include the use of in vitro and 3D models, which are useful for mechanistic studies on cellular physiology but do not fully replicate the human skin environment. Interestingly, as solar lentigo and post-inflammatory hyperpigmentation share some features with melasma pathology [49,50], the combination of these compounds might be used to treat other types of skin hyperpigmentation. Given the observed results on the compounds' safety, their mechanisms of action, and their efficacy in the 3D epidermal model, we expect that the use of this combination of compounds in daily-use creams, masks, or similar topical formulations can boost the efficacy of the current skin whitening treatments.

5. Conclusions

Our research shows that apigenin and phloretin have synergistic effects on the inhibition of melanin accumulation in melanocytes by targeting some of the main features of melasma pathology, including the Wnt pathway, autophagy impairment, miRNA regulation, and paracrine factors released by the surrounding cells. In addition, the topical application of these compounds significantly decreased melanin accumulation in a phototype VI 3D epidermis model. Thus, this combination of compounds shows promising results for the treatment of skin hyperpigmentation and aging. Further studies should be conducted to confirm the depigmenting effect in physiological skin conditions.

Supplementary Materials: The following supporting information can be downloaded at: https://www.mdpi.com/article/10.3390/cosmetics11040128/s1, Table S1: Primer sequences used for qPCR assay.

Author Contributions: Conceptualization, A.M.-G. and M.C.G.; methodology, J.S. and T.N.; software: A.M.-G., J.S. and T.N.; validation, A.M.-G. and M.C.G.; formal analysis, A.M.-G. and J.S.; investigation, A.M.-G., J.S. and T.N.; resources, A.M.-G. and M.C.G.; data curation, A.M.-G. and J.S.; writing—original draft preparation, A.M.-G. writing—review and editing, A.M.-G. and M.C.G.; supervision, A.M.-G. and M.C.G.; project administration, M.C.G.; funding acquisition, M.C.G. All authors have read and agreed to the published version of the manuscript.

Funding: The research has been funded by Mesoestetic Pharma Group S.L.

Data Availability Statement: All data related to this research has been included in this paper.

Acknowledgments: We would like to thank all the collaborators and company departments that supported our research in its different stages.

Conflicts of Interest: The authors are employees of Mesoestetic Pharma Group S.L. The authors declare that the research was conducted in the absence of any commercial or financial relationships that could be construed as a potential conflict of interest. The funders had no role in the design of the study; in the collection, analyses or interpretation of data; in the writing of the manuscript; or in the decision to publish the results.

References

1. Espósito, A.C.C.; Cassiano, D.P.; da Silva, C.N.; Lima, P.B.; Dias, J.A.F.; Hassun, K.; Bagatin, E.; Miot, L.D.B.; Miot, H.A. Update on Melasma—Part I: Pathogenesis. *Dermatol. Ther.* **2022**, *12*, 1967–1988. [CrossRef] [PubMed]
2. Yuan, X.H.; Jin, Z.H. Paracrine regulation of melanogenesis. *Br. J. Dermatol.* **2018**, *178*, 632–639. [CrossRef]
3. Yamaguchi, Y.; Itami, S.; Watabe, H.; Yasumoto, K.; Abdel-Malek, Z.A.; Kubo, T.; Rouzaud, F.; Tanemura, A.; Yoshikawa, K.; Hearing, V.J. Mesenchymal–epithelial interactions in the skin: Increased expression of dickkopf1 by palmoplantar fibroblasts inhibits melanocyte growth and differentiation. *J. Cell Biol.* **2004**, *165*, 275–285. [CrossRef] [PubMed]
4. Yamaguchi, Y.; Passeron, T.; Hoashi, T.; Watabe, H.; Rouzaud, F.; Yasumoto, K.; Hara, T.; Tohyama, C.; Katayama, I.; Miki, T.; et al. Dickkopf 1 (DKK1) regulates skin pigmentation and thickness by affecting Wnt/ β-catenin signaling in keratinocytes. *FASEB J.* **2008**, *22*, 1009–1020. [CrossRef]
5. Passeron, T.; Picardo, M. Melasma, a photoaging disorder. *Pigment. Cell Melanoma Res.* **2018**, *31*, 461–465. [CrossRef]
6. Kim, J.C.; Park, T.J.; Kang, H.Y. Skin-Aging Pigmentation: Who Is the Real Enemy? *Cells* **2022**, *11*, 2541. [CrossRef]
7. Espósito, A.C.C.; Brianezi, G.; Miot, L.D.B.; Miot, H.A. Fibroblast morphology, growth rate and gene expression in facial melasma. *An. Bras. Dermatol.* **2022**, *97*, 575–582. [CrossRef] [PubMed]

8.   Ni, C.; Narzt, M.-S.; Nagelreiter, I.-M.; Zhang, C.F.; Larue, L.; Rossiter, H.; Grillari, J.; Tschachler, E.; Gruber, F. Autophagy deficient melanocytes display a senescence associated secretory phenotype that includes oxidized lipid mediators. *Int. J. Biochem. Cell Biol.* **2016**, *81*, 375–382. [CrossRef]

9.   Artzi, O.; Horovitz, T.; Bar-Ilan, E.; Shehadeh, W.; Koren, A.; Zusmanovitch, L.; Mehrabi, J.N.; Salameh, F.; Nelkenbaum, G.I.; Zur, E.; et al. The pathogenesis of melasma and implications for treatment. *J. Cosmet. Dermatol.* **2021**, *20*, 3432–3445. [CrossRef]

10.   Majma Sanaye, P.; Mojaveri, M.R.; Ahmadian, R.; Jahromi, M.S.; Bahrasoltani, R. Apigenin and its dermatological applications: A comprehensive review. *Phytochemistry* **2022**, *203*, 113390. [CrossRef]

11.   Anunciato Casarini, T.P.; Frank, L.A.; Pohlmann, A.R.; Guterres, S.S. Dermatological applications of the flavonoid phloretin. *Eur. J. Pharmacol.* **2020**, *889*, 173593. [CrossRef] [PubMed]

12.   Vichai, V.; Kirtikara, K. Sulforhodamine B colorimetric assay for cytotoxicity screening. *Nat. Protoc.* **2006**, *1*, 1112–1116. [CrossRef] [PubMed]

13.   Martínez-Gutiérrez, A.; Asensio, J.A.; Aran, B. Effect of the combination of different depigmenting agents in vitro. *J. Cosmet. Sci.* **2014**, *65*, 365–375. [PubMed]

14.   Meijer, L.; Skaltsounis, A.-L.; Magiatis, P.; Polychronopoulos, P.; Knockaert, M.; Leost, M.; Ryan, X.P.; Vonica, C.A.; Brivanlou, A.; Dajani, R. GSK-3-Selective Inhibitors Derived from Tyrian Purple Indirubins. *Chem. Biol.* **2003**, *10*, 1255–1266. [CrossRef] [PubMed]

15.   Alfredo, M.G.; Maribel, P.M.; Eloy, P.R.; Susana, G.E.; Luis, L.G.S.; Carmen, G.M. Depigmenting topical therapy based on a synergistic combination of compounds targeting the key pathways involved in melasma pathophysiology. *Exp. Dermatol.* **2023**, *32*, 611–619. [CrossRef] [PubMed]

16.   Espósito, A.C.C.; Brianezi, G.; de Souza, N.P.; Miot, L.D.B.; Miot, H.A. Exploratory Study of Epidermis, Basement Membrane Zone, Upper Dermis Alterations and Wnt Pathway Activation in Melasma Compared to Adjacent and Retroauricular Skin. *Ann. Dermatol.* **2020**, *32*, 101–108. [CrossRef] [PubMed]

17.   Lee, A.Y. Recent progress in melasma pathogenesis. *Pigment. Cell Melanoma Res.* **2015**, *28*, 648–660. [CrossRef] [PubMed]

18.   Chen, J.; Li, Q.; Ye, Y.; Huang, Z.; Ruan, Z.; Jin, N. Phloretin as both a substrate and inhibitor of tyrosinase: Inhibitory activity and mechanism. *Spectrochim. Acta A Mol. Biomol. Spectrosc.* **2020**, *226*, 117642. [CrossRef]

19.   Li, Y.; Zhao, Z.; Luo, J.; Jiang, Y.; Li, L.; Chen, Y.; Zhang, L.; Huang, Q.; Cao, Y.; Zhou, P.; et al. Apigenin ameliorates hyperuricemic nephropathy by inhibiting URAT1 and GLUT9 and relieving renal fibrosis via the Wnt/β-catenin pathway. *Phytomedicine* **2021**, *87*, 153585. [CrossRef]

20.   Ozbey, U.; Attar, R.; Romero, M.A.; Alhewairini, S.S.; Afshar, B.; Sabitaliyevich, U.Y.; Hanna-Wakim, L.; Ozcelik, B.; Farooqi, A.A. Apigenin as an effective anticancer natural product: Spotlight on TRAIL, WNT/β-catenin, JAK-STAT pathways, and microRNAs. *J. Cell Biochem.* **2019**, *120*, 1060–1067. [CrossRef]

21.   Pan, F.F.; Shao, J.; Shi, C.J.; Li, Z.P.; Fu, W.M.; Zhang, J.F. Apigenin promotes osteogenic differentiation of mesenchymal stem cells and accelerates bone fracture healing via activating Wnt/β-catenin signaling. *Am. J. Physiol. Endocrinol. Metab.* **2021**, *320*, E760–E771. [CrossRef]

22.   Lin, C.M.; Chen, H.H.; Lin, C.A.; Wu, H.C.; Sheu, J.J.; Chen, H.J. Apigenin-induced lysosomal degradation of β-catenin in Wnt/β-catenin signaling. *Sci. Rep.* **2017**, *7*, 372. [CrossRef]

23.   Kern, M.; Pahlke, G.; Ngiewih, Y.; Marko, D. Modulation of Key Elements of the Wnt Pathway by Apple Polyphenols. *J. Agric. Food Chem.* **2006**, *54*, 7041–7046. [CrossRef]

24.   Casado-Díaz, A.; Rodríguez-Ramos, Á.; Torrecillas-Baena, B.; Dorado, G.; Quesada-Gómez, J.M.; Gálvez-Moreno, M.Á. Flavonoid Phloretin Inhibits Adipogenesis and Increases OPG Expression in Adipocytes Derived from Human Bone-Marrow Mesenchymal Stromal-Cells. *Nutrients* **2021**, *13*, 4185. [CrossRef]

25.   Kim, U.; Kim, C.Y.; Lee, J.M.; Oh, H.; Ryu, B.; Kim, J.; Park, J.H. Phloretin Inhibits the Human Prostate Cancer Cells Through the Generation of Reactive Oxygen Species. *Pathol. Oncol. Res.* **2020**, *26*, 977–984. [CrossRef]

26.   Kapoor, S.; Padwad, Y.S. Phloretin induces G2/M arrest and apoptosis by suppressing the β-catenin signaling pathway in colorectal carcinoma cells. *Apoptosis* **2023**, *28*, 810–829. [CrossRef]

27.   Kim, N.H.; Choi, S.H.; Kim, C.H.; Lee, C.H.; Lee, T.R.; Lee, A.Y. Reduced MiR-675 in Exosome in H19 RNA-Related Melanogenesis via MITF as a Direct Target. *J. Investig. Dermatol.* **2014**, *134*, 1075–1082. [CrossRef] [PubMed]

28.   Nimal, S.; Kumbhar, N.; Rathore, S.; Naik, N.; Paymal, S.; Gacche, R.N. Apigenin and its combination with Vorinostat induces apoptotic-mediated cell death in TNBC by modulating the epigenetic and apoptotic regulators and related miRNAs. *Sci. Rep.* **2024**, *14*, 9540. [CrossRef] [PubMed]

29.   Husain, K.; Villalobos-Ayala, K.; Laverde, V.; Vazquez, O.A.; Miller, B.; Kazim, S.; Blanck, G.; Hibbs, M.L.; Krystal, G.; Elhussin, I.; et al. Apigenin Targets MicroRNA-155, Enhances SHIP-1 Expression, and Augments Anti-Tumor Responses in Pancreatic Cancer. *Cancers* **2022**, *14*, 3613. [CrossRef]

30.   Li, L.; Li, M.; Xu, S.; Chen, H.; Chen, X.; Gu, H. Apigenin restores impairment of autophagy and downregulation of unfolded protein response regulatory proteins in keratinocytes exposed to ultraviolet B radiation. *J. Photochem. Photobiol. B* **2019**, *194*, 84–95. [CrossRef] [PubMed]

31.   Yang, J.; Pi, C.; Wang, G. Inhibition of PI3K/Akt/mTOR pathway by apigenin induces apoptosis and autophagy in hepatocellular carcinoma cells. *Biomed. Pharmacother.* **2018**, *103*, 699–707. [CrossRef] [PubMed]

32. Kim, T.W.; Lee, H.G. Apigenin Induces Autophagy and Cell Death by Targeting EZH2 under Hypoxia Conditions in Gastric Cancer Cells. *Int. J. Mol. Sci.* **2021**, *22*, 13455. [CrossRef] [PubMed]
33. Fan, C.; Zhang, Y.; Tian, Y.; Zhao, X.; Teng, J. Phloretin enhances autophagy by impairing AKT activation and inducing JNK-Beclin-1 pathway activation. *Exp. Mol. Pathol.* **2022**, *127*, 104814. [CrossRef] [PubMed]
34. Dierckx, T.; Haidar, M.; Grajchen, E.; Wouters, E.; Vanherle, S.; Loix, M.; Boeykens, A.; Bylemans, D.; Hardonnière, K.; Kerdine-Römer, S.; et al. Phloretin suppresses neuroinflammation by autophagy-mediated Nrf2 activation in macrophages. *J. Neuroinflammation* **2021**, *18*, 148. [CrossRef]
35. He, C.; Klionsky, D.J. Regulation mechanisms and signaling pathways of autophagy. *Annu. Rev. Genet.* **2009**, *43*, 67–93. [CrossRef]
36. Yamaguchi, Y.; Morita, A.; Maeda, A.; Hearing, V.J. Regulation of Skin Pigmentation and Thickness by Dickkopf 1 (DKK1). *J. Investig. Dermatol. Symp. Proc.* **2009**, *14*, 73–75. [CrossRef]
37. Fang, J.; Zhou, Q.; Liu, L.Z.; Xia, C.; Hu, X.; Shi, X.; Jiang, B.H. Apigenin inhibits tumor angiogenesis through decreasing HIF-1α and VEGF expression. *Carcinogenesis* **2006**, *28*, 858–864. [CrossRef]
38. Osada, M.; Imaoka, S.; Funae, Y. Apigenin suppresses the expression of VEGF, an important factor for angiogenesis, in endothelial cells via degradation of HIF-1α protein. *FEBS Lett.* **2004**, *575*, 59–63. [CrossRef]
39. Tu, F.; Pang, Q.; Chen, X.; Huang, T.; Liu, M.; Zhai, Q. Angiogenic effects of apigenin on endothelial cells after hypoxia-reoxygenation via the caveolin-1 pathway. *Int. J. Mol. Med.* **2017**, *40*, 1639–1648. [CrossRef]
40. Tuli, H.S.; Rath, P.; Chauhan, A.; Ramniwas, S.; Vashishth, K.; Varol, M.; Jaswal, V.S.; Haque, S.; Sak, K. Phloretin, as a Potent Anticancer Compound: From Chemistry to Cellular Interactions. *Molecules* **2022**, *27*, 8819. [CrossRef]
41. Hytti, M.; Ruuth, J.; Kanerva, I.; Bhattarai, N.; Pedersen, M.L.; Nielsen, C.U.; Kauppinen, A. Phloretin inhibits glucose transport and reduces inflammation in human retinal pigment epithelial cells. *Mol. Cell Biochem.* **2023**, *478*, 215–227. [CrossRef]
42. Matsugami, H.; Harada, Y.; Kurata, Y.; Yamamoto, Y.; Otsuki, Y.; Yaura, H.; Inoue, Y.; Morikawa, K.; Yoshida, A.; Shirayoshi, Y.; et al. VEGF secretion by adipose tissue-derived regenerative cells is impaired under hyperglycemic conditions via glucose transporter activation and ROS increase. *Biomed. Res.* **2014**, *35*, 397–405. [CrossRef]
43. Zhang, Y.; Wang, J.; Cheng, X.; Yi, B.; Zhang, X.; Li, Q. Apigenin induces dermal collagen synthesis via smad2/3 signaling pathway. *Eur. J. Histochem.* **2015**, *59*. [CrossRef]
44. Martel, J.; Ojcius, D.M.; Wu, C.Y.; Peng, H.H.; Voisin, L.; Perfettini, J.L.; Ko, Y.F.; Young, J.D. Emerging use of senolytics and senomorphics against aging and chronic diseases. *Med. Res. Rev.* **2020**, *40*, 2114–2131. [CrossRef] [PubMed]
45. Perrott, K.M.; Wiley, C.D.; Desprez, P.Y.; Campisi, J. Apigenin suppresses the senescence-associated secretory phenotype and paracrine effects on breast cancer cells. *Geroscience* **2017**, *39*, 161–173. [CrossRef] [PubMed]
46. Malavolta, M.; Bracci, M.; Santarelli, L.; Sayeed, M.A.; Pierpaoli, E.; Giacconi, R.; Costarelli, L.; Piacenza, F.; Basso, A.; Cardelli, M.; et al. Inducers of Senescence, Toxic Compounds, and Senolytics: The Multiple Faces of Nrf2-Activating Phytochemicals in Cancer Adjuvant Therapy. *Mediat. Inflamm.* **2018**, *2018*, 1–32. [CrossRef]
47. Zolghadri, S.; Beygi, M.; Mohammad, T.F.; Alijanianzadeh, M.; Pillaiyar, T.; Garcia-Molina, P.; Garcia-Canovas, F.; Munoz-Munoz, J.; Saboury, A.A. Targeting tyrosinase in hyperpigmentation: Current status, limitations and future promises. *Biochem. Pharmacol.* **2023**, *212*, 115574. [CrossRef]
48. Qian, W.; Liu, W.; Zhu, D.; Cao, Y.; Tang, A.; Gong, G.; Su, H. Natural skin-whitening compounds for the treatment of melanogenesis (Review). *Exp. Ther. Med.* **2020**, *20*, 173–185. [CrossRef] [PubMed]
49. Cardinali, G.; Kovacs, D.; Picardo, M. Mechanisms underlying post-inflammatory hyperpigmentation: Lessons from solar lentigo. *Ann. Dermatol. Venereol.* **2012**, *139*, S148–S152. [CrossRef]
50. Sadick, N.; Pannu, S.; Abidi, Z.; Arruda, S. Topical Treatments for Melasma and Post-inflammatory Hyperpigmentation. *J. Drugs Dermatol.* **2023**, *22*, 1118–1123. [CrossRef]

Article

Glossogyne tenuifolia Essential Oil Prevents Forskolin-Induced Melanin Biosynthesis via Altering MITF Signaling Cascade

Wan-Teng Lin [1,†], Yi-Ju Chen [2,3,†], Hsin-Ning Kuo [1], Cheng-Yeh Yu [4], Mosleh Mohammad Abomughaid [5] and K. J. Senthil Kumar [4,6,*]

[1] Department of Hospitality Management, College of Agriculture and Health, Tunghai University, Taichung 407224, Taiwan; 040770@thu.edu.tw (W.-T.L.); shirleykuo@thu.edu.tw (H.-N.K.)
[2] Department of Surgery, Taichung Veterans General Hospital, Taichung 40705, Taiwan; chenyiju5668@gmail.com
[3] Department of Animal Science and Biotechnology, College of Agriculture and Health, Tunghai University, Taichung 407224, Taiwan
[4] Bachelor Program of Biotechnology, National Chung Hsing University, Taichung 402, Taiwan; apertureronald@gmail.com
[5] Department of Medical Laboratory Sciences, College of Applied Medical Sciences, University of Bisha, Bisha 67714, Saudi Arabia; moslehali@ub.edu.sa
[6] Center for General Education, National Chung Hsing University, Taichung 402, Taiwan
* Correspondence: zenkumar@dragon.nchu.edu.tw; Tel.: +886-04-2284 (ext. 537)
[†] These authors contributed equally to this work.

Abstract: *Glossogyne tenuifolia* (Labill.) Cass. ex Cass (Compositae) is a herbaceous plant that is endemic to Taiwan. Traditional Chinese Medicine has utilized it as a treatment for fever, inflammation, and liver preservation. Recent research has unveiled its bioactivities, including anti-inflammation, anti-cancer, antiviral, antioxidant, anti-fatigue, hepatoprotection, and immune modulation elements. Nevertheless, its effect on skin health remains to be investigated. Thus, we investigated the impact of *G. tenuifolia* essential oil (GTEO) on forskolin (FRK)-induced melanin biosynthesis and its mechanisms in B16-F10 murine melanoma in vitro. Treatment of GTEO resulted in a substantial decrease in FRK-induced melanin production, accompanied by a significant decrease in tyrosinase mRNA and protein expression levels. Additionally, our data demonstrated that the decrease in tyrosinase expression resulted from the suppression of MITF, as indicated by the reduced movement of MITF into the cell nucleus. Moreover, GTEO prompted a prolonged ERK1/2 activation, leading to the decline of MITF through proteasomal degradation, and it was verified that GTEO had no inhibitory impact on MITF activity in ERK1/2 inhibitor-treated cells. Additional studies demonstrated that α-pinene and D-limonene, which are the primary components in GTEO, showed strong melanin and tyrosinase inhibitory effects, indicating that α-pinene and D-limonene may contribute to its anti-melanogenic effects. Collectively, these data presented compelling proof that GTEO, along with its primary components α-pinene and D-limonene, show great potential as natural sources for developing innovative skin-whitening agents in the field of cosmetics.

Keywords: *Glossogyne tenuifolia*; α-pinene; D-limonene; anti-melanogenesis; tyrosinase inhibition; MITF

Citation: Lin, W.-T.; Chen, Y.-J.; Kuo, H.-N.; Yu, C.-Y.; Abomughaid, M.M.; Senthil Kumar, K.J. *Glossogyne tenuifolia* Essential Oil Prevents Forskolin-Induced Melanin Biosynthesis via Altering MITF Signaling Cascade. *Cosmetics* **2024**, *11*, 142. https://doi.org/10.3390/cosmetics11040142

Academic Editor: Enzo Berardesca

Received: 12 June 2024
Revised: 22 July 2024
Accepted: 12 August 2024
Published: 20 August 2024

1. Introduction

Melanin is involved in pigmentation in the skin of a diverse range of animal species. Melanin is an inherent pigment that gives color to the skin, eyes, and hair while also playing a crucial function in safeguarding the skin from potential harm, including from damaging light like ultraviolet rays [1,2]. Despite its beneficial effects, abnormalities in melanin synthesis and regulation give rise to pigment-related conditions like albinism, vitiligo, melasma, freckles, and lentigo [3]. Various elements like intracellular pH shifts, prolonged exposure to UV radiation, and the aging process can disturb the usual synthesis of melanin, resulting in its buildup within the epidermis [1]. As an example, exposure to UV radiation activates melanocytes, leading to irregular melanin production and distribution [4].

The basal layer of the epidermis contains special cells called melanocytes, which are responsible for producing melanin. This progression entails a sequence of enzyme-driven actions, predominantly facilitated by enzymes like tyrosinase, TYRP-1, and TYRP-2 [5]. Among these, tyrosinase is crucial in the two fundamental stages of melanin production; first L-tyrosine is hydroxylated into L-3,4-dihydroxyphenylalanine (L-DOPA), which is oxidized into dopaquinone and transformed into dopachrome. Furthermore, TYRP-2 carries out the transformation of L-dopachrome into DHICA, and TYRP-1 facilitates the conversion of DHICA into IQCA. It is via this sequence of processes that eumelanin is synthesized [6]. It is possible to reduce skin pigmentation in several ways with the use of a depigmentation agent, including hindering the functioning of the tyrosinase family and (2) also preventing ROS generation when exposed to UV radiation [3,7]. Among them, inhibiting tyrosinase activity is a promising strategy for controlling excessive melanin production [8].

Plant volatiles, commonly referred to as essential oils, are commonly used in modern cosmetic products because they contain various active ingredients, potent fragrances, and appealing marketing images. Recently, essential oils and their constituents have garnered significant public interest, owing to their broad consumer acceptance and versatile functional applications [9,10]. Increasing scientific evidence supports the utilization of essential oils for addressing various skin disorders, including acne, premature skin aging, and hyperpigmentation, and offering a defense against UV radiation [11,12]. Essential oils have been identified for their strong in vitro anti-melanogenic effects [13,14].

Glossogyne tenuifolia (Labill.) Cass. ex Cass is a herb that is indigenous to Penghu Island in Taiwan, and it is also present in various regions across Asia and Australia. Herbal teas made from this herb have been traditionally utilized by Penghu Islanders to treat conditions such as fever, hepatitis, and inflammation [15]. Recent scientific research has uncovered the diverse range of bioactive attributes present in different extracts of *G. tenuifolia*. These attributes include antimicrobial, anti-inflammatory, anti-fatigue, antioxidant, antiviral, anti-angiogenesis, anti-cancer, hepatoprotective, and immune-modulating effects [16–22]. Additionally, Chyau et al. [23] employed concurrent steam distillation and solvent extraction methods to isolate essential oil from *G. tenuifolia*, yielding 62 distinct compounds. The primary components of *G. tenuifolia* essential oil are primarily terpenes. In one of our recent studies, we extracted essential oil from *G. tenuifolia* using the steam distillation method, yielding 21 compounds which make up 95.95% of the entire oil. Among them, *p*-cymene, β-myrcene, β-cedrene, β-ocimene, α-pinene, and D-limonene were index compounds of GTEO with concentrations of 35.5%, 14.68%, 9.8%, 8.49%, 6.69%, and 5.17%, respectively. Furthermore, we demonstrated that *G. tenuifolia* essential oil (GTEO) possesses strong anti-inflammatory activity against lipopolysaccharide-induced inflammatory responses in macrophage cells [19]. Based on our current understanding, the effect of *G. tenuifolia* essential oil (GTEO) on skin health, including skin protection and hypo-pigmentary effects, are unexplored. Consequently, we aimed to examine the influence of GTEO on melanin synthesis induced by forskolin in murine melanoma (B16-F10) cells and to uncover the underlying mechanism.

2. Materials and Methods

2.1. Chemicals and Reagents

Roswell Park Memorial Institute (RPMI) 1640 medium, fetal bovine serum (FBS), and penicillin–streptomycin were obtained from the Life Technologies Corporation (Grand Island, NY, USA). Forskolin (FRK) was acquired from Selleckchem (Houston, TX, USA. Tyrosinase (EC 1.14.18.1), L-tyrosine, L-3,4-dihydroxyphenylalanine (L-DOPA), 4′-6-diamidino-2-phenylindole (DAPI), and 3-(4,5-dimethyl-thiazol-2-yl)-2,5-diphenyl tetrazolium bromide (MTT), resveratrol, gallic acid, kojic acid (KA), and ascorbic acid (AA) were procured from Sigma–Aldrich, St. Louis, CA, USA. PD98059, an ERK1/2 inhibitor was obtained from Calbiochem (La Jolla, CA, USA). Antibodies targeting tyrosinase and GAPDH were acquired from Santa Cruz Biotechnology, Dallas, TX, USA. Cell Signalling Technology (Danvers, MA, USA) provided horseradish peroxidase (HRP)-linked antibodies against mouse IgG and rabbit IgG. All chemicals, except for those specified, were of high quality and provided by either Sigma–Aldrich or Merck (Darmstadt, Germany).

2.2. Cell Culturing and Viability Testing

A B16-F10 cell line was obtained from the American Type Culture Collection (ATCC) in Manassas, VA, USA. The cells were cultivated in RPMI-1640 and supplemented with 10% FBS, penicillin, and streptomycin and incubated in a humidified environment with 5% CO_2 at a temperature of 37 °C. Cell viability was evaluated using the MTT colorimetric test. Briefly, B16-F10 cells were placed in a 96-well culture plate at a concentration of 1×10^4 cells/well. Following a 24 h incubation period, cells were exposed to increasing doses of GTEO (12.5–200 µg/mL) or pure compounds (25–200 µM) for an additional 24 h. The culture media was removed, and MTT (1 mg/mL) in 200 µL of fresh culture media was introduced. The MTT farmazon crystals were dissolved in 200 µL of DMSO. The optical density of the test samples was measured at a wavelength of 570 nm (A_{570}) using an ELISA microplate reader. The cell viability percentage (%) was determined by multiplying the ratio of the absorbance of treated cells to the absorbance of untreated cells by 100.

2.3. Quantification of Melanin Formation and Fontana-Masson Staining

A melanin formation assay and Fontana-Masson staining were conducted following previously established procedures [24]. Concisely, cells were placed in a 6-well cell culture plate with a density of 1×10^5 cells/well. Following a 24 h incubation period, cells were exposed to FRK (20 µM) for 48 h either with or without the presence of GTEO, KA, α-pinene, *p*-cymene, and D-limonene. Following the treatment, the cells were gathered and washed two times with PBS and the melanin was dissolved in 1 N NaOH. The mixture was then heated at a temperature of 68 °C for 20 min. Cellular melanin content was determined by quantifying the optical density at a wavelength of 475 nm, measured using an ELISA microplate reader. To detect melanin pigment, B16F10 cells were cultured on an 8-well chamber slide (ibidi GmbH, Gräfelfing, Germany). Following 48 h of treatment, the cells were fixed with formalin. Subsequently, they were stained using the Fontana-Masson ammoniacal silver staining method. In summary, the cells were immersed in an ammoniacal silver solution overnight, followed by fixation in an acid-fixing solution and staining with Kernechtrot. Images were captured using an inverted microscope (Motic Electric Group, Xiamen, Fujian, China) at 20× magnification.

2.4. Determination of Cellular Tyrosinase Activity

To assess the activity of tyrosinase in cells, cells were seeded at 1×10^5 per dish in 6 cm dishes. Following a 24 h incubation period, cells were exposed to FRK (20 µM) for 48 h, either with or without the presence of GTEO, KA, α-pinene, *p*-cymene, and D-limonene. Following treatment, the cells were lysed using a lysis buffer. After centrifugation at 16,000× *g* for 10 min, a clear liquid (supernatant) was obtained. A 96-well plate was filled with 90 µL of each lysate, which contained 100 µg of protein in total. Following that, 10 µL of 15 mM L-DOPA was added to every well. An ELISA microplate reader was used to quantify the production of dopachrome at 475 nm after a 20 min incubation time at 37 °C.

2.5. Mushroom Tyrosinase Inhibition Assay

L-tyrosine and L-DOPA were used as substrates for mushroom tyrosinase assay. We investigated the inhibitory activity of GTEO against the oxidation of L-tyrosine catalyzed by tyrosinase, as described in our previous research [24]. Concisely, 1.5 mM L-tyrosine substrate (40 µL) was mixed with 120 µL of 100 mM phosphate-buffer solution. In this mixture, 20 µL of various concentrations of GTEO (25–100 µg/mL) or kojic acid (KA, 40 µM) were introduced. Subsequently, mushroom tyrosinase 2000 U/mL in 20 µL was added to initiate the reaction. After that, the mixture was allowed to incubate for a duration of 15 min at 37 °C, and then the absorbance was measured at 475 nm, which indicates the formation of dopachrome. Furthermore, the effect of GTEO and KA on the activity of mushroom tyrosinase in the oxidation of L-DOPA was evaluated. A mixture containing 100 μL of 0.1 M phosphate buffer and 20 μL of various concentrations of GTEO (25–100 μg/mL) or KA (40 μM) was prepared. This mixture was incubated with 20 μL of mushroom tyrosinase (2000 U/mL in phosphate buffer). After five minutes of incubation at 37 °C, 40 μL of L-DOPA (4 mM in 0.1 M of phosphate buffer) was added to the mixture. The mixture was then incubated for a further 10 min at 37 °C, and the absorbance of the reaction mixture was measured at 475 nm. Using the following formula, the percentage of the inhibition of L-tyrosine or L-DOPA oxidation was determined: percentage inhibition = 100 − (B/A × 100), where A = $^{\Delta}OD_{475}$ over 10 min without the sample, and B = $^{\Delta}OD_{475}$ over 10 min with the tested sample.

2.6. Protein Extraction and Immunoblotting

Cellular protein was extracted using RIPA buffer (Pierce Biotechnology, Rockford, IL, USA), and the protein concentration was measured using the Bradford dye-binding method with Bio-Rad reagents. Afterward, to separate protein samples (100 μg), 8–12% SDS-PAGE was used, followed by transfer onto a PVDF membrane. Following the transfer, protein membranes were blocked with 5% non-fat skim milk at room temperature for 30 min, then left to incubate overnight with particular primary antibodies at 4 °C. The membranes were then probed for 2 h using HRP-conjugated anti-rabbit or anti-mouse antibodies. Enhanced chemiluminescence (ECL) reagents (Advansta Inc., San Jose, CA, USA) were used to visualize immunoblots, and the ChemiDoc XRS^{+} docking system was utilized to capture pictures. Image lab software version 6.0.1 from Bio-Rad Laboratories was used to perform a quantitative analysis of the protein bands.

2.7. RNA Extraction and Q-PCR Analyses

The GeneMark Total RNA Purification Kit (GeneMark, New Taipei City, Taiwan) was used to extract total RNA. A First-Strand Synthesis Kit (SuperScript^{TM} IV; Invitrogen, Waltham, MA, USA) was used to transcribe 2 μg of isolated total RNA into cDNA. Then, employing the Applied Biosystems Real-Time PCR System (Applied Biosystems, Waltham, MA, USA) and Power SYBR Green Master Mix (Applied Biosystems), the levels of mRNA expression were measured. The qPCR reaction was carried out as follows: initial denaturation at 96 °C for 3 min was followed by 40 cycles of denaturation at 96 °C for 1 min, annealing at 50 °C for 30 s, and extension at 72 °C for 90 s. The qPCR primer sequences for each gene were as follows: *Tyrosinase*—forward primer (F), 5′-TATTGAGCCTTACTTGGAAC-3′; reverse primer (R), 5′-AAATAGGTCGAGTGAGGTAA-3′, and *GAPDH*—forward primer (F), 5′-TCAACGGCACAGTCAAGG-3′; reverse primer (R), 5′-ACTCCACGACATACTCAGC-3′. The quantity of each transcript was assessed by computing the relative copy number, which was adjusted based on the GAPDH copy number. Using the $^{2\Delta}Ct$ cycle threshold method, the relative abundance of the target mRNA in each sample was determined by calculating the $^{\Delta}Ct$ values of the target and the endogenous reference gene GAPDH.

2.8. Statistical Analysis

The results are shown as the mean ± standard deviation. Statistical analysis was performed using GraphPad Prism version 6.0 for Windows (GraphPad Software, La Jolla, CA, USA). Statistical assessment was conducted using a one-way ANOVA followed by Dunnett's test for multiple comparisons. p values of less than 0.05 *, 0.01 **, and 0.001 *** were considered statistically significant for the FRK vs. sample treatment groups. Additionally, p values of less than 0.001 $^{\$\$\$}$ were considered statistically significant when evaluating the FRK in comparison to the control group.

3. Results

3.1. The Effect of GTEO on Melanin Biosynthesis

Before examining GTEO's anti-melanogenic properties, the cytotoxicity of murine melanoma (B16-F10) was examined. Cells were treated with various doses of GTEO (12.5, 25, 50, 100, and 200 µg/mL) for 48 h and cell viability was determined by MTT testing. GTEO did not show any detectable reduction in cell viability with doses of 100 µg/mL, while 200 µg/mL demonstrated a substantial decrease in cell viability (Figure 1A). Therefore, the non-cytotoxic concentrations of GTEO (12.5, 25, 50, and 100 µg/mL) were subjected to further studies. Next, the melanin inhibitory effect of GTEO was examined through FRK-induced melanogenesis in B16-F10 cells.

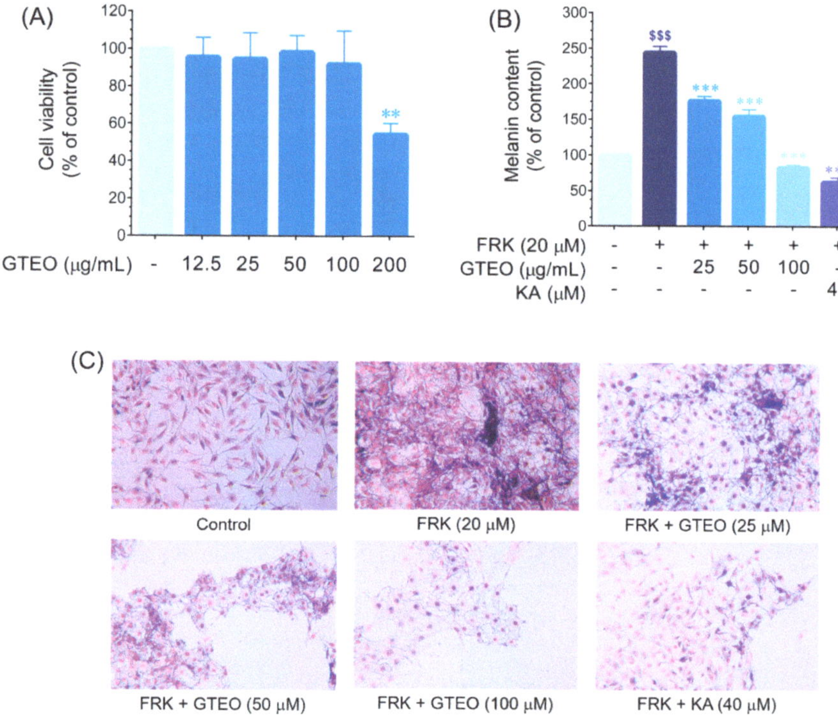

Figure 1. Effect of GTEO on FRK-induced melanin biosynthesis. (A) Cells were treated with GTEO (ranging from 12.5 to 200 µg/mL) for 48 h, after which their viability was determined. The findings were depicted as the average ± SD from three separate experiments. Statistical significance (** $p < 0.01$) was control vs. GTEO-treated group. (**B**) Cells were co-treated with FRK and GTEO (25–100 µg/mL) or KA (40 µM) for 48 h. Cellular melanin formation was determined calorimetrically. (**C**) Following treatment with FRK in the presence or absence of GTEO or KA for 48 h, cells were stained using Fontana–Masson staining. The images were captured with an inverted light microscope. The data, presented as the mean ± SD of three independent experiments, shows statistical significance: $$$ $p < 0.001$ of the control vs. FRK and *** $p < 0.001$ FRK vs. GTEO or KA.

The cells were treated with various concentrations of GTEO (25–100 µg/mL) or 40 µM KA in combination with FRK (20 µM) for 48 h. The intracellular melanin content was determined by the colorimetric method. Figure 1B demonstrates the cellular melanin content in FRK-stimulated cells, which exhibited a significant increase from 3.1 µg/mL (baseline 100%) to 8.54 µg/mL (246%). However, when treated in combination with GTEO, there was a significant decrease in the melanin content in B16-F10 cells caused by FRK, which was reduced in a dose-dependent manner. The reduction was observed at concentrations of 5.22 µg/mL (172%), 3.84 µg/mL (165%), and 2.56 µg/mL (82%) with GTEO doses of 25 µg/mL, 50 µg/mL, and 100 µg/mL, respectively. Notably, the melanin inhibition produced by GTEO (100 µg/mL) was highly comparable with KA (40 µM), a known melanin inhibitor. Additionally, microscopic observation of B16-F10 cells stained with the Fontana-Masson (FM) method highlights increased intracellular melanin accumulation in FRK-induced cells, as indicated by unstained dark regions. However, melanin content in B16-F10 cells was considerably and dose-dependently decreased by co-treatment with GTEO. This result further supports the notion that GTEO inhibits melanin biosynthesis and dendrite formation stimulated by FRK (Figure 1C).

3.2. GTEO Inhibits Mushroom Tyrosinase Activity

Tyrosinase serves as a crucial enzyme controlling the pace of melanin biosynthesis. Thus, we proceeded to investigate whether GTEO influences the activity of mushroom tyrosinase in a system devoid of cells, using L-tyrosine and L-DOPA as substrates. As depicted in Figure 2A, when mushroom tyrosinase was incubated with L-tyrosine, it resulted in high enzyme activity, as indicated by rapid browning of the mixture, whereas co-incubation with GTEO significantly reduced the activity of mushroom tyrosinase in a dosage-dependent manner. Additionally, mushroom tyrosinase with L-DOPA produced high enzyme activity that was significantly and dose-dependently inhibited by GTEO (Figure 2B). Indeed, the mushroom tyrosinase inhibitory activity of GTEO was highly comparable with ascorbic acid (AA). Taken together, these data suggest that GTEO could inhibit tyrosinase during both monophenolase and diphenolase activities. This finding indicates that GTEO possessed a strong tyrosinase inhibitory effect.

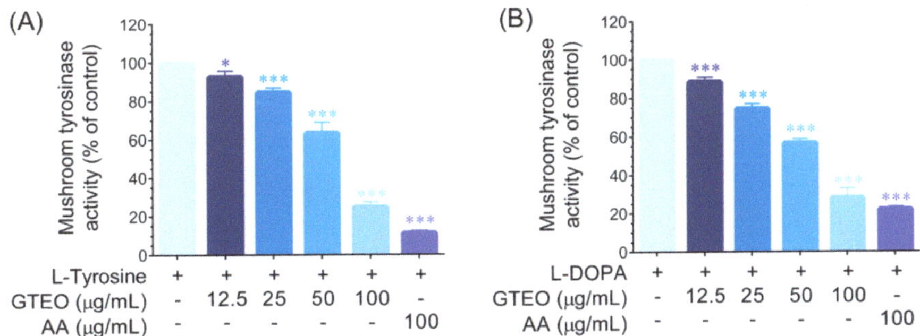

Figure 2. Effect of GTEO on mushroom tyrosinase activity. GTEO (12.5–100 µg/mL or AA (100 µg/mL) were incubated with 2000 U/mL of mushroom tyrosinase for 10 min. Then, the mixture was incubated with either 15 mM of L-tyrosine (**A**) or L-DOPA (**B**) for 30 min. L-DOPA and dopachrome formation were measured using an ELISA microplate reader. The data, presented as the mean ± SD of three independent experiments. * $p < 0.05$, and *** $p < 0.001$ were the statistical significances between the control and GTEO treatment groups.

3.3. GTEO Inhibits Cellular Tyrosinase Activity and Expression

Several anti-melanogenic agents that inhibit melanin production have been found to directly inhibit tyrosinase and disrupt melanin production in melanocytes. To further elucidate the mechanism of inhibition, GTEO was tested on FRK-stimulated cellular tyrosinase activity. Figure 3A demonstrates that the cellular tyrosinase activity increased significantly to 302.5% after FRK stimulation. Nevertheless, co-administration of GTEO caused a reduction in cellular tyrosinase activity to 230.0%, 167.9%, and 127.7% at doses of 25, 50, and 100 µg/mL, respectively. Moreover, GTEO exhibited a similar inhibitory effectiveness as kojic acid (KA). These findings suggest that GTEO effectively inhibits intracellular tyrosinase activity. To further comprehend how GTEO inhibits melanin production and tyrosinase activity, we sought to evaluate its impact on protein expression level of tyrosinase. As depicted in Figure 3B,C, after FRK stimulation, tyrosinase expression notably surged by 7.6-fold compared to control cells. Nonetheless, co-treatment of GTEO substantially decreased tyrosinase expression, nearly returning to basal levels (1.9-fold) at a concentration of 100 µg/mL. To delve deeper into understanding how GTEO reduces tyrosinase protein expression, we continued by examining the mRNA expression levels of tyrosinase using q-PCR. Figure 3D shows that the tyrosinase mRNA level was markedly increased after FRK-stimulation to 5.1-fold. Interestingly, co-treatment with GTEO altered the mRNA expression level of tyrosinase. Indeed, the positive drug control galic acid (GA) significantly inhibited tyrosinase in FRK-stimulated cells, which is highly comparable with GTEO. These findings indicate that GTEO could disrupt tyrosinase at both transcriptional and translational stages.

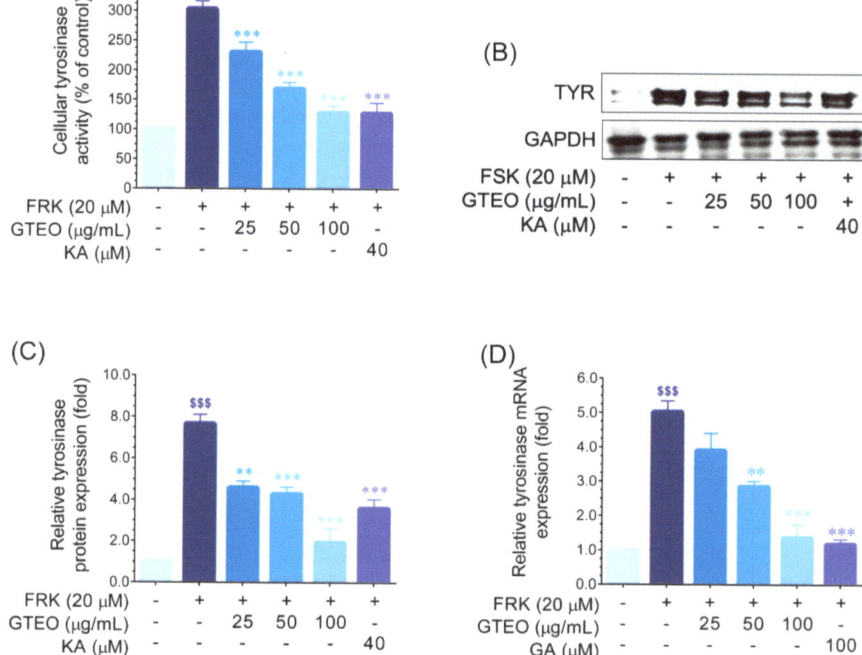

Figure 3. GTEO inhibits cellular tyrosinase and melanin biosynthesis regulators in B16-F10 cells. Cells were treated with FRK and GTEO (25–100 µg/mL) or KA (40 µM) or 100 µM GA for 48 h. (**A**) Cellular tyrosinase activity was measured. (**B**) Tyrosinase protein level was determined by immunoblotting. (**C**) The histogram shows relative protein expression. Target proteins were normalized with internal control GAPDH. (**D**) Tyrosinase mRNA level was determined by q-PCR, where GAPDH was served as an internal control. The data, presented as the mean ± SD of three independent experiments. $^{\$\$\$}$ $p < 0.001$ significance between control vs. FRK. ** $p < 0.01$, and *** $p < 0.001$ significance between FRK vs. GTEO.

3.4. GTEO Inhibits Melanin Biosynthesis via MITF Signaling Pathway

MITF, a transcription factor, is responsible for transcribing tyrosinase and its associated genes [25]. Thus, our subsequent objective was to investigate whether GTEO influences MITF at transcriptional and translational levels. Our study revealed that increased MITF mRNA expression was observed in FRK-stimulated cells. However, GTEO treatment resulted in a significant and dose-dependent decrease in MITF expression (Figure 4A). Additionally, treatment with GTEO significantly and dose-dependently reduced the FRK-induced MITF protein levels (Figure 4B). These data suggest that GTEO could affect MITF transcriptional activity. Therefore, we sought to examine whether GTEO modulates the nuclear translocation of MITF through the use of immunofluorescence. Figure 4C demonstrates that FRK-stimulated cells exhibited enhanced nuclear exportation of MITF, as indicated by the increased nuclear accumulation of MITF in FRK-stimulated cells. Remarkably, the concurrent administration of GTEO effectively prevented the FRK-mediated nuclear traslocation of MITF. To clarify the impact of GTEO-mediated suppression of MITF and melanin biosythesis, MITF siRNA was transiently transfected into cells for 6 h. Subsequently, the cells were exposed to FRK for 48 h, either with or without the presence of GTEO. FRK-treated cells had a substantial reduction in melanin synthesis following MITF siRNA transfection. Cells treated with GTEO also showed a comparable inhibition, which was further reduced when combined with GTEO and MITF-specific siRNA (Figure 4D). These findings indicate that GTEO hinders the process of melanogenesis by inhibiting the MITF signaling pathway.

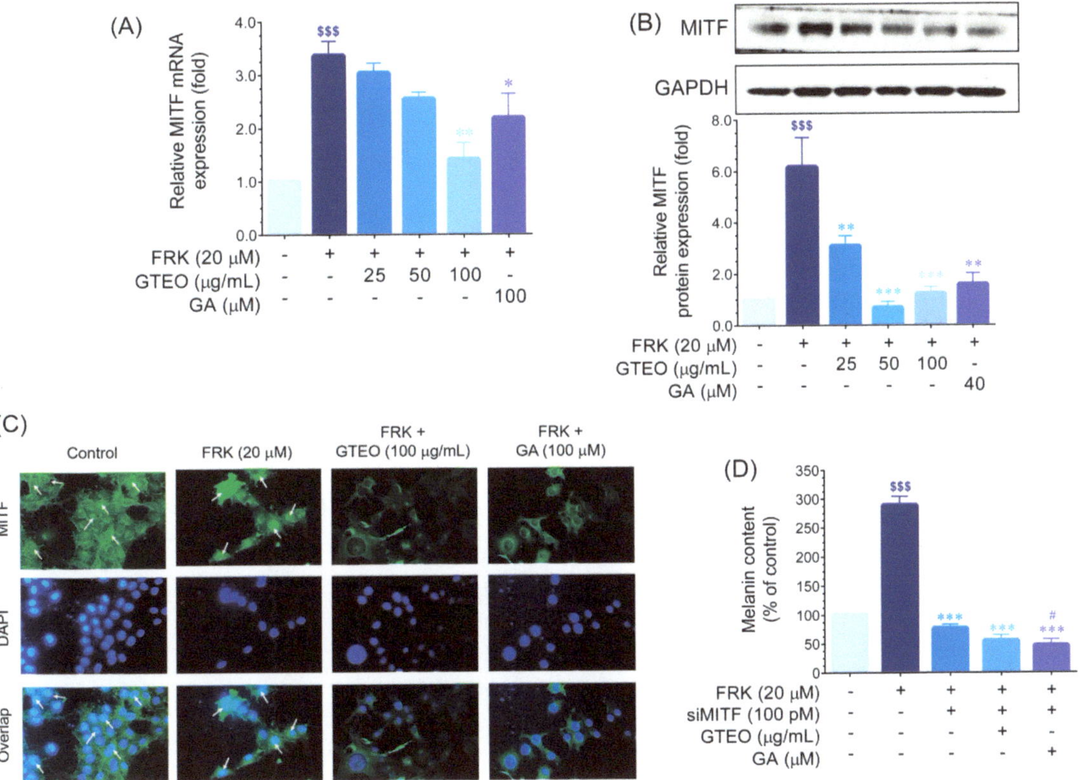

Figure 4. Effect of GTEO on MITF transcriptional activity. (**A**) MITF's mRNA expression level was evaluated using Q-PCR following a 6 h treatment with FRK and GTEO or GA. (**B**) Immunoblotting was used to measure MITF protein expression, and the histogram was standardized against GAPDH. (**C**) The intracellular location of MITF was determined using immunofluorescence, utilizing a secondary antibody labeled with fluorescein isothiocyanate (FITC). (**D**) The melanin formation was assessed in cells that were transiently transfected with either MITF siRNA or control siRNA and then treated with FRK and GTEO (100 µg/mL) or GA (100 µM). The data, presented as the mean ± SD of three independent experiments. $^{\$\$\$}$ $p < 0.001$ significance between control vs. FRK. * $p < 0.05$, ** $p < 0.01$, and *** $p < 0.001$ significance between FRK vs. GTEO. $^{\#}$ $p < 0.05$ significance between control vs. GTEO or GA.

3.5. GTEO Suppress MITF Activity through ERK1/2 Activation

ERK1/2 is a crucial factor in melanogenesis, as it regulates the activation of MITF. ERK1/2 activation results in the degradation of MITF through ubiquitination, which is crucial for the suppression of melanogenesis. Consequently, we tried to investigate whether GTEO controls MITF activity via the ERK1/2 pathway. The application of GTEO resulted in a dose-dependent and substantial increase in ERK1/2 phosphorylation, as demonstrated by Western blot analysis (Figure 5A). Based on these observations, we speculate that MITF degradation by proteasomes may be triggered by GTEO-mediated ERK1/2 phosphorylation. As anticipated, GTEO failed to inhibit FRK-stimulated MITF activity in PD98059, a pharmacological inhibitor of ERK1/2 pre-treated cells (Figure 5B). To further clarify, we investigated the impact of GTEO on FRK-induced melanin synthesis under ERK1/2 inhibition. ERK1/2 inhibition exhibited a notable augmentation in melanin synthesis, surpassing that of FRK-treated cells. Surprisingly, the addition of GTEO did not prevent the increase in melanin synthesis caused by FRK in cells treated with an inhibitor of ERK1/2 (Figure 5C).

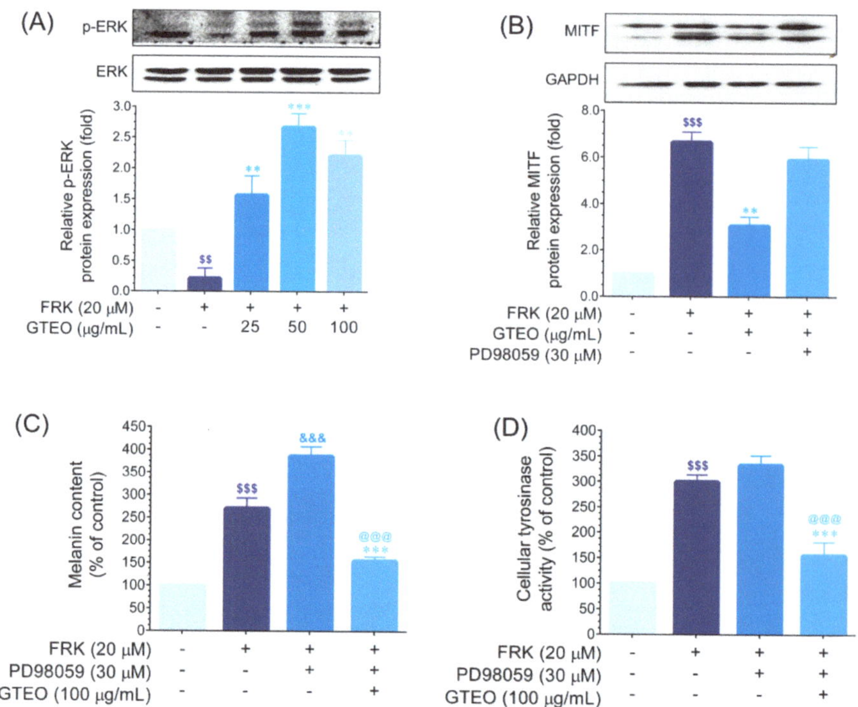

Figure 5. GTEO activates ERK1/2 in B16-F10 cells. (**A**) Cells were treated with FRK and GTEO for 15 min. Phosphorylated levels of ERK1/2 were examined by immunoblotting using a corresponding antibody. The histogram displays the relative phos-ERK1/2 protein levels, which were normalized with total levels of ERK1/2 protein. (**B**) Cells were treated with FRK and GTEO (100 µg/mL) or ERK1/2 inhibitor PD98059 for 6 h. MITF protein levels were measured by immunoblotting. (**C**) The cellular melanin content was measured after treatment with GTEO, ERK1/2 inhibitor, or their combination. (**D**) Cellular tyrosinase activity was determined after treatment with GTEO, ERK1/2 inhibitor, or their combination. The data, presented as the mean ± SD of three independent experiments. $^{\$\$}$ $p < 0.01$ and $^{\$\$\$}$ $p < 0.001$ significance between control vs. FRK. ** $p < 0.01$, and *** $p < 0.001$ significance between FRK vs. GTEO. $^{\&\&\&}$ $p < 0.001$ significance between control vs. ERK1/2 inhibitor. $^{@@@}$ $p < 0.001$ significance between ERK1/2 inhibitor vs. ERK1/2 inhibitor + GTEO.

3.6. Effect of Major Compounds of GTEO on Melanin Production and Cellular Tyrosinase Activity

Initially, the cytotoxic effects of α-pinene, β-cedrene, β-myrcene, β-ocimene, *p*-cymene, and D-limonene were quantified by MTT assay. Indeed, *p*-cymene or α-pinene did not display a cytotoxicity against B16-F10 cells with a maximum tested dose (100 μM) for 48 h, whereas β-cedrene and β-ocimene displayed cytotoxicity starting from 25 μM. At the same time, β-myrcene and D-limonene exhibited cytotoxicity after 50 μM (Figure 6A). Due to the strong cytotoxicity, β-myrcene, β-cedrene, and β-ocimene were excluded for further studies. In order to assess the effect of GTEO's bioactive components on melanogenesis, we investigated cellular tyrosinase activity and melanin biosynthesis in B16-F10 cells stimulated with FRK. It is interesting to note that D-limonene (100 μM) and α-pinene (100 μM) greatly prevented the FRK-mediated rise in melanin formation. Indeed, these compounds decreased the melanin formation from 307.2% (FRK-treatment) to 111.5% and 74.2%, respectively (Figure 6B,C), However, *p*-cymene treatment did not change the FRK-stimulated melanin production. Further, we studied the effect of *p*-cymene, α-pinene, and D-limonene on cellular tyrosinase activity. The cellular tyrosinase activities exhibited a resemblance to the melanin content (Figure 6D,E) while *p*-cymene did not inhibit cellular tyrosinase activity.

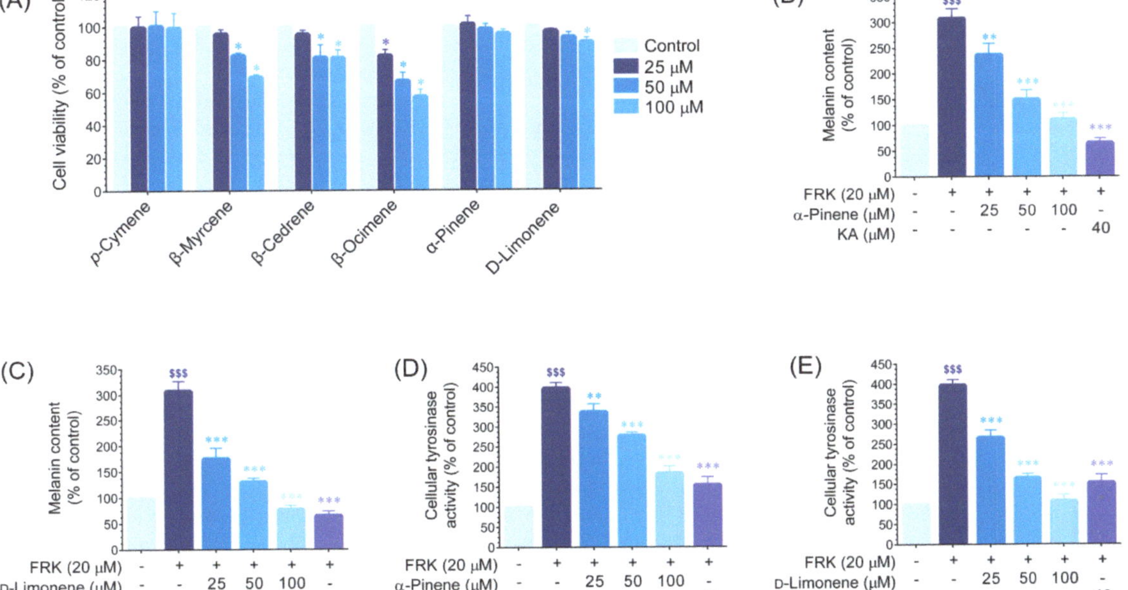

Figure 6. Effect of GTEO's principal compounds on tyrosinase activity and melanin biosynthesis in FRK-stimulated cells. (**A**) Cells were incubated with increasing concentrations of index compounds for 48 h. The MTT assay was used to assess cell viability. The findings were depicted as the average ± SD from three separate experiments. Statistical significance (* $p < 0.01$) was measured by examining the control vs. index compounds-treated groups. (**B,C**) Cellular melanin content was determined by colorimetric assay. (**D,E**) Cellular tyrosinase activity was quantified using an ELISA microplate reader. The data, presented as the mean ± SD of three independent experiments. $^{\$\$\$}$ $p < 0.001$ significance between control vs. FRK. ** $p < 0.01$, and *** $p < 0.001$ significance between FRK vs. index compounds.

4. Discussion

There are several chemically synthesized compounds employed in the cosmetic sector for skin whitening purposes. Many of these compounds directly inhibit tyrosinase enzyme activity; however, these agents have severe drawbacks, such as high cytotoxicity, limited dermal absorbance, and reduced efficacy. One example is kojic acid, which is a synthetic substance that is commonly used in cosmetics to suppress the enzyme tyrosinase and achieve skin lightening effects, which has been linked to various adverse effects, including erythema and contact dermatitis. Hence, there is a pressing need to identify safe, biocompatible, and naturally derived skin-whitening agents [26]. Recently, public attention has significantly increased towards essential oils and their components. This is due to their widespread acceptance among consumers and their ability to be used in various functional applications. Their biological activities span a wide range, including antibacterial, antifungal, antiviral, antiseptic, analgesic, anti-inflammatory, and dermato-protective effects [27]. Moreover, their distinct and pleasing aromas make them suitable for use in cosmetic products. Mounting scientific data substantiates the use of essential oils in treating diverse skin conditions such as acne, premature skin aging, hyperpigmentation, and protecting against UV radiation [28]. Multiple essential oils have been researched for their anti-melanogenic properties, demonstrating robust inhibition of melanin biosynthesis by either suppressing tyrosinase activity or modifying melanogenic signaling pathways [13,14]. Previously, we conducted a study where we recorded the anti-inflammatory properties of essential oils derived from *G. tenuifolia* [19]. Furthermore, other researchers have observed the powerful antioxidant properties of GTEO. Based on this information, we suggest that the essential oils derived from *G. tenuifolia* may also possess anti-melanogenic characteristics.

B16-F10, a murine melanoma cell line that is well regarded for its melanin-producing capabilities, serves as a common model for investigating the anti-melanogenic effects of both synthetic and natural agents. Various substances can induce melanogenesis in these melanoma cells, including α-MSH and FRK, which are identified as cAMP activators, stimulating melanin biosynthesis [29]. In the present study, FRK was employed to stimulate melanin biosynthesis and evaluate the effect of GTEO and its major bioactive compounds on melanogenesis. GTEO exhibited the ability to hinder melanin biosynthesis in B16-F10 cells. The melanin and cell viability assays revealed that treating B16-F10 cells with 25–100 µM GTEO resulted in a reduction in melanin biosynthesis while maintaining a cell viability of above 90%. Indeed, the GTEO's potency was highly comparable with a known skin-lightening agent, kojic acid. The process of melanogenesis is tightly controlled by specific enzymes, including tyrosinase, TRP-1, and TRP-2. Hence, we aimed to investigate whether GTEO has the potential to regulate the activity of the tyrosinase enzyme. The mushroom tyrosinase inhibitory assay is commonly employed to assess the skin-whitening potential of candidate substances in a cell-free system, as tyrosinase serves as the key enzyme in melanin biosynthesis. Employing this assay, we examined the direct tyrosinase inhibitory properties of GTEO, where L-tyrosine and L-DOPA served as substrates. Indeed, our results revealed a significant inhibition of mushroom tyrosinase enzyme activity upon co-incubation with GTEO, notably observed in the presence of either L-tyrosine or L-DOPA substrates. L-tyrosine and L-DOPA are commonly used as substrates in mushroom tyrosinase assays because they serve as precursors in the biosynthesis of melanin. Tyrosinase catalyzes the conversion of L-tyrosine to L-DOPA (monophenolase) and further converts L-DOPA to dopaquinone (diphenolase), which is a key step in melanin synthesis [30].

Numerous natural and synthetic compounds demonstrate dual capabilities in directly inhibiting tyrosinase and altering cellular signaling pathways, effectively suppressing melanin production [31,32]. As anticipated, the treatment with GTEO demonstrated a potent tyrosinase inhibitor, which aligns with others' findings indicating that various essential oils possess the ability to inhibit cellular tyrosinase activity. This extended to the inhibition of tyrosinase both transcriptionally and translationally by GTEO.

The primary transcription factor responsible for directly regulating the transcription of tyrosinase, TRP-1, and TRP-2 is MITF. At the transcriptional level, CREB, a cAMP-dependent transcription factor, induces the transcription of MITF upon stimulation [33]. We found that FRK, functioning as a cAMP activator, elevated MITF at both transcriptional and translational levels, whereas GTEO treatment notably suppressed this elevation. The nuclear translocation of MITF signifies a hallmark event in MITF's transcriptional activities. The administration of GTEO effectively hinders the FRK-induced nuclear translocation of MITF and subsequently suppresses its transcriptional activation. In a previous study, it was noted that the majority of melanoma cells exhibit hyper-activated MAPKs, resulting in MITF phosphorylation and subsequently leading to ubiquitin-mediated proteasomal degradation [34]. Additionally, we previously reported that essential oils can impede the protein stability of MITF by activating ERK [29]. To confirm this hypothesis, cells were treated with ERK1/2 inhibitor prior to FRK and GTEO treatment, after which melanin production was assessed. In this study, we found that the application of GTEO did not hinder FRK-induced melanin formation in cells treated with the ERK inhibitor (PD98059). However, notable reductions in melanin formation were observed in cells treated solely with GTEO. These data align with a notable increase in ERK phosphorylation induced by GTEO. Other researchers have also reported a comparable anti-melanogenic mechanism [35].

Chyau et al. [23] utilized a combined approach of steam distillation and solvent extraction methods to isolate essential oil from *G. tenuifolia*, resulting in the identification of 62 distinct compounds, predominantly terpenes, which are the primary constituents of *G. tenuifolia* essential oil. Similarly, in our previous work, we employed the steam distillation method to extract essential oils from *G. tenuifolia*, identifying 21 compounds. Monoterpenoids and sesquiterpenoids comprise the majority of these constituents. Among these, *p*-cymene, β-myrcene, β-cedrene, β-ocimene, α-pinene, and D-limonene were identified as key compounds with concentrations of 35.5%, 14.68%, 9.8%, 8.49%, 6.69%, and 5.17%, respectively [19]. The cell cytotoxicity assay conducted with these primary compounds indicated that the treatment with β-myrcene, β-cedrene, and β-ocimene resulted in significant cytotoxicity toward B16-F10 melanoma cells. This observation aligns with our earlier findings, where these compounds demonstrated cytotoxic effects on the murine macrophage (RAW 264.7) cell line [19]. Subsequent investigations showed that the primary compound, *p*-cymene, did not exhibit any significant effect on melanin inhibition. This finding is consistent with our previous study, which indicated that *p*-cymene isolated from the essential oil of *Alpinia nantoensis* did not demonstrate notable melanin inhibition even at a concentration of 100 μM [29]. Indeed, α-pinene and D-limonene markedly suppressed FRK-induced melanin formation, subsequently inhibiting cellular tyrosinase activity. This finding was strongly supported by previous observations indicating that α-pinene exhibits potent melanin inhibitory properties [36]. Furthermore, our earlier research also documented that D-limonene impedes FRK-induced melanin formation in B16-F10 cells [29]. These findings indicate that α-pinene and D-limonene present in GTEO may be accountable for its potent anti-melanogenic properties.

5. Conclusions

In this study, we have illustrated that the essential oils derived from *G. tenuifolia*, as well as their bioactive constituents α-pinene and D-limonene, diminish melanin biosynthesis in cultured B16-F10 melanoma cells. We propose a dual mechanism of action for GTEO. Firstly, through a competitive inhibition model, GTEO binds to tyrosinase, the enzyme responsible for catalyzing its oxidation. This binding impedes the oxidation of tyrosine, slowing down melanogenesis. Secondly, GTEO alters cellular signaling pathways, thereby intercepting melanin biosynthesis. Presumably, GTEO reduces tyrosinase at both the transcriptional and translational levels by limiting MITF transcriptional activity and subsequently inducing protein instability, achieved through ERK activation. Subsequent investigations revealed that the anti-melanogenic properties of GTEO were attributed to the presence of α-pinene and D-limonene. Also, other secondary metabolites present in smaller proportions may indeed act as adjuvants to the main metabolites. To our knowledge, this is the first in vitro examination of the essential oils of *G. tenuifolia* regarding their impact on tyrosinase and melanin biosynthesis in melanocytes. These essential oils exhibit potential in regard to the treatment of skin hyperpigmentation disorders; however, further exploration through in vivo testing is essential to ascertain their efficacy and potential side effects. Furthermore, these natural products may provide an alternative to synthetic compounds that have demonstrated significant adverse effects.

Author Contributions: Conceptualization, W.-T.L. and K.J.S.K.; methodology K.J.S.K.; validation, K.J.S.K., H.-N.K. and Y.-J.C.; formal analysis K.J.S.K., Y.-J.C. and M.M.A.; investigation, K.J.S.K. and C.-Y.Y.; writing—original draft preparation, K.J.S.K. and Y.-J.C.; writing—review and editing K.J.S.K and W.-T.L.; supervision, K.J.S.K. and W.-T.L.; project administration, K.J.S.K. and W.-T.L.; funding acquisition, K.J.S.K. and W.-T.L. All authors have read and agreed to the published version of the manuscript.

Funding: This study was supported by grants from the National Science and Technology Council, Taiwan (NSTC 111-2313-B-005-052-MY3) and Tunghai University (THU111611). The funding body did not have any role in the design of the study and collection, analysis, and interpretation of data and in writing the manuscript, and K.J.S.K. funded the APC.

Institutional Review Board Statement: Not applicable.

Informed Consent Statement: Not applicable.

Data Availability Statement: The data presented in this study are available on request from the corresponding author.

Conflicts of Interest: The authors declare no conflicts of interest.

References

1. Miyamura, Y.; Coelho, S.; Wolber, R.; Miller, S.; Wakamatsu, K.; Zmudzka, B.; Ito, S.; Smuda, C.; Passeron, T.; Choi, W.; et al. Regulation of human skin pigmentation and response to ultraviolet radiation. *Pigment. Cell Res.* **2007**, *20*, 2–13. [CrossRef] [PubMed]

2. Solano, F. Melanins: Skin pigments and much more—Types, structural models, biological functions, and formation routes. *New J. Sci.* **2014**, *2014*, 498276. [CrossRef]

3. Ebanks, J.P.; Wickett, R.R.; Boissy, R.E. Mechanisms regulating skin pigmentation: The rise and fall of complexion coloration. *Int. J. Mol. Sci.* **2009**, *10*, 4066–4087. [CrossRef]

4. Virador, V.M.; Muller, J.; Wu, X.; Abdel-Malek, Z.A.; Yu, Z.X.; Ferrans, V.J.; Kobayashi, N.; Wakamatsu, K.; Ito, S.; Hammer, J.A.; et al. Influence of alpha-melanocyte-stimulating hormone and ultraviolet radiation on the transfer of melanosomes to keratinocytes. *FASEB J.* **2002**, *16*, 105–107. [CrossRef]

5. Maranduca, M.A.; Branisteanu, D.; Serban, D.N.; Branisteanu, D.C.; Stoleriu, G.; Manolache, N.; Serban, I.L. Synthesis and physiological implications of melanic pigments. *Oncol. Lett.* **2019**, *17*, 4183–4187. [CrossRef]

6. Kim, Y.J.; Uyama, H. Tyrosinase inhibitors from natural and synthetic sources: Structure, inhibition mechanism and perspective for the future. *Cell Mol. Life Sci.* **2005**, *62*, 1707–1723. [CrossRef] [PubMed]

7. Shin, J.W.; Park, K.C. Current clinical use of depigmenting agents. *Dermatol. Sin.* **2014**, *32*, 205–210. [CrossRef]

8. Logesh, R.; Prasad, S.R.; Chipurupalli, S.; Robinson, N.; Mohankumar, S.K. Natural tyrosinase enzyme inhibitors: A path from melanin to melanoma and its reported pharmacological activities. *Biochem. Biophys. Acta* **2023**, *1878*, 188968. [CrossRef]

9. Sarkic, A.; Stappen, I. Essential oils and their single compounds in cosmetics—A critical review. *Cosmetics* **2018**, *5*, 11. [CrossRef]

10. Sharmeen, J.B.; Mahomoodally, F.M.; Zengin, G.; Maggi, F. Essential oils as natural sources of fragrance compounds for cosmetics and cosmeceuticals. *Molecules* **2021**, *26*, 666. [CrossRef]

11. Bungau, A.F.; Radu, A.-F.; Bungau, S.G.; Vesa, C.M.; Tit, D.M.; Purza, A.L.; Endres, L.M. Emerging insights into the applicability of essential oils in the management of acne vulgaris. *Molecules* **2023**, *28*, 6395. [CrossRef] [PubMed]

12. Kashyap, N.; Kumari, A.; Raina, N.; Zakir, F.; Gupta, M. Prospects of essential oil loaded nanosystems for skin care. *Phytomed. Plus* **2022**, *2*, 100198. [CrossRef]

13. Wijayadi, L.; Kelvin, K. The effect of natural essential oil depigmenting agent for alternative treatment of melasma. *J. Food Pharm. Sci.* **2023**, *11*, 770–779. [CrossRef]

14. Yang, J.; Lee, S.Y.; Jang, S.K.; Kim, K.J.; Park, M.J. Inhibition of melanogenesis by essential oils from the citrus cultivars' peels. *Int. J. Mol. Sci.* **2023**, *24*, 4207. [CrossRef] [PubMed]

15. Chen, C.C.; Chang, H.C.; Kuo, C.L.; Agrawal, D.C.; Wu, C.R.; Tsay, H.S. In vitro propagation and analysis of secondary metabolites in *Glossogyne tenuifolia* (Hsiang-Ju)—A medicinal plant native to Taiwan. *Bot. Stud.* **2014**, *55*, 45. [CrossRef]

16. Asokan, S.M.; Wang, R.Y.; Hung, T.H.; Lin, W.T. Hepato-protective effects of *Glossogyne tenuifolia* in streptozotocin-nicotinamide-induced diabetic rats on high-fat diet. *BMC Complement. Altern. Med.* **2019**, *19*, 117. [CrossRef] [PubMed]

17. Chen, Y.J.; Baskaran, R.; Shibu, M.A.; Lin, W.T. Anti-fatigue and exercise performance improvement effect of *Glossogyne tenuifolia* extract in mice. *Nutrients* **2022**, *14*, 1011. [CrossRef] [PubMed]

18. Ha, C.L.; Weng, C.Y.; Wang, L.; Lian, T.W.; Wu, M.J. Immunomodulatory effect of *Glossogyne tenuifolia* in murine peritoneal macrophages and splenocytes. *J. Ethnopharmacol.* **2006**, *107*, 116–125. [CrossRef] [PubMed]
19. Lin, W.T.; He, Y.H.; Lo, Y.H.; Chiang, Y.T.; Wang, S.Y.; Bezirganoglu, I.; Kumar, K.J.S. Essential oil from *Glossogyne tenuifolia* inhibits lipopolysaccharide-induced inflammation-associated genes in macrophage cells via suppression of NF-κB signaling pathway. *Plants* **2023**, *12*, 1241. [CrossRef]
20. Wu, M.J.; Huang, C.L.; Lian, T.W.; Kou, M.C.; Wang, L. Antioxidant activity of *Glossogyne tenuifolia*. *J. Agric. Food Chem.* **2005**, *53*, 6305–6312. [CrossRef]
21. Wu, M.J.; Weng, C.Y.; Ding, H.Y.; Wu, P.J. Anti-inflammatory and antiviral effects of *Glossogyne tenuifolia*. *Life Sci.* **2005**, *76*, 1135–1146. [CrossRef] [PubMed]
22. Yang, T.S.; Chao, L.K.; Liu, T.T. Antimicrobial activity of the essential oil of *Glossogyne tenuifolia* against selected pathogens. *J. Sci. Food Agric.* **2014**, *94*, 2965–2971. [CrossRef] [PubMed]
23. Chyau, C.C.; Tsai, S.Y.; Yang, J.H.; Weng, C.C.; Han, C.M.; Shih, C.C.; Mau, J.L. The essential oil of *Glossogyne tenuifolia*. *Food Chem.* **2007**, *100*, 808–812. [CrossRef]
24. Kumar, K.J.S.; Vani, M.G.; Chinnasamy, M.; Lin, W.T.; Wang, S.Y. Patchouli alcohol: A potent tyrosinase inhibitor derived from patchouli essential oil with potential in the development of a skin-lightening agent. *Cosmetics* **2024**, *11*, 38. [CrossRef]
25. Cheli, Y.; Ohanna, M.; Ballotti, R.; Bertolotto, C. Fifteen-year quest for microphthalmia-associated transcription factor target genes. *Pigment Cell Melanoma Res.* **2010**, *23*, 27–40. [CrossRef] [PubMed]
26. Phasha, V.; Senabe, J.; Ndzotoyi, P.; Okole, B.; Fouche, G.; Chuturgoon, A. Review on the use of kojic acid—A skin-lightening ingredient. *Cosmetics* **2022**, *9*, 64. [CrossRef]
27. Pezantes-Orellana, C.; German Bermúdez, F.; Matías De la Cruz, C.; Montalvo, J.L.; Orellana-Manzano, A. Essential oils: A systematic review on revolutionizing health, nutrition, and omics for optimal well-being. *Front. Med.* **2024**, *11*, 1337785. [CrossRef] [PubMed]
28. Maddheshiya, S.; Ahmad, A.; Ahmad, W.; Zakir, F.; Aggarwal, G. Essential oils for the treatment of skin anomalies: Scope and potential. *S. Afr. J. Bot.* **2022**, *151*, 187–197. [CrossRef]
29. Kumar, K.J.S.; Vani, M.G.; Wu, P.C.; Lee, H.J.; Tseng, Y.H.; Wang, S.Y. Essential oils of *Alpinia nantoensis* retard forskolin-induced melanogenesis via ERK1/2-mediated proteasomal degradation of MITF. *Plants* **2020**, *9*, 1672. [CrossRef]
30. Goldfeder, M.; Kanteev, M.; Adir, N.; Fishman, A. Influencing the monophenolase/diphenolase activity ratio in tyrosinase. *Biochim. Biophys. Acta* **2013**, *1834*, 629–633. [CrossRef]
31. Pillaiyar, T.; Manickam, M.; Namasivayam, V. Skin whitening agents: Medicinal chemistry perspective of tyrosinase inhibitors. *J. Enzym. Inhib. Med. Chem.* **2017**, *32*, 403–425.
32. Tobin, D.J. How to design robust assays for human skin pigmentation: A "Tortoise and Hare challenge". *Exp. Dermatol.* **2021**, *30*, 624–627. [CrossRef] [PubMed]
33. Hsiao, J.J.; Fisher, D.E. The roles of microphthalmia-associated transcription factor and pigmentation in melanoma. *Arch. Biochem. Biophys.* **2014**, *563*, 28–34. [CrossRef] [PubMed]
34. Wu, M.; Hemesath, T.J.; Takemoto, C.M.; Horstmann, M.A.; Wells, A.G.; Price, E.R.; Fisher, D.Z.; Fisher, D.E. c-Kit triggers dual phosphorylations, which couple activation and degradation of the essential melanocyte factor Mi. *Genes Dev.* **2000**, *14*, 301–312. [CrossRef]
35. Huang, H.C.; Chang, S.J.; Wu, C.Y.; Ke, H.J.; Chang, T.M. [6]-Shogaol inhibits α-MSH-induced melanogenesis through the acceleration of ERK and PI3K/Akt-mediated MITF degradation. *BioMed Res. Int.* **2014**, *2014*, 842569. [CrossRef] [PubMed]
36. Chao, W.W.; Su, C.C.; Peng, H.Y.; Chou, S.T. *Melaleuca quinquenervia* essential oil inhibits α-melanocyte-stimulating hormone-induced melanin production and oxidative stress in B16 melanoma cells. *Phytomedicine* **2017**, *34*, 191–201. [CrossRef] [PubMed]

MDPI AG
Grosspeteranlage 5
4052 Basel
Switzerland
Tel.: +41 61 683 77 34

Cosmetics Editorial Office
E-mail: cosmetics@mdpi.com
www.mdpi.com/journal/cosmetics

www.ingramcontent.com/pod-product-compliance
Lightning Source LLC
LaVergne TN
LVHW071937160726
843515LV00011B/2632